Predictive Filtering for Microsatellite Control System

Predictive Filtering for Microsatellite Control System

Lu Cao
National Innovation Institute of Defense Technology
Chinese Academy of Military Sciences
Beijing, China

Xiaoqian Chen
National Innovation Institute of Defense Technology
Chinese Academy of Military Sciences
Beijing, China

Bing Xiao
School of Automation
Northwestern Polytechnical University
Xi'an, China

ELSEVIER

ACADEMIC PRESS
An imprint of Elsevier

Academic Press is an imprint of Elsevier
125 London Wall, London EC2Y 5AS, United Kingdom
525 B Street, Suite 1650, San Diego, CA 92101, United States
50 Hampshire Street, 5th Floor, Cambridge, MA 02139, United States
The Boulevard, Langford Lane, Kidlington, Oxford OX5 1GB, United Kingdom

Notices

Knowledge and best practice in this field are constantly changing. As new research and experience broaden our understanding, changes in research methods, professional practices, or medical treatment may become necessary.

Practitioners and researchers must always rely on their own experience and knowledge in evaluating and using any information, methods, compounds, or experiments described herein. In using such information or methods they should be mindful of their own safety and the safety of others, including parties for whom they have a professional responsibility.

To the fullest extent of the law, neither the Publisher nor the authors, contributors, or editors, assume any liability for any injury and/or damage to persons or property as a matter of products liability, negligence or otherwise, or from any use or operation of any methods, products, instructions, or ideas contained in the material herein.

Library of Congress Cataloging-in-Publication Data
A catalog record for this book is available from the Library of Congress

British Library Cataloguing-in-Publication Data
A catalogue record for this book is available from the British Library

ISBN: 978-0-12-821865-5

For information on all Academic Press publications
visit our website at https://www.elsevier.com/books-and-journals

Publisher: Mara Conner
Acquisitions Editor: Sonnini R. Yura
Editorial Project Manager: Amy Moone
Production Project Manager: Kamesh Ramajogi
Designer: Mark Rogers

Typeset by VTeX

To our great country and our family

Contents

List of figures

List of tables

List of algorithms

Biography

Lu Cao

Dr. Cao received the B.S., the M.S., and the Ph.D. degrees in navigation, guidance, and control of spacecraft from the National University of Defense Technology, Changsha, China, in 2007, 2010, and 2014, respectively. From 2013 to 2014, he was a visiting Ph.D. student with the Department of Mechanical Engineering, McGill University, Canada. From 2014 to 2017, he was a senior engineer of the State Key Laboratory of Astronautic Dynamics, China Xi'an Satellite Control Center, China. Since 2018, he has been an associate research fellow with the National Innovation Institute of Defense Technology, Beijing, China. He has more than ten years' engineering experience in nanosatellite, and designed attitude determination and control systems for eight in-orbital micro/nano satellites. His research interests include the areas of satellite navigation, control and dynamics, and state estimation.

Xiaoqian Chen

Dr. Chen received the M.Sc. and Ph.D. degrees in aerospace engineering from the National University of Defense Technology, Changsha, China, in 1997 and 2001, respectively. From 2002 to 2016, he was a professor with the National University of Defense Technology, Changsha, China. Since 2017, he has been a professor with the National Institute Innovation of Defense Technology, Beijing, China. He has been elected as an academician of the International Academy of Astronautics (IAA), a member of the International Astronautical Federation (IAF), and a member of the robotics committee of the Chinese Society of Astronautics. Dr. Chen has published more than 200 journal and conference papers. His research interests include spacecraft systems engineering, advanced digital design methods of space systems, and multidisciplinary design optimization.

Bing Xiao

Dr. Xiao received the B.S. degree in mathematics from Tianjin Polytechnic University, Tianjin, China, in 2007, and the M.S. and Ph.D. degrees in control science and engineering from Harbin Institute of Technology, Harbin, China, in

2010 and 2014, respectively. From May 2014 to 2017, he was a Professor with the College of Engineering, Bohai University, China. He was also a postdoctoral researcher with the School of Automation, Nanjing University of Science and Technology, Nanjing, China. Since 2018, he has been an associate professor with the School of Automation, Northwestern Polytechnical University. His research interests include spacecraft attitude control and its fault-tolerant control design.

Preface

Microsatellite with its weight less than 100 kilograms and featured by low-cost is exploding in the civil and the military sector. More than 300 microsatellites have been launched into space by the end of 2019. Attitude and orbit control system should be designed for microsatellite to accomplish planned missions. Because the sensors of microsatellite are usually designed by using commercial off-the-shelf components or micro-electro-mechanical systems, their measurement data are with low accuracy and some states may be unmeasurable. Therefore, precisely estimating all the states of microsatellite control system by merging all on-board measurement data is quite essential. For general linear systems, the study of the state estimation problem was date back to the 1960s, in which the well-known Kalman filtering theory was developed. However, the implementation of the Kalman filter necessitates a precise system dynamic model and the assumption of Gaussian white noise for measurement. Moreover, it is only applicable to linear control systems.

Almost all the control systems in practice are unfortunately characterized by nonlinearity. Meanwhile, it is impossible to establish a mathematical model for nonlinear systems to fully and accurately describe their dynamics. There is always a discrepancy or modeling error between the actual system dynamics and its mathematical model. These discrepancies mostly come from external disturbances, unknown system parameters, and unmodeled dynamics. The measurement provided by sensors is also with non-Gaussian white noise, which is widely seen in microsatellite's sensors. These problems let the Kalman filter be unable to achieve a desirable state estimation. In fact, designing state estimation methods that provide desired performance in the presence of modeling error, non-Gaussian white noise, and system nonlinearity is a very challenging task for a control engineer. This has led to intense interest in the development of the so-called nonlinear filtering methods, which are supposed to solve this problem.

Based on the standard Kalman filter theory, the extended Kalman filter, the unscented Kalman filter, and the cubature Kalman filter have been presented for nonlinear systems. However, an accurate system model and Jacobians of the system are required for those filters. This drives researchers to assort to other theories outside the framework of the Kalman filter. As a result, some filters namely particle filters, robust filters, and predictive filters are reported in the literature. Although those filters could achieve good state estimation for nonlinear systems with modeling error and non-Gaussian white noise, their application to state estimation of the microsatellite control system is severely limited. That is because those filters demand expensive computation resources, while the embedded computer of microsatellite has limited computation capability.

This book addresses the state estimation problem of nonlinear control systems-in particular, the microsatellite control system. Despite the extensive and successful development of nonlinear filters, the predictive filter remains, probably, the most successful approach in handling modeling errors. Hence, the predictive filtering approach is integrated with other mathematical tools to develop several advanced predictive filters in this book. More specifically, the unscented transformation, the cubature rules, and the Stirling interpolation technique are invoked to design three sigma-point predictive filters for a general class of nonlinear systems with modeling error and measurement noise. These filters successfully eliminate the drawbacks of the Kalman filters and the classical predictive filters. Then, the adaptive control theory and the variable structure control theory are applied to synthesize several predictive filters, which are summarized as the predictive variable structure filters. Those developed filtering approaches guarantee high-accuracy state estimation and great robustness to modeling error. Any measurement noise but not limited to Gaussian noise can be handled. Moreover, those filters have the superior capability of ensuring real-time estimation with few computation resources demanded. With the application of the developed filters to the microsatellite control system, its high-accuracy state estimation problem is successfully solved with simulation results shown.

The book itself provides the reader with the current state of the art in the state estimation area of the general nonlinear control system and the microsatellite control system. Moreover, it also contains the six-degree-of-freedom control system model of a single microsatellite and the relative position as well as the relative attitude control system model of the microsatellite formation flying. The book has two readers: the first is primarily interested in the attitude and position determination system design engineering or theory of satellite and other unmanned flying systems such as unmanned aerial/underwater vehicles. The other is primarily interested in signal processing and state estimation of nonlinear control systems. Prerequisites for understanding the book are a sound of knowledge of satellite attitude and orbit dynamics, as well as knowledge of some basic random concepts related to the probability theory.

Lu Cao
Xiaoqian Chen
Bing Xiao
Beijing, China
April 28, 2020

Acknowledgments

The authors would like to thank everyone who has supported us along the way, including our family, collaborators, and co-workers. Some parts of the research presented in this book are based on collaborations with Arun K. Misra, Hengnian Li, Dechao Ran, Qiang Zhang, and Yong Zhao. Our gratitude first goes to them. As a special note, the first author would also like to dedicate this modest work to his life, academic, and work career's advisor, Prof. Xiaoqian Chen, a genius scholar, hardware expert, and research leader, and most important, a genuine person, who left us prematurely. The second author would like to express his sincere thanks to all the students with whom he has worked over the past twenty years. The third author is grateful to his friends, Xing Huo, Youmin Zhang, Dong Ye, Yanning Guo, and Yaohong Qu for their encouragement and enthusiastic help. A special thanks to the China Scholarship Council (CSC) for supporting the first and the third author's visit to McGill University, Canada, Georgia Institute of Technology, USA, respectively.

Most parts of this work would not have been possible without the financial support from the National Natural Science Foundation of China under Grants 11972373, 11902359, 11725211, and 61873207, and in part by the China Postdoctoral Science Foundation under Grants 2018M631360, 2018T110052, and 2017T100369.

Finally, the authors want to thank Sonnini Yura, Amy Moone, and Emma Hayes from Elsevier Limited for their patience throughout this book.

Symbols and abbreviations

$\mathbb{R}$	Set of all real numbers
$\mathbb{R}_+$	Set of all positive real numbers
$\mathbb{N}$	Set of all positive integers
$\mathbb{N}_0$	Set of all non-negative integers
$\mathbb{R}^n$	n dimensional real space
$\mathbb{R}^{n\times m}$	Set of all $n \times m$ real matrices
$\forall$	For all
$\in$	Belong(s) to
$\triangleq$	Be defined as
$\square$	End of proof
$\to$	Tends to
$\|\cdot\|$	Euclidean norm or its induced norm
$\vert\cdot\vert$	Scalar-valued or vector-valued abs function
$i!$	Multiply all whole integers from $i \in \mathbb{N}$ down to 1
$\boldsymbol{I}_n$	The $n \times n$ identity matrix
$\mathbf{0}$	Zero vector or matrix with appropriate dimension
$\boldsymbol{A}^{\mathrm{T}}$	Transpose of real matrix $\boldsymbol{A}$
$\boldsymbol{A}^{-1}$	Inverse or pseudo inverse of matrix $\boldsymbol{A}$
$\boldsymbol{A} >, \geq 0$	Symmetric positive-definite, semi-definite matrix $\boldsymbol{A}$
$\boldsymbol{x} > \boldsymbol{y}$	Each element of vector $\boldsymbol{x} - \boldsymbol{y}$ is positive, $\boldsymbol{x} \in \mathbb{R}^n$, $\boldsymbol{y} \in \mathbb{R}^n$
$\boldsymbol{a}^{\times}$	$\boldsymbol{a}^{\times} = \begin{bmatrix} 0 & -a_3 & a_2 \\ a_3 & 0 & -a_1 \\ -a_2 & a_1 & 0 \end{bmatrix}, \forall \boldsymbol{a} = [a_1 \quad a_2 \quad a_3]^{\mathrm{T}} \in \mathbb{R}^3$
$\boldsymbol{x}^2$	$\boldsymbol{x}^2 = \boldsymbol{x}\boldsymbol{x}^{\mathrm{T}}, \forall \boldsymbol{x} \in \mathbb{R}^n$ or $\forall \boldsymbol{x} \in \mathbb{R}^{n\times m}$
lim	Limit
min	Minimum
sup	Supremum
$\sum$	Sum
$\prod$	Matrix multiplication
$\mathbb{P}(\cdot)$	Mathematical probability of a random signal
$\mathbb{E}(\cdot)$	Mathematical expectation of a random signal
$\mathrm{Cov}(\boldsymbol{X}, \boldsymbol{Y})$	Covariance matrix of two random signals $\boldsymbol{X}$ and $\boldsymbol{Y}$
$\mathrm{Var}(\boldsymbol{X})$	Covariance matrix of a random signal $\boldsymbol{X}$
$w.p.$	With probability
$\frac{\partial f}{\partial \boldsymbol{x}}$ or $\frac{\partial}{\partial \boldsymbol{x}} f$	The derivative of the function f with respect to $\boldsymbol{x}$
$\int$	Integral
δ_{kl}	Kronecker delta, $\delta_{kl} = 1$ if $k = l$, otherwise $\delta_{kl} = 0$
$\det(\boldsymbol{A})$	Determinant of matrix $\boldsymbol{A}$
$\mathrm{rank}(\boldsymbol{A})$	Rank of matrix $\boldsymbol{A}$

$\mathrm{tr}(\boldsymbol{A})$	Trace of square matrix $\boldsymbol{A}$
$\exp(\cdot)$	Scalar-valued or vector-valued exponential function
$\mathrm{sgn}(\cdot)$	Signum function of a real number
$\mathrm{sat}(\cdot)$	Saturation function with $\mathrm{sat}(x) = \begin{cases} x, & \text{if } \lvert a\rvert \le 1 \\ \mathrm{sgn}(x), & \text{if } \lvert x\rvert > 1 \end{cases}, x \in \mathbb{R}$
$[a_{ij}]_{n\times m}$	Matrix with n rows and m columns, a_{ij} is the entry in the ith row and the jth column
$[a_i]_n$	n dimensional vector with its ith element as a_i
$\mathbf{diag}(\boldsymbol{y})$	Diagonal matrix, $\mathbf{diag}(\boldsymbol{y}) = \begin{bmatrix} y_1 & 0 & \cdots & 0 \\ 0 & y_2 & \cdots & 0 \\ \vdots & \vdots & \vdots & \vdots \\ 0 & 0 & \cdots & y_n \end{bmatrix}, \forall \boldsymbol{y} = [y_i]_n$
$\vec{A}$	$\vec{A} = \mathrm{diag}([a_{11} \quad a_{22} \quad \cdots \quad a_{nn}]^{\mathrm{T}}), \forall \boldsymbol{A} = [a_{ij}]_{n\times n}$
$\mathbf{sgn}(\boldsymbol{y})$	Signum vector $\mathbf{sgn}(\boldsymbol{y}) = [\mathrm{sgn}(y_1) \quad \mathrm{sgn}(y_2) \quad \cdots \quad \mathrm{sgn}(y_n)]^{\mathrm{T}} \in \mathbb{R}^n, \forall \boldsymbol{y} = [y_i]_n$
$\mathbf{Sat}(\boldsymbol{y})$	Saturation vector $\mathbf{Sat}(\boldsymbol{y}) = [\mathrm{sat}(y_1) \quad \mathrm{sat}(y_2) \quad \cdots \quad \mathrm{sat}(y_n)]^{\mathrm{T}} \in \mathbb{R}^n, \forall \boldsymbol{y} = [y_i]_n$
$\mathbf{Sat}(\boldsymbol{A})$	Saturation matrix $\mathbf{Sat}(\boldsymbol{A}) = [\mathrm{sat}(a_{ij})]_{n\times m} \in \mathbb{R}^n, \forall \boldsymbol{A} = [a_{ij}]_{n\times m}$
$\mathbb{C}(\boldsymbol{q})$	Rotation matrix with the respect to the unit quaternion $\boldsymbol{q} = [q_i]_4$, $\mathbb{C}(\boldsymbol{q}) = [c_{ij}]_{3\times 3}$, where $c_{11} = q_1^2 + q_2^2 - q_3^2 - q_4^2$, $c_{22} = q_1^2 - q_2^2 + q_3^2 - q_4^2$, $c_{33} = q_1^2 - q_2^2 - q_3^2 + q_4^2$, $c_{12} = c_{21} = 2q_2q_3 + 2q_1q_4$, $c_{13} = c_{31} = 2q_2q_4 - 2q_1q_3$, and $c_{23} = c_{32} = 2q_3q_4 + 2q_1q_2$
AODS	Attitude and Orbit Determination Subsystem
COTS	Commercial Off-The-Shelf
MEMS	Micro-Electro-Mechanical Systems
EKF	Extended Kalman Filter
UKF	Unscented Kalman Filter
CKF	Cubature Kalman Filter
GHF	Gauss-Hermite Filter
CDF	Central Difference Filter
DDF	Divided Difference Filter
CDKF	Central Difference Kalman Filter
SPKF	Sigma Point Kalman Filter
PF	Predictive Filter
MME	Minimum Modeling Error
SIS	Sequence Important Sampling
MIMO	Multi-Input-and-Multi-Output
SFF	Satellite Formation Flying
LVLH	Local-Vertical-Local-Horizontal
CW	Clohessy-Wiltshire
T-H	Tschauner-Hempel
UT	Unscented Transformation
UPF	Unscented Predictive Filter
VDSP	Vision-based Digital Signal Processor
PSD	Position-Sensing-Diode
CDPF	Central Difference Predictive Filter
IGRF	International Geomagnetic Reference Field
CPF	Cubature Predictive Filter
VSC	Variable Structure Control
PVSF	Predictive Variable Structure Filter
MPVSF	Modified Predictive Variable Structure Filter
PSVSF	Predictive Smooth Variable Structure Filter
CCD	Charge Couple Device

PAVSF	Predictive Adaptive Variable Structure Filter
ST-PAVSF	Strong Tracking Predictive Adaptive Variable Structure Filter
ST-SP-PAVSF	Strong Tracking Sigma Point Predictive Adaptive Variable Structure Filter
PHVSF	Predictive High-order Variable Structure Filter
OPHVSF	Orthogonal Predictive High-order Variable Structure Filter
HPHVSF	Huber Predictive High-order Variable Structure Filter
RSME	Root Square Mean Error

Chapter 1

Overview

1.1 Introduction

During the past six decades, thousands of satellites were launched. Unlike the first satellite launched in 1957, modern satellites are fixed with payloads such as camera or antenna to accomplish missions including high-resolution imaging, internet communication services, environmental monitoring, and military surveillance. Those missions demand satellites to control their attitude or orbit to provide payloads with normal working conditions. Control system design is, therefore, necessary for any satellite. This drives the development of control system theory and design methodology. Different from the classical transfer function-based control schemes, the modern control approach developed in the state-space is widely applied in satellite control system design. In this approach, the dynamics of a satellite is described by a state-space mathematical model. The states determine how the satellite control system is dynamically evolving. Hence, the evolution with time of one or more of the satellite control system's states should be controlled in aerospace engineering with satisfactory performance including the control accuracy and the steady-state behavior guaranteed. To achieve this objective, almost all controllers in use today for satellite engineering employ the form of negative feedback of system states.

Unfortunately, the states of the satellite control system may not always accessible. That is induced by cost limitation or sensor faults due to the aging of components. Even the sensor for measuring states is healthy, the real state information can not be directly obtained due to measurement noise. At this time, a state estimation method is necessary for controller design. Someone would claim that the mathematical model and the input of the satellite control system can be used to work out its states. However, this idea is impractical, because the satellite control system can not be precisely modeled and described by using the current finite modeling techniques. Actually, unknown external disturbance inevitably acts on satellites. Moreover, unknown parameters are met in the satellite control system. These issues lead to modeling error or a discrepancy between the actual system dynamics and the established mathematical model. Therefore, it definitely necessitates state estimation methods for the satellite control system design in practice.

Predictive Filtering for Microsatellite Control System. https://doi.org/10.1016/B978-0-12-821865-5.00013-9

1.2 Microsatellite control system and its design

In the past two decades, the development of satellite is toward two directions: one is the large satellite and the other is the microsatellite. The weight of a microsatellite is usually less than 100 kilograms. Multiple microsatellites flying in a formation can achieve large-scale distributed sensing, while their larger counterparts cannot economically achieve. Comparing with large satellites, microsatellite has many advantages including low launch cost, low design cost, low-risk, and short developing period. Those advantages let microsatellite be an enabling technology for new space missions. Moreover, microsatellite has also become a low-cost way of testing new equipment, new ideas, and new concepts before going through a more expensive or commercial satellite. In recent five years, microsatellite has entered an outbreak. Up to December 2019, more than 300 microsatellites were launched such as the "Tian Tuo 1" and the "Tian Tuo 3" microsatellite developed by the National University of Defense Technology, China, which are shown in Fig. 1.1 and Fig. 1.2, respectively. The mechanical dimension of "Tian Tuo 1" is 425 mm (length) × 410 mm (width) × 80 mm (height) (Ran et al., 2014), while its mass is 9.3 kilograms only. Given the preceding benefits, the main content of this book focuses on the microsatellite, in particular, its state estimation method design.

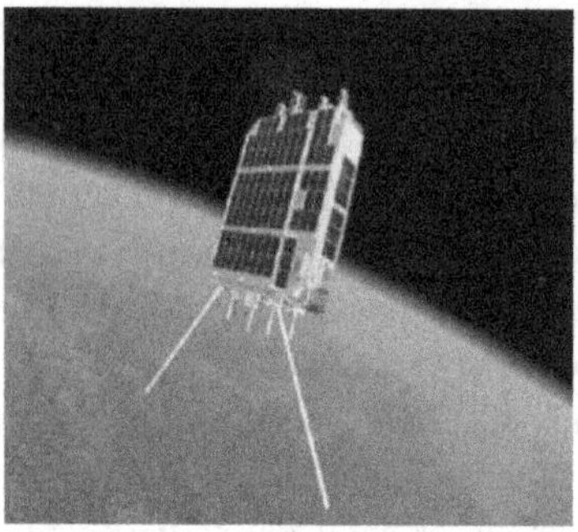

FIGURE 1.1 "Tian Tuo 1" microsatellite.

FIGURE 1.2 "Tian Tuo 3" microsatellite.

1.2.1 Microsatellite control system

Like most control systems in practice, the microsatellite control system is a Multi-Input-and-Multi-Output (MIMO) nonlinear system. Aiming to analyze the dynamic motion of microsatellite and design a desired control system for it, a mathematical model to describe the relationship between its input and its output should be established first.

In the continuous-time state space, by introducing a set of appropriate state variables and applying the Newton's second law, the mathematical model or the dynamic behavior of the microsatellite control system can be generally represented by the following nonlinear equations.

$$\begin{cases} \dot{\boldsymbol{x}} = \boldsymbol{f}(\boldsymbol{x}, \boldsymbol{u}) + \boldsymbol{f}_u(\boldsymbol{x}, \boldsymbol{u}) \\ \boldsymbol{z} = \boldsymbol{h}(\boldsymbol{x}, \boldsymbol{u}) \end{cases} \tag{1.1}$$

where $\boldsymbol{x} \in \mathbb{R}^n$ is the system state, $\boldsymbol{z} \in \mathbb{R}^m$ is the real output of the system, and $\boldsymbol{u}$ is the system input. $\boldsymbol{f}(\cdot) \in \mathbb{R}^n$ and $\boldsymbol{h}(\cdot) \in \mathbb{R}^m$ are known, and they are the process and the output functions, respectively. $\boldsymbol{f}_u(\cdot) \in \mathbb{R}^n$ is unknown and denotes the system uncertainty induce by unmodeling dynamics, uncertain parameters, and external disturbance, etc.

Fig. 1.3 illustrates a general diagram of the microsatellite control system (1.1) with the input $\boldsymbol{u}$ and the real output $\boldsymbol{z}$ in the presence of system uncertainty.

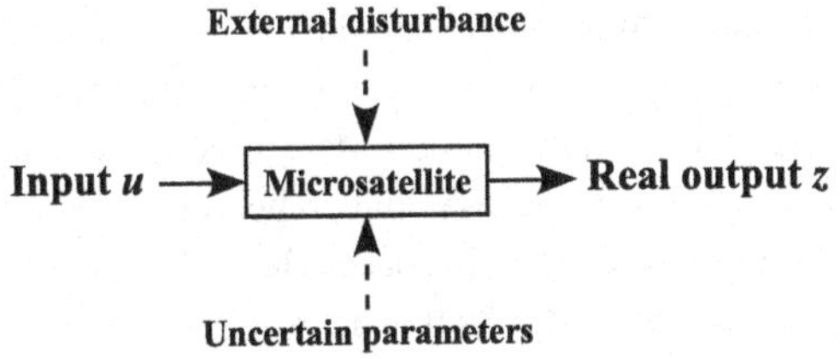

FIGURE 1.3 Open-loop microsatellite control system.

1.2.2 Microsatellite control system design

Due to external disturbance and system uncertainty, the microsatellite control system (1.1) is open-loop unstable. This is not desirable for microsatellite. To solve this problem, a closed-loop control system design is necessary. Hence, sensors are needed to measure the controlled variables and compare their behavior with the reference signals. Based on these error signals, a controller is developed to generate the desired inputs to the microsatellite system with satisfactory control performance achieved even in the presence of external disturbance and system uncertainty. In correspondence with a controller, the closed-loop microsatellite control system described in a general framework is shown in Fig. 1.4 with the real output $\boldsymbol{y}$ contaminated by noise.

Because sensors are embedded into the closed-loop microsatellite system, measurement noise exists for any sensor. Then, it could not be appropriate for applying (1.1) to describe the dynamic behavior of a microsatellite. In contrast to (1.1), its mathematical model is represented as

$$\begin{cases} \dot{\boldsymbol{x}} = \boldsymbol{f}(\boldsymbol{x}, \boldsymbol{u}) + \boldsymbol{f}_u(\boldsymbol{x}, \boldsymbol{u}) + \boldsymbol{w} \\ \boldsymbol{y} = \boldsymbol{h}(\boldsymbol{x}, \boldsymbol{u}) + \boldsymbol{v} \end{cases} \tag{1.2}$$

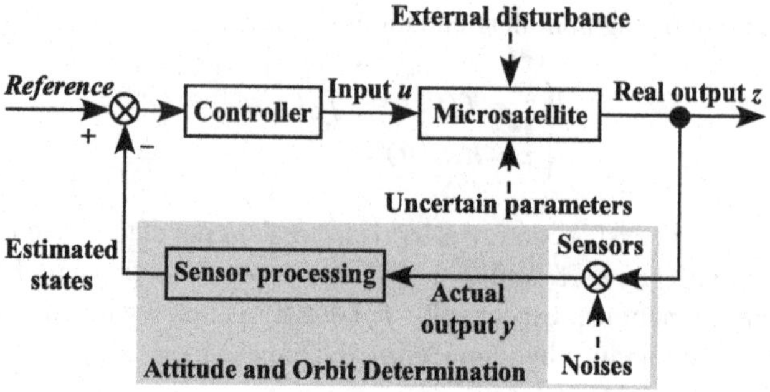

FIGURE 1.4 Closed-loop microsatellite control system with state estimation.

where $\boldsymbol{w} \in \mathbb{R}^n$ is the process noise and $\boldsymbol{v} \in \mathbb{R}^m$ is the measurement noise. $\boldsymbol{y} \in \mathbb{R}^m$ is the measurement output of the microsatellite control system, *i.e.*, $\boldsymbol{y} = \boldsymbol{z} + \boldsymbol{v}$.

The dynamic model (1.2) is given in the continuous-time state space. In aerospace engineering, the control system design is implemented on the embedded on-board computer in microsatellite. The controller is developed and implemented in the discrete-time state space. Hence, it needs to discretize the model (1.2) into the discrete-time equation as:

$$\begin{cases} \dot{\boldsymbol{x}}_k = \boldsymbol{f}(\boldsymbol{x}_{k-1}, \boldsymbol{u}_{k-1}) + \boldsymbol{f}_u(\boldsymbol{x}_{k-1}, \boldsymbol{u}_{k-1}) + \boldsymbol{w}_{k-1} \\ \boldsymbol{y}_k = \boldsymbol{h}(\boldsymbol{x}_k, \boldsymbol{u}_k) + \boldsymbol{v}_k \end{cases} \tag{1.3}$$

where $k \in \mathbb{N}$ is the time index.

The controller design seen in Fig. 1.4 is one of the main work that should be done in the microsatellite control system design. Knowing that the controllers (such as the proportional-integral-derivative controller) developed by using the classical control theory are unable to achieve high-performance control in the presence of external disturbance and system uncertainty, many controllers developed by using nonlinear control theory have been reported. Table 1.1 lists the most common nonlinear control methods that are applied to the microsatellite control system design in the literature.

1.3 Attitude and orbit determination subsystem design

For any microsatellite, its Attitude and Orbit Determination Subsystem (AODS) usually consist of sensors and a sensor processing module. Sensors measure the states or the output of the microsatellite control system. The core of the sensor processing module is an embedded state estimation method to fuse multi-sensor measurement data. It generates the estimation value of the real states.

TABLE 1.1 Nonlinear control techniques used to design microsatellite controller.

Control technique	Example of recent results using this technique
Backstepping control	Singh and Yim (2005); Kristiansen et al. (2009b); Ali et al. (2010); Xiao et al. (2010); Sun and Zheng (2017)
Sliding mode control	Chen and Lo (1993); Crassidis and Markley (1996a); Varma and Kumar (2009); Giri and Sinha (2016); Gui and Vukovich (2015); Crassidis et al. (2000b)
Optimal control	Carrington and Junkins (1986); Sharma and Tewari (2004); Yang and Wu (2007); Pukdeboon (2013); Zhang et al. (2013); Park (2013)
Adaptive control	Parlos and Sunkel (1992); Egeland and Godhavn (1994); Costic et al. (2001); Maganti and Singh (2007); Pisu and Serrani (2007); Zhao et al. (2018)
$\mathcal{H}_\infty$ control	Yang and Sun (2002); Luo et al. (2005); Zheng and Wu (2009); Nagashio et al. (2011); Chen et al. (2000)
Finite-time control	Du and Li (2012); Xiao et al. (2013); Zou et al. (2017); Guo et al. (2019); Shi et al. (2017)

1.3.1 Sensors for microsatellite control system

Sensors widely used in satellite engineering are listed in Table 1.2. The measurement principle of those sensors can be referred to Sidi (1997). Usually, several types of the sensors listed in Table 1.2 are selected out and integrated to design the attitude and orbit determination subsystem for microsatellite. Moreover, due to limitations of cost, size, mass, and power, those sensors for microsatellite are designed using Commercial Off-The-Shelf (COTS) component or Micro-Electro-Mechanical Systems (MEMS) that cost tenths of the expensive radiation-resistant and space qualified components. Figs. 1.5–1.8 show four typical MEMS-based sensors used in the AODS design of microsatellite.

TABLE 1.2 Sensors used in AODS design.

Sensors for satellite/microsatellite	Measurement output
Gyroscope	Angular velocity
GPS Array	Position and or attitude
Earth Sensor	Attitude
Sun Sensor	Attitude
Star Tracker	Attitude
Magnetometer	Attitude
Stereo Camera	Relative attitude
Lidar	Relative position
Differential GPS (DGPS)	Relative position

FIGURE 1.5 Star sensor.

FIGURE 1.6 Sun sensor.

FIGURE 1.7 MEMS-based digital GPS.

FIGURE 1.8 Magnetometer with MEMS.

1.3.2 State estimation method design

After acquiring measurement outputs from sensors, the next work of the attitude and orbit determination subsystem design is the development of the state estimation method. State estimation is a process of estimating the unmeasured states by using the noised measurement outputs, control input along with process, and measurement model. This issue has attracted significant attention in the academic and engineering area in the past several decades. The state estimation methods in the literature can be classified into two categories. The first category is the observer-based state estimation. The other category is the filtering-based state estimation.

1.3.2.1 Observer-based state estimation

For any observer-based state estimation method, an observer is developed to construct the model of the control system. The constructed model is driven by the same input signals applied to the control system and the difference between the states of the observer and the available measurement of the control system. By choosing appropriate observer gains, the difference between the observer states and the system states is governed to converge to zero or into a small set containing zero. Correspondingly, the states of the observer can be taken as a good estimation of the system states. The Luenberger observer presented for linear systems in 1966 was considered as the first result of the observer-based state estimation methods (Luenberger, 1966). Thereafter, significant development has been witnessed in this area. The common approaches in the literature to the state observer design problem are listed in Table 1.3.

TABLE 1.3 Observers for state estimation.

Observer type	References
High-gain observer	Farza et al. (2011); Hammouri et al. (2010); Mercorelli (2015); Lin and Cheng (2010); Veluvolu and Soh (2009)
Sliding mode observer	Davila et al. (2005); Ahmed-Ali and Lamnabhi-Lagarrigue (1999); Floquet et al. (2007); Daly and Wang (2009); Feng et al. (2013)
Adaptive observer	Pisu and Serrani (2007); Liu et al. (2014); Singh and Zhang (2004)
$\mathcal{H}_\infty$ observer	Pertew et al. (2006); Wang et al. (1999); Ibaraki et al. (2005); Li et al. (2016)
Finite-time observer	Hu and Jiang (2018); Orlov et al. (2011); Gui and Vukovich (2017); Hu et al. (2018)
Lyapunov stability-based observer	Tayebi (2008); Kristiansen et al. (2009a); Schlanbusch et al. (2012); Berkane et al. (2018); Hu and Zhang (2013)

Note that the observer-based approach is only appropriate for the state estimation problem of deterministic systems. Moreover, measurement noise is usually not considered in the observer design. However, all practical control systems are stochastic and with measurement noise. Therefore, the practical application of the observer-based state estimation approaches is limited.

1.3.2.2 Filtering-based state estimation

The filtering-based state estimation method is developed for stochastic systems. Its development has been a long history. It is widely known that Karl Friedrich Gauss was the earliest scientist to develop a filter to estimate state for astronomical studies (Gauss, 1809). In 1795, Gauss applied the deterministic least square method with the squared discrepancies minimized to predict the orbit and the motion of the planet and comet by using the telescopic measurements only. The measurement model was required only to implement that method. Any statistical property of the telescopic measurements was not needed. After more than a century of Gauss's invention, Ronald Aylmer Fisher invented the maximum likelihood estimation theory based on probability density (Fisher, 1912). In the 1940s, Norbert Weiner (Wiener, 1942) and Andrey Nikolaevich Kolmogorov (Kolmogoroff, 1941) invented the Wiener filtering theory independently to solve the least square estimation problem for continuous stochastic and discrete stochastic systems, respectively. Although the Wiener filter provides an important foundation for modern filtering theory, it can handle the stationary processes only. To circumvent this drawback, the modern Wiener filtering method was invented for processing multidimensional and non-stationary stochastic signals (Kucera, 1980). However, this method consumes expensive computation and has a non-recursive form. Its practical application is thus limited.

In 1960, Rudolf Emil Kalman proposed the well-known Kalman filter by introducing a state-space model into the optimal estimation theory (Kalman, 1960). The limitations of the Wiener filter were addressed. Key differences between the Wiener and the Kalman filter can be referred to Chandra and Gu (2019). In 1961, Rudolf Emil Kalman extended his work for continuous-time systems and completely established the Kalman filtering theory (Kalman and Bucy, 1961). The Kalman filter was developed with the variance of the estimation error minimized to have a recursive form for more generic time-varying, non-stationary, and multidimensional processes. Its computation burden and data restoration are not large. It is thus user friendly for real-time online computation and implementation. Since its successful application to aerospace engineering in the 1960s, the Kalman filter has been widely applied in practical control systems.

Since the 1960s, many methods have been presented to improve the Kalman filtering theory. For example, to solve the filtering divergence problem caused by round-off errors and accumulated truncation errors and to ensure that the propagated error covariance matrix is always positive-definite (otherwise, numerical instability may be induced during the online implementation), the square-root Kalman filter was presented in Battin (1964). It was practically applied to achieve state estimation for the Apollo lunar module. This result was extended in Bierman (1974); Carlson (1973); Schmidt (1970); Oshman and Bar-Itzhack (1986). It should be stressed that the implementation of the Kalman filters necessitates the prior statistical properties of measurement noises. However, it is difficult to acquire those statistical properties in practice. Moreover, the statistical properties of noise will change along with internal or external uncertainty. Such change results in the degradation of estimation accuracy and even divergence. To overcome this limitation, adaptive Kalman filters were proposed (Sage and Andrew, 1969; Yoshimura and Soeda, 1978). Since the real-time requirement will not be satisfied when the Kalman filter is applied to estimate states of systems with high dimension, the decentralized Kalman filtering method was developed in Speyer (1979) and Kerr (1987) via the parallel computing technique to address this challenge. Based on this method, the federated Kalman filter was proposed in Carlson (1990). Although those extended results have significantly promoted the development of the Kalman filtering theory, its natural drawbacks such as its ineffectiveness of estimating nonlinear systems' states can not be solved.

1.4 Review of nonlinear filtering

The Kalman filtering approach was initially developed to the state estimation problem of linear control systems. In contrast, almost all the control systems in practice are nonlinear. Hence, many filtering methods have been presented for nonlinear systems. This type of method is called as nonlinear filter, and has been extensively studied. In this section, some nonlinear filters widely seen in the literature are reviewed.

1.4.1 Nonlinear Kalman filtering

1.4.1.1 Extended Kalman filtering

To extend the Kalman filter to nonlinear continuous systems, Stanley F. Schmidt made the first successful attempt and designed the so-called extended Kalman filter (EKF) (McGee and Schmidt, 1985). The Kalman filter was also extended to be an EKF for nonlinear discrete stochastic systems (Sunahara, 1969; Bucy and Senne, 1971). The basic idea of the EKF is to locally linearize the nonlinear state functions and the measurement functions first and then applies the Kalman filter to those linearized functions. The EKF is a nonlinear Gaussian filter requiring the first-order approximation of the nonlinear systems. When it is applied to systems subject to strong nonlinearity, low state estimation accuracy is led. The EKF also requires that the states of stochastic systems have Gaussian distribution. It is suitable for "mild" nonlinearities only, where the first-order approximation of state functions is accurate enough. However, those requirements and assumptions are rarely satisfied for many physical systems. Besides, the state Jacobians of nonlinear functions are required at every iteration, *i.e.*, the continuous differentiability must be ensured. The calculation of the Jacobian matrix for high dimensional systems is rather cumbersome. This leads to low numerical stability and even divergence. To address those challenges of the EKF, the truncated second-order EKF (Maybeck, 1982) and the iterated EKF (Jazwinski, 1970) have been reported. Nevertheless, their applications are limited due to large computational burdens.

1.4.1.2 Sigma-point Kalman filtering

Because approximating the probability density distribution of nonlinear functions is easier than approximating nonlinear functions themselves, the nonlinear distribution approximations have drawn considerable attention. To overcome the theoretical limitations of the EKF, an unscented Kalman filter (UKF) was proposed in Julier et al. (2000) and Julier (2002). Its algorithm for state estimation is almost the same as the Kalman filters. The only difference is that the unscented transformation instead of the local linearization of the EKF is used in the UKF. However, the Gaussian distribution assumption of the stochastic systems' states should still be satisfied. Fortunately, the state functions and the measurement functions are not needed to be continuously differentiable. The calculation of the Jacobian matrix is eliminated. It is proved that the unscented transformation can capture the posterior mean and covariance accurately to the 3rd order Taylor series expansion for states of any nonlinear Gaussian systems. This ensures better performance than the EKF in the state estimation problem (Julier, 2002). Based on Julier et al. (2000) and Julier (2002), several other unscented transformation strategies including the symmetric sampling, the simplex sampling, and the third-moment skew sampling were developed (Julier and Uhlmann, 2002; Julier, 2003).

Besides the UKF, two suboptimal Gaussian filters, *i.e.*, the Gauss-Hermite filter (GHF) and the central difference filter (CDF), were designed (Ito and Xiong, 2000). The GHF utilizes Gauss-Hermite quadrature formula to approximate and predict the multiple integrals for nonlinear function covariance. It regards multiple integrals as a nested sequence of single integrals, and then applies a single integration formula to each variable in the proper sequence. Hence, the computational burden of the GHF increases exponentially along with state dimensions, which hinders its practical application. The CDF calculates multiple integrals by using polynomial interpolation, whose precision is lower than the GHF. However, the CDF has simple computation and is featured by easy implementation (Simandl and Dunik, 2009; Subrahmanya and Shin, 2009). Meanwhile, a divided difference filter (DDF) by adopting Stirling's interpolation formula to approximate the multiple integrals of nonlinear functions was also presented (Nørgaard et al., 2000). It is rigorously proved that the DDF can capture the posterior mean and covariance accurately to the 2nd order Taylor series expansion for any nonlinear Gaussian systems' states. Because the DDF and the CDF almost have the same theoretical basis, both were generalized as the central difference Kalman filter (CDKF) with its general recursive format given in Merwe and Wan (2004).

The polynomial interpolation used by the CDKF is equivalent to the unscented transformation with the symmetric sampling strategy in the UKF. The only difference between them is the expressions of sampling weights and covariance predictions. A unified filter framework was therefore given in Werve (2004) to describe the UKF and the CDKF. This framework is known as the sigma-point Kalman filter (SPKF). In Werve and Wan (2001), a square-root SPKF was designed by introducing the QR decomposition and the Cholesky decomposition into the prediction and the updating step of the filter. This significantly improved the estimation stability and the computational efficiency of the SPKF. In 2009, a cubature Kalman filter (CKF) was presented in Arasaratnam and Haykin (2009). The 3rd order spherical-radial integration rule was applied to approximate the multidimensional integrals. Its algorithm to achieve state estimation is similar to the UKF, while $2n$ sampling points are used for numerical integration in the CKF. It is also proved that the CKF ensures better state estimation accuracy than the EKF and the UKF. The work in Arasaratnam and Haykin (2009) was extended by Bhaumik and Swati in 2013 to develop a cubature quadrature Kalman filter with the CKF improved

(Bhaumik and Swati, 2013). Since the filtering structures of the UKF and the CDKF are almost the same, the CKF could be unified into the SPKF framework.

Although the SPKF has better performance than the EKF in the state estimation accuracy, a rigorous proof of the convergence of the SPKF is currently unavailable. Inspired by the proof of the EKF, a sufficient condition for an error bound on the UKF was given in Xiong et al. (2006). That is, the UKF's estimation errors are bounded when the nonlinearities and the prior noise statistical properties conditions are satisfied. This conclusion was extended to the GHF and the CDKF (Wu et al., 2007). In Xiong et al. (2007), the state estimation stability of the GHF, the UKF, and the CDKF was further proved in the presence of nonlinear state function and measurement functions. Note that when designing the SPKF, the system noise and the measurement noise are assumed to be independent; moreover, their prior statistical properties are assumed to be known. Those two assumptions may not be satisfied in a practical control system. Degradation of state estimation or filtering accuracy and even divergence may induce. Therefore, the improvement of the SPKF has attracted more and more attention during the past several decades.

1.4.2 Predictive filtering

In addition to extend the Kalman filter for nonlinear systems, a number of other nonlinear filtering methods developed outside the Kalman filtering framework have been seen in the literature. The nonlinear predictive filter (PF) has attracted many attention (Crassidis and Markley, 1997c). This filtering approach is developed based on the criterion of Minimum Modeling Error (MME). The PF does not require the modeling error to be a Gaussian process. The MME criterion is a novel optimal estimation index. It aims to minimize the loss function expressed by the squared weighted sum of measurement error and the modeling error, and to satisfy the necessary condition of the covariance constraint. The criterion was originally proposed in Mook and Junkins (1988). They proposed a minimum model error estimation method based on the approach beforementioned. It could achieve a high-accuracy state estimation even in the presence of severe modeling error, while the statistical properties of the modeling error are not required. However, the MME algorithm is not easy to implement. To avoid this drawback, based on the MME criterion, several predictive filters were proposed by Crassidis and Markley with the prediction tracking technique in Lu (1994) introduced. This lets the PF have the following significant advantages:

- *Unknown modeling error without any assumption and constraint can be identified in real-time.*
- *The state estimation and the identification of the modeling error are achieved simultaneously.*
- *High computational efficiency is guaranteed because the PF does not need the propagation of the state variance matrix.*

The PF is found to have the real-time advantage of the Kalman filter and the modeling error identification advantage of the MME algorithm. It needs to point out that when implementing the PF in practice, it is difficult to select out the optimal modeling error weights matrix. This matrix is usually chosen based on the engineers' experience. Moreover, the low convergence rate of state estimation is seen in the PF.

1.4.3 Particle filtering

The problem of designing nonlinear filters for state estimation can also be studied in the Bayesian framework (Bayes, 1763). That is because the Bayesian stochastic theory can provide an optimal and accurate framework for nonlinear filtering, which is referred as the Bayesian filter. The Kalman filter, the EKF, the GHF, and the SPKF, can be regarded as an optimal Bayesian filter simplified under certain conditions. In Ho and Lee (1964), the iterative Bayesian filter was first seen, while the Kalman filter was a special case of this Bayesian filter. It should be stressed that it is difficult to obtain an analytical solution of a Bayesian filter for nonlinear and non-Gaussian control systems. To circumvent this difficulty, approximation methods including the analytical approximation methods and the simulation-based methods should be adopted.

In the 1950s and the 1960s, particle filter was introduced in the fields of statistics, physics (Hammersley and Morton, 1954), and automatic control (Handschin and Mayne, 1969). Thereafter, significant development has been witnessed for particle filter, but particle degradation and computational complexity were not considered. Until 1993, a bootstrap particle filter was proposed (Gordon et al., 1993), which provided a theoretical basis for the particle filtering methodologies. In 2000, a general description of particle filtering was given based on the sequence-importance-sampling (SIS) (Doucet et al., 2000). For that description, the multiple integrals in the process of Bayesian estimation are solved by the Monte Carlo method. A set of particles in the dynamic state-space is then obtained by using the SIS. Each particle has an assigned likelihood weight that represents the probability of the particle being sampled from the probability density function. These weighted sums could be used to approximate the posterior probability density. The distribution with minimum variance is finally acquired based on the sample mean instead of the original integrals. It is proved in Crisan and Doucet (2002) that the particle filter can be guaranteed to be stable when enough particles are chosen; moreover, and the convergence rate is not constrained to the states' dimensions.

The particle filter is an integration of the Bayesian optimal filtering and the Monte Carlo sampling. It eliminates the assumption that the control system is linear and Gaussian. However, this filtering approach has the following three drawbacks, which have a negative effect on its estimation performance.

- *Particle depletion problem due to resampling process*: Particle degradation indicates that particle loses diversity as the iterations increase.

- *Real-time problem*: Compared with the Kalman filter, the particle filter has inferior performance in real-time. For MIMO control systems, as the state dimension increases, the number of sample points chosen to describe the posterior probability density will increase greatly. This finally affects the calculation efficiency of the particle filter.
- *Selection problem of important density function*: It needs to estimate the posterior probability density function. The posterior probability density function of states can not be directly sampled.

To avoid the first drawback, many improved filters have been proposed, including the regularized particle filter (Musso et al., 2001) and the intelligence optimization resampling-based particle filter (Haug, 2005). To solve the real-time problem, a series of improvements have been investigated, such as the marginalized particle filter (Schon et al., 2005), the Rao-Blackwellised particle filter (Baziw, 2005), and the adaptive particle filter (Fox, 2002). Motivated by overcoming the third drawback, a re-sampling approach can be adopted. This approach includes two steps: The first is to define a new density function that can cover all property of the original function, which is usually called the importance density function. The second step is to re-sample that importance density function. Generally, the appropriate selection of that density function contributes to reduce particle depletion and improve estimation accuracy. Hence, several filters such as the auxiliary variable-based particle filtering, the extended Kalman particle filter, the sigma-point Kalman particle filters, the Gaussian particle filter, have been developed to improve the probability density function (Kotecha and Djuric, 2003a,b).

1.4.4 Robust filtering

Another alternative to the state estimation problem of nonlinear systems is the development of nonlinear filters using the concepts of robust control. This type of filters is classed as robust filters including the $\mathcal{H}_\infty$ filter (Souza et al., 1995; Li and Fu, 1997; Zhang et al., 1999), the $\mathcal{H}_2/\mathcal{H}_\infty$ filter (Wang et al., 1997; Wang and Unbehauen, 1999; Wang and Huang, 2000), the $\mathcal{L}_1$ filter (Fialho and Georgiou, 1995; Kim and Won, 2000b,a), and the $\mathcal{L}_2/\mathcal{L}_\infty$ filter (Gao and Wang, 2003b,a; Gao et al., 2003). In 1989, A. Elsayed investigated the $\mathcal{H}_\infty$ filtering problem by assuming the inputs as energy-bounded signals (Elsayed and Grimble, 1989). This assumption mainly aims to minimize the upper bounds on the $\mathcal{H}_\infty$ norm of the filtering error of dynamic systems. The investigation in Banavar and Speyer (1991); Grimble (1988); Nagpal and Khargonekar (1991); Shaked and Theodor (1992); Yaesh and Shaked (1989) can be referred as more recent development in the $\mathcal{H}_\infty$ filter. During the $\mathcal{H}_\infty$ estimation, the $\mathcal{L}_2$ gain derived from noise and estimation errors must be guaranteed to be less than a given index. The noise handled only needs to be an energy-bounded signal. The key advantages of the $\mathcal{H}_\infty$ filter are that it does not need to impose any assumption on the statistical properties of input signals, and it is robust to system uncertainty including observation noise, time-delay, and uncertain parameters (Theodor et al., 1994; Shaked, 1990).

The robust $\mathcal{H}_2$ filter is a special class of state estimators to ensure that the estimation error covariance is bounded in the presence of acceptable system uncertainties. This filter can be viewed as the extension of the Kalman filter to uncertain systems (Shaked and Souza, 1994; Xie and Soh, 1994; Souza and Shaked, 1999; Mahmoud et al., 1999; Fridman et al., 2002). If the statistics of partial noise are known and the rests are energy-bounded, the $\mathcal{H}_2/\mathcal{H}_\infty$ filter can be developed (Haddad et al., 1991). For the $\mathcal{H}_2/\mathcal{H}_\infty$ filters, the $\mathcal{H}_2$ estimation errors are minimized by guaranteeing that the errors are satisfied with the assumption of the given disturbance attenuation level.

Another robust filter, *i.e.*, the $\mathcal{L}_2/\mathcal{L}_\infty$ filter, is called as the energy-to-peak filter. The feature of this filtering strategy is that all the input and output signals are assumed to be energy-bounded and peak-bounded. A $\mathcal{L}_2/\mathcal{L}_\infty$ filter was first proposed in Wilson (1989). It was extended in Rotea (1993) with practical application considered. In Grigoriadis and Watson (1997) as well as Watson and Grigoriadis (1998), the $\mathcal{L}_2/\mathcal{L}_\infty$ performance index were studied during filter design. Another robust $\mathcal{L}_2/\mathcal{L}_\infty$ filter was seen in Palhares and Peres (2000), while the system considered was subject to uncertainty. Assuming that disturbance and filtering errors were both peak-bounded, the $\mathcal{L}_1$ filter was deduced to the peak-to-peak filter. The maximum peak-to-peak gain was defined as the performance index (Abedor et al., 1996; Vincent et al., 1996; Voulgaris, 1996).

1.4.5 Nonlinear filters applied to satellite control system

State estimation is a prerequisite of achieving high-precision and high-reliability control for in-orbital satellites. The filtering theory is therefore widely applied in the relative/absolute attitude and orbit determination or state estimation of satellite (Wu et al., 2016; Psiaki, 2004; Adnane et al., 2018). From the 1960s to the 1970s, the EKF was utilized in satellite attitude and orbit determination subsystem design (White et al., 1975; Lefferts et al., 1982; Schmidt, 2012; Chen and Liu, 2017; Xing et al., 2012). Since mathematical tools for attitude representation vary, the corresponding filters vary from the EKF-based QUEST method (Shuster, 1990), the modified Rodrigues parameters-based EKF filter (Idan, 1996; Crassidis and Markley, 1996b), the MEKF/AEKF described by the unit quaternion (Pittelkau, 2003; Bar-Itzhack et al., 1991), the backward-smoothing EKF (Psiaki, 2005) to the deterministic EKF-like estimator (Markley et al., 1994; Smith, 1995). Among them, the EKF-based state estimation algorithm is the most prevailing method and its famous applications include the Hubble space telescope.

During the recent two decades, many nonlinear filtering methods have been applied in many space missions. The UKF is one of the widely applied methods in aerospace engineering (Yokoyama, 2011; Crassidis and Markley, 2003; Soken and Hajiyev, 2013; Hajiyev and Soken, 2014; Soken and Hajiyev, 2010). It was analyzed in Crassidis and Markley (2003) that when the initial attitude errors are large, the UKF had better robustness in attitude determination than the EKF. The Monte Carlo method-based particle filter was also employed for satellite state estimation (Cheng and Crassidis, 2004, 2010; Oshman and Carmi, 2006, 2004; Carmi and Oshman, 2007). Considering the predictive filter has great capability of estimating unknown models with high accuracy, it was regarded as an important state estimation method for the satellite control system (Crassidis and Markley, 1997b; Lin and Deng, 2002). On the other hand, the robust filters have gradually gained attention since the 1980s (Xie et al., 1994). In 1993, the robust filter was first adopted for the attitude determination of the RADCAL satellite (Markley et al., 1993). In Xiong et al. (2008), the robust filtering theory was applied to calibrate sensors and correct their errors. Some robust CKF were also applied to solve the attitude estimation problem, as suggested in Garcia et al. (2019) and Huang et al. (2016). Because the Kalman filtering theory has been successfully applied to estimate satellite's attitude in practice, applying this theory to the attitude estimation problem in the presence of system uncertainty is still attracting significant attention. For example, robust Kalman filters were presented in Soken et al. (2014), Cilden-Guler et al. (2019), and Chang et al. (2017) to achieve a high-accuracy attitude estimation even in the presence of uncertainty and non-Gaussian noise.

1.5 Motivations for predictive filtering for microsatellite control systems

Uncertain parameters and external disturbance are seen in the dynamics of any microsatellite. In contrast to large satellites, actuators or sensors of microsatellite are not precisely configurated. They have a misalignment with low calibration accuracy. Moreover, due to the use of COTS or MEMS components, fault or failure would occur in actuators and sensors. As a result, a microsatellite and its dynamics, as well as measurement are characterized by

- *Severe system uncertainty is introduced by external disturbance, uncertain parameters, and unknown torque or force due to actuator misalignment and faults.*
- *Measurement noise is non-Gaussian white noise.*
- *The embedded computer has few onboard computation resources and capability.*

Due to cost limitations, it is impossible and not worth to spend a huge amount of money on establishing an accurate model for microsatellite. The following challenges will be raised in microsatellite state estimation.

- *The above characteristics let the Kalman and its extended version such as the EKF, the UKF, and the CDKF, be inapplicable for state estimation of the microsatellite control system. That is because the Kalman-type filters require an accurate system model and they can handle Gaussian white measurement noise only. Moreover, they are inefficient to deal with system uncertainty.*

- *Although the existing predictive filters have high calculation efficiency and can achieve high-performance state estimation for microsatellite even in the presence of system uncertainty, its implementation severely depends on the estimation weighting matrix. Moreover, the predictive filters ensure a slow convergence rate of state estimation.*
- *Particle filters are available for state estimation of nonlinear non-Gaussian systems. However, their implementation consumes expensive computational resources. Moreover, they are difficult to satisfy the real-time requirement on state estimation. Particle filters are therefore not applicable for the microsatellite control system to achieve state estimation.*

Motivated by solving the above three challenges, this book deals with the design of predictive filters for the microsatellite control system. This book focuses on designing advanced techniques to estimate the microsatellite control system's states including the attitude, the angular velocity, the position or relative position, and the velocity or relative velocity. Real-time and robustness capability, estimation accuracy, as well as less computation consumption are three main requirements for the filtering algorithms developed.

1.6 What is in this book

The book is organized as ten chapters, including three parts of technical results (totally eight chapters). Part I includes Chapters 2–3 with the classical predictive filter briefly introduced and the microsatellite control system mathematically modeled. Part II and Part III concentrate on the development of new predictive filtering approaches together with the application of these approaches to the microsatellite control system. A graphical representation of this book is shown in Fig. 1.9.

Chapter 1 gives an overview of the microsatellite control system and the state estimation. It briefly introduces a literature review about filter-based state estimation methodologies already developed for linear and nonlinear systems as well as microsatellite control system.

Part I, consisting of two chapters, is intended for providing preliminary for the filtering approaches design in the rest of the book. Chapter 2 begins with some basic mathematical definitions related to the probability theory. Then, it briefly introduces the basic theory of the classical predictive filter. Chapter 3 establishes a mathematical model for the microsatellite control system, which is used for filter design and verification.

Part II focuses on developing new predictive filtering approaches for a general class of nonlinear systems by using sigma point techniques. Chapter 4 presents an unscented predictive filtering algorithm. Chapter 5 is concerned with developing a central difference predictive filter by using Stirling's polynomial interpolation technique. Chapter 6 discusses another filtering approach design by using cubature rules. The filters synthesized in Chapter 4, Chapter 5, and

Chapter 6 are applied in a microsatellite control system with simulation results shown to demonstrate their effectiveness.

Part III is dedicated to solve the drawbacks of the filters in Chapter 4, Chapter 5, and Chapter 6 by integrating the classical predictive filter with the variable structure control theory. In Chapter 7, this integration is carried out first to develop a predictive variable structure filter. Chapter 8 focuses on improving the filter presented in Chapter 7 by synthesizing a predictive adaptive variable structure filtering method. In Chapter 9, a predictive high-order variable structure filtering approach is designed with the filtering methods in Chapter 7 and Chapter 8 improved. In those three chapters, the developed filter is validated to be applicable to solve the state estimation problem of the microsatellite control system.

Chapter 10 concludes this book with the presented filtering methods and some future work summarized.

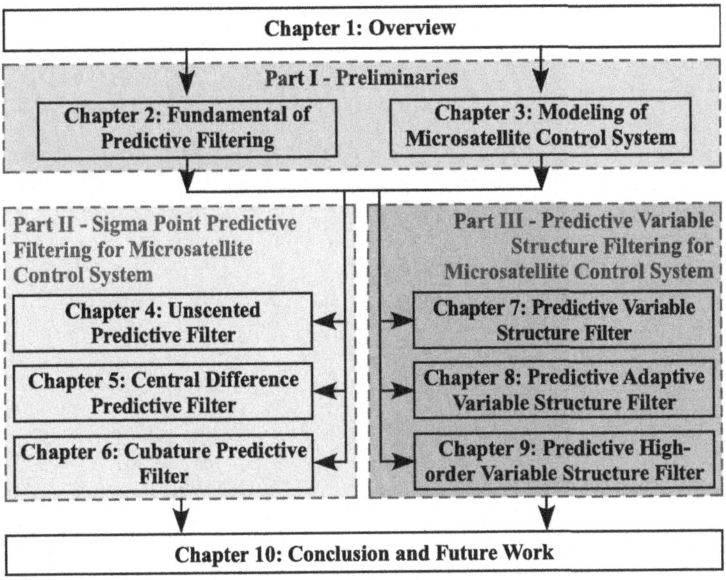

FIGURE 1.9 Block diagram of this book.

Part I

Preliminaries

Chapter 2

Fundamental of predictive filtering

2.1 Introduction

The goal of filtering is to extract valued information or signals from measured output signals, which are contaminated by noise. The early signal filters are designed in the frequency domain. The signal and the noise spectra are assumed to be non-overlapping, *i.e.*, those two signals are in different frequency bands. In this case, the filter is to let the desired signals to pass, while the unwanted noise or some frequency bands are attenuated. The classical frequency-domain filtering methodologies for signal processing could be the Chebyshev filter (Oppenheim et al., 1999), the band pass or stop filter, the low-pass or high-pass filter (Porat, 1997), and other advanced filters such as the wavelet-based denoising method (Hussein et al., 2015, 2007).

If the measurement noise and the valued signals are overlapped in the spectrum, the frequency-domain filters are unable to separate the valued signals from the noise. In such a situation, the statistical ideas dated back to the 1940s are available to design time-domain filters. In contrast to the frequency-domain filters, the time-domain filters focus on the differences in the statistical properties of the valued signals and the noise. Since the well-known time-domain Kalman filtering theory was founded, many time-domain filters have been presented. Recognized in Section 1.4.2 that the predictive filter has many advantages for state estimation of nonlinear non-Gaussian systems despite modeling error, its design has witnessed significant development (Lu, 1995, 1994). In the predictive filter, the modeling error is estimated and compensated online. This method ensures its state estimation performance to be robust to modeling error. Driven by this, the main focus of this book is the advanced predictive filters design for estimating the microsatellite control system's states. It is necessary to preliminarily introduce some basic concepts of this filtering method.

2.2 Basic probability of random vector

Because the predictive filtering method is developed based on the statistics, it is useful to concisely give the basic probability of random vectors in this book. Readers could find details in standard textbooks on probability and stochastic systems such as Leon-Garcia (2008) and Papoulis and Pillai (2002).

Predictive Filtering for Microsatellite Control System. https://doi.org/10.1016/B978-0-12-821865-5.00015-2

2.2.1 Random vector

In Probability Theory, a random vector is defined as $\boldsymbol{X} = [X_1 \quad X_2 \quad \cdots \quad X_n]^{\mathrm{T}}$, where X_i is a random variable, $i = 1, 2, \cdots, n$. Its joint cumulative distribution function is defined as the probability of an n-dimensional semi-infinite rectangle associated with the sample point $\boldsymbol{x} = [x_1 \quad x_2 \quad \cdots \quad x_n]^{\mathrm{T}}$. It is given by

$$\boldsymbol{F}_{\boldsymbol{X}}(\boldsymbol{x}) = \boldsymbol{F}_{\boldsymbol{X}}(x_1, x_2, \cdots, x_n) = \mathbb{P}(X_1 \leq x_1, X_2 \leq x_2, \cdots, X_n \leq x_n) \quad (2.1)$$

If $\boldsymbol{X}$ is continuous random vector, then its joint probability density function is defined as

$$\boldsymbol{f}_{\boldsymbol{X}}(\boldsymbol{x}) = \frac{\partial^n}{\partial x_1 \cdots \partial x_n} \boldsymbol{F}_{\boldsymbol{X}}(x_1, x_2, \cdots, x_n) \quad (2.2)$$

For a discrete vector $\boldsymbol{X}$, its joint probability mass function is defined as

$$\mathcal{P}_{\boldsymbol{X}}(\boldsymbol{x}) = \mathbb{P}(X_1 = x_1, X_2 = x_2, \cdots, X_n = x_n) \quad (2.3)$$

The joint probability density function and the joint probability mass function satisfy

$$\int_{-\infty}^{\infty} \cdots \int_{-\infty}^{\infty} \boldsymbol{f}_{\boldsymbol{X}}(\boldsymbol{x}) dx_1 \cdots dx_n = 1 \quad (2.4)$$

$$\sum_{x_1} \cdots \sum_{x_n} \mathcal{P}_{\boldsymbol{X}}(\boldsymbol{x}) = 1 \quad (2.5)$$

Moreover, for all continuous random vectors, $\boldsymbol{f}_{\boldsymbol{X}}(\boldsymbol{x})$ is such that

$$\boldsymbol{F}_{\boldsymbol{X}}(\boldsymbol{x}) = \int_{-\infty}^{x_1} \int_{-\infty}^{x_2} \cdots \int_{-\infty}^{x_n} \boldsymbol{f}_{\boldsymbol{X}}(\tau_1, \tau_2, \cdots, \tau_n) d\tau_1 d\tau_2 \cdots d\tau_n \quad (2.6)$$

Let $\boldsymbol{X} = [X_1 \quad X_2 \quad \cdots \quad X_n]^{\mathrm{T}}$ and $\boldsymbol{Y} = [Y_1 \quad Y_2 \quad \cdots \quad Y_m]^{\mathrm{T}}$ be two random vectors having n and m components, respectively. The joint cumulative distribution function of $\boldsymbol{X}$ and $\boldsymbol{Y}$ at $\boldsymbol{x} = [x_1 \quad x_2 \quad \cdots \quad x_n]^{\mathrm{T}}$ and $\boldsymbol{y} = [y_1 \quad y_2 \quad \cdots \quad y_m]^{\mathrm{T}}$ is defined as

$$\boldsymbol{F}_{\boldsymbol{X},\boldsymbol{Y}}(\boldsymbol{x}, \boldsymbol{y}) = \mathbb{P}(X_1 \leq x_1, \cdots, X_n \leq x_n, Y_1 \leq y_1, \cdots, Y_m \leq y_m) \quad (2.7)$$

When $\boldsymbol{X}$ and $\boldsymbol{Y}$ are discrete, we define the joint probability mass function of $\boldsymbol{X}$ and $\boldsymbol{Y}$ as

$$\mathcal{P}_{\boldsymbol{X},\boldsymbol{Y}}(\boldsymbol{x}, \boldsymbol{y}) = \mathbb{P}(X_1 = x_1, \cdots, X_n = x_n, Y_1 = y_1, \cdots, Y_m = y_m) \quad (2.8)$$

and their marginal probability functions as

$$\mathcal{P}_{\boldsymbol{X},\boldsymbol{X}}(\boldsymbol{x}) = \sum_{y_1} \cdots \sum_{y_m} \mathcal{P}_{\boldsymbol{X},\boldsymbol{Y}}(\boldsymbol{x}, \boldsymbol{y}) \quad (2.9)$$

$$\mathcal{P}_{\boldsymbol{Y},\boldsymbol{Y}}(\boldsymbol{y}) = \sum_{x_1} \cdots \sum_{x_n} \mathcal{P}_{\boldsymbol{X},\boldsymbol{Y}}(\boldsymbol{x}, \boldsymbol{y}) \quad (2.10)$$

For continuous $\boldsymbol{X}$ and $\boldsymbol{Y}$, we define their joint probability density function as

$$f_{X,Y}(\boldsymbol{x}, \boldsymbol{y}) = \frac{\partial^{n+m}}{\partial x_1 \cdots \partial x_n \partial y_1 \cdots \partial y_m} F_{X,Y}(\boldsymbol{x}, \boldsymbol{y}) \tag{2.11}$$

and their marginal probability functions as

$$f_{X,X}(\boldsymbol{x}) = \int_{-\infty}^{\infty} \cdots \int_{-\infty}^{\infty} f_{X,Y}(\boldsymbol{x}, \boldsymbol{y}) dy_1 dy_2 \cdots dy_m \tag{2.12}$$

$$f_{Y,Y}(\boldsymbol{y}) = \int_{-\infty}^{\infty} \cdots \int_{-\infty}^{\infty} f_{X,Y}(\boldsymbol{x}, \boldsymbol{y}) dx_1 dx_2 \cdots dx_n \tag{2.13}$$

Then, the random vectors $\boldsymbol{X}$ and $\boldsymbol{Y}$ are *independent* if and only if

$$\begin{cases} \mathcal{P}_{X,Y}(\boldsymbol{x}, \boldsymbol{y}) = \mathcal{P}_{X,X}(\boldsymbol{x}) \mathcal{P}_{Y,Y}(\boldsymbol{y}), & \text{if } \boldsymbol{X} \text{ and } \boldsymbol{Y} \text{ are discrete} \\ f_{X,Y}(\boldsymbol{x}, \boldsymbol{y}) = f_{X,X}(\boldsymbol{x}) f_{Y,Y}(\boldsymbol{y}), & \text{if } \boldsymbol{X} \text{ and } \boldsymbol{Y} \text{ are continuous} \end{cases} \tag{2.14}$$

2.2.2 Mean vector

The expectation or mean of a random vector $\boldsymbol{X} = [X_1 \quad X_2 \quad \cdots \quad X_n]^{\mathrm{T}}$ is denoted as $\mathbb{E}(\boldsymbol{X})$. It is defined as the vector of expectations of the corresponding random variable, *i.e.*,

$$\mathbb{E}(\boldsymbol{X}) = [\mathbb{E}(X_1) \quad \mathbb{E}(X_2) \quad \cdots \quad \mathbb{E}(X_n)]^{\mathrm{T}} \in \mathbb{R}^n \tag{2.15}$$

where

$$\mathbb{E}(X_i) = \begin{cases} \int_{-\infty}^{\infty} \cdots \int_{-\infty}^{\infty} \tau_i f_X d\tau_1 d\tau_2 \cdots d\tau_n, & \text{if } \boldsymbol{X} \text{ is continuous} \\ \sum x_i \mathcal{P}_X(x_1, x_2, \cdots, x_n), & \text{if } \boldsymbol{X} \text{ is discrete} \end{cases} \tag{2.16}$$

2.2.3 Covariance matrix

Suppose that $\boldsymbol{X} = [X_1 \quad X_2 \quad \cdots \quad X_n]^{\mathrm{T}}$ and $\boldsymbol{Y} = [Y_1 \quad Y_2 \quad \cdots \quad Y_m]^{\mathrm{T}}$ are two random vectors, the covariance of those two random vectors is defined as

$$\mathrm{Cov}(\boldsymbol{X}, \boldsymbol{Y}) = \mathbb{E}((\boldsymbol{X} - \mathbb{E}(\boldsymbol{X}))(\boldsymbol{Y} - \mathbb{E}(\boldsymbol{Y}))^{\mathrm{T}}) \tag{2.17}$$

which is a matrix with a dimension of $n \times m$.

The variance matrix of the random vector $\boldsymbol{X} = [X_1 \quad X_2 \quad \cdots \quad X_n]^{\mathrm{T}}$ is the $n \times n$ matrix defined by

$$\mathrm{Var}(\boldsymbol{X}) = \mathrm{Cov}(\boldsymbol{X}, \boldsymbol{X}) = \mathbb{E}((\boldsymbol{X} - \mathbb{E}(\boldsymbol{X}))((\boldsymbol{X} - \mathbb{E}(\boldsymbol{X}))^{\mathrm{T}}) \tag{2.18}$$

The covariance of any two variables X_i and Y_j are defined as

$$\mathrm{Cov}(X_i, Y_j) = \mathbb{E}((X_i - \mathbb{E}(X_i))(Y_j - \mathbb{E}(Y_j))) \tag{2.19}$$

which reflects how much X_i and Y_j change together.

From the definitions given in (2.17) and (2.19), it can be further calculated that

$$\begin{aligned}\mathrm{Cov}(\boldsymbol{X}, \boldsymbol{Y}) &= \mathbb{E}(\boldsymbol{X}\boldsymbol{Y}^{\mathrm{T}} - \boldsymbol{X}\mathbb{E}(\boldsymbol{Y})^{\mathrm{T}} - \mathbb{E}(\boldsymbol{X})\boldsymbol{Y}^{\mathrm{T}} + \mathbb{E}(\boldsymbol{X})\mathbb{E}(\boldsymbol{Y})^{\mathrm{T}}) \\ &= \mathbb{E}(\boldsymbol{X}\boldsymbol{Y}^{\mathrm{T}}) - 2\mathbb{E}(\boldsymbol{X})\mathbb{E}(\boldsymbol{Y})^{\mathrm{T}} + \mathbb{E}(\boldsymbol{X})\mathbb{E}(\boldsymbol{Y})^{\mathrm{T}} \\ &= \mathbb{E}(\boldsymbol{X}\boldsymbol{Y}^{\mathrm{T}}) - \mathbb{E}(\boldsymbol{X})\mathbb{E}(\boldsymbol{Y})^{\mathrm{T}}\end{aligned} \tag{2.20}$$

$$\mathrm{Cov}(X_i, Y_j) = \mathbb{E}(X_i Y_j) - \mathbb{E}(X_i)\mathbb{E}(Y_j) \tag{2.21}$$

2.2.4 Normal distribution random vector

A random vector $\boldsymbol{X} = [X_1 \quad X_2 \quad \cdots \quad X_n]^{\mathrm{T}}$ is said to be the normal distribution random vector if its joint probability density function is given by

$$f_{\boldsymbol{X}}(\boldsymbol{x}) = \frac{1}{(2\pi)^{\frac{n}{2}}\sqrt{\det(\boldsymbol{\Sigma}_{\boldsymbol{X}})}} \exp\left(-\frac{1}{2}(\boldsymbol{x} - \boldsymbol{X}_0)^{\mathrm{T}}\boldsymbol{\Sigma}_{\boldsymbol{X}}^{-1}(\boldsymbol{x} - \boldsymbol{X}_0)\right) \tag{2.22}$$

where $\boldsymbol{x} = [x_1 \quad x_2 \quad \cdots \quad x_n]^{\mathrm{T}}$, $\boldsymbol{X}_0 = \mathbb{E}(\boldsymbol{X})$, and $\boldsymbol{\Sigma}_{\boldsymbol{X}} = \mathrm{Cov}(\boldsymbol{X})$. It is known that $\boldsymbol{\Sigma}_{\boldsymbol{X}} = \boldsymbol{\Sigma}_{\boldsymbol{X}}^{\mathrm{T}}$ and $\boldsymbol{\Sigma}_{\boldsymbol{X}} > \boldsymbol{0}$.

The normal distribution random vector is also called the jointly Gaussian random vector and written as $\boldsymbol{X} \sim \mathcal{N}(\boldsymbol{X}_0, \boldsymbol{\Sigma}_{\boldsymbol{X}})$. It is completely specified by the individual mean $\boldsymbol{X}_0$ and the pairwise covariance $\boldsymbol{\Sigma}_{\boldsymbol{X}}$. $\boldsymbol{X}_0$ indicates where the peak of the density $f_{\boldsymbol{X}}(\boldsymbol{x})$ lies at. $\boldsymbol{\Sigma}_{\boldsymbol{X}}$ indicates the spreading distribution. In the one dimensional case, *i.e.*, $n = 1$, $\boldsymbol{X}$ is known as the Gaussian distribution, which is a very important class of statistical distribution.

In particular, suppose that $\boldsymbol{A} \in \mathbb{R}^{m\times n}$ is a constant matrix with $\mathrm{rank}(\boldsymbol{A}) = m \leq n$ and $\boldsymbol{b} \in \mathbb{R}^m$ is constant vector, given an n-dimensional random vector $\boldsymbol{X} \sim \mathcal{N}(\boldsymbol{X}_0, \boldsymbol{\Sigma}_{\boldsymbol{X}})$, then $\boldsymbol{Y} = \boldsymbol{A}\boldsymbol{X} + \boldsymbol{b}$ is also a random vector and $\boldsymbol{Y} \sim \mathcal{N}(\boldsymbol{A}\boldsymbol{X}_0 + \boldsymbol{b}, \boldsymbol{A}\boldsymbol{\Sigma}_{\boldsymbol{X}}\boldsymbol{A}^{\mathrm{T}})$.

2.2.5 White noise process

Let $\{\boldsymbol{w}(t), t \in \mathbb{T}\}$ be a stochastic process, which is family of random variables with the discrete index set $\mathbb{T} = \mathbb{N}_0$ or the continuous index set $\mathbb{T} = [0, \infty)$. Mathematically, a stochastic process $\boldsymbol{w}(t)$ is said to an white noise process, if it is continuous and satisfies

$$\mathbb{E}(\boldsymbol{w}(t)) = \boldsymbol{0} \tag{2.23}$$

$$\mathbb{E}(\boldsymbol{w}(t_{k_1})\boldsymbol{w}(t_{k_2})^{\mathrm{T}}) = \frac{\boldsymbol{C}_0}{2}\delta_{k_1 k_2} \tag{2.24}$$

where $k_1 \in \mathbb{N}_0$, $k_2 \in \mathbb{N}_0$, and $\boldsymbol{C}_0$ is constant matrix with appropriate dimension.

It is implied from (2.23) and (2.24) that the process $\boldsymbol{w}(t)$ has zero mean everywhere and infinite power at zero time shift. The white noise process is therefore purely a theoretical and an ideal signal with its power spectrum distributed uniformly in the whole frequency domain. In practice, it is impossible to generate such a process, because this process is actually a specific signal whose total power is infinite. However, a white signal with a flat spectrum could be reasonably assumed within a finite frequency range.

2.3 Preliminary definitions and lemma

In this section, some mathematical definitions and lemmas are introduced, which are useful for analyzing the state estimation performance of the filters presented in the rest of this book.

Definition 2.1. (Agniel and Jury, 1971) The stochastic process $\boldsymbol{\xi}_k$ is exponentially bounded in the mean square sense, if there are three real numbers $\eta \in \mathbb{R}_+$, $\upsilon \in \mathbb{R}_+$, $\vartheta \in \mathbb{R}_+$, and $0 < \vartheta < 1$ such that

$$\mathbb{E}(||\boldsymbol{\xi}_k||^2) \leq \eta||\boldsymbol{\xi}_0||^2\vartheta^k + \upsilon \tag{2.25}$$

is satisfied for any $k \geq 0$.

Definition 2.2. (Tarn and Rasis, 1976) The stochastic process $\boldsymbol{\xi}_k$ is said to be bounded with probability one if

$$\mathbb{P}\left(\sup_{k\geq 0}||\boldsymbol{\xi}_k|| < \infty\right) = 1 \tag{2.26}$$

Lemma 2.1. *(Reif et al., 1999) Suppose that there exist a stochastic process $V_k(\boldsymbol{\xi}_k)$ and four real numbers $\underline{\upsilon} \in \mathbb{R}_+$, $\bar{\upsilon} \in \mathbb{R}_+$, $\mu \in \mathbb{R}_+$, $\alpha \in \mathbb{R}_+$, and $0 < \alpha < 1$ such that*

$$\underline{\upsilon}||\boldsymbol{\xi}_k||^2 \leq V_k(\boldsymbol{\xi}_k) \leq \bar{\upsilon}||\boldsymbol{\xi}_k||^2 \tag{2.27}$$

and

$$\mathbb{E}(V_{k+1}(\boldsymbol{\xi}_{k+1})|\boldsymbol{\xi}_k) - V_k(\boldsymbol{\xi}_k) \leq \mu - \alpha V_k(\boldsymbol{\xi}_k) \tag{2.28}$$

are satisfied. Then, the stochastic process $\boldsymbol{\xi}_k$ is exponentially bounded in mean square sense, and the following inequality holds for $k \geq 0$.

$$\mathbb{E}(||\boldsymbol{\xi}||^2) \leq \frac{\bar{\upsilon}}{\underline{\upsilon}}\mathbb{E}(||\boldsymbol{\xi}_0||^2)(1-\alpha)^k + \frac{\mu}{\underline{\upsilon}}\sum_{i=1}^{k-1}(1-\alpha)^i \tag{2.29}$$

Moreover, the stochastic process $\boldsymbol{\xi}_k$ is bounded with probability one.

Remark 2.1. Following Tarn and Rasis (1976) and applying the inequality:

$$\sum_{i=1}^{k-1}(1-\alpha)^i \leq \sum_{i=1}^{\infty}(1-\alpha)^i = \frac{1}{\alpha} \tag{2.30}$$

we can rewrite the inequality (2.29) as

$$\mathbb{E}(||\boldsymbol{\xi}||^2) \leq \frac{\bar{\upsilon}}{\underline{\upsilon}}\mathbb{E}(||\boldsymbol{\xi}_0||^2)(1-\alpha)^k + \frac{\mu}{\underline{\upsilon}\alpha} \tag{2.31}$$

2.4 Desired properties of filter

The aim of developing filters for nonlinear control systems including the microsatellite control system is to estimate the real information of their states. Let $\hat{\boldsymbol{x}}(t)$ generated by the developed filter be the estimated value of the system states $\boldsymbol{x}(t)$, then $\boldsymbol{x}_e(t) = \hat{\boldsymbol{x}}(t) - \boldsymbol{x}(t)$ represents the state estimation error. In this book, the following four indices or desired properties are introduced to evaluate the state estimation performance of the developed filters.

- *Stability*: This is the basic performance index that any filter should guarantee. It is also the basic desired property that any filter should have. It means that $\lim_{t\to\infty} \boldsymbol{x}_e(t)$ should exit, *i.e.*, $\boldsymbol{x}_e(t)$ is stable in the sense of the Lyapunov stability theory. Otherwise, the filter is viewed to diverge, which lets the filter be inapplicable.
- *Estimation accuracy*: This performance index is used to evaluate how well the estimation $\hat{\boldsymbol{x}}(t)$ can approximate the real states $\boldsymbol{x}(t)$. The smaller $\boldsymbol{x}_e(t)$ is, the higher estimation accuracy is achieved by the developed filter. If $\lim_{t\to\infty} \boldsymbol{x}_e(t) = \mathbf{0}$, then the estimation $\hat{\boldsymbol{x}}(t)$ will fully estimate or reconstruct the real system states.
- *Robustness*: This desired property is to evaluate how well the filter can handle system uncertainty, unmodeling error, and measurement noise. If the filter developed has more capability of dealing with those issues, then we call that the filter has more robustness.
- *Real-time property*: Almost all the practical control systems run in real-time. This requires the developed filter to estimate the real system states in real-time rather than off-line.
- *Consumption of computation resource*: In practice, filters are constantly implemented in an embedded computer with limited computation resources. Hence, it demands the developed filters to be user friendly with high computation efficiency. Then, few computation resources are consumed. For example, the computation capability of the embedded computer in a microsatellite is very limited. If a filtering approach needs expensive computation resources, then its on-line implementation in the microsatellite will be out of the embedded computer's memory.

2.5 Basic theory of predictive filtering

Consider a class of nonlinear stochastic discrete-time systems with their mathematical model described by

$$\boldsymbol{x}_{k+1} = \boldsymbol{f}(\boldsymbol{x}_k) + \boldsymbol{g}(\boldsymbol{x}_k)\boldsymbol{d}_k \tag{2.32}$$

$$\boldsymbol{y}_k = \boldsymbol{h}(\boldsymbol{x}_k) + \boldsymbol{v}_k \tag{2.33}$$

where $k \in \mathbb{N}_0$ is the discrete time index, $\boldsymbol{x}_k \in \mathbb{R}^n$ is the system state at t_k with mean $\underline{\boldsymbol{x}}_k \in \mathbb{R}^n$ and covariance $\boldsymbol{P}_{x,k} \in \mathbb{R}^{n\times n}$, $\boldsymbol{y}_k \in \mathbb{R}^m$ is the system measurement output at t_k, $\boldsymbol{f}(\boldsymbol{x}_k) = [f_1(\boldsymbol{x}_k) \quad f_2(\boldsymbol{x}_k) \quad \cdots \quad f_n(\boldsymbol{x}_k)]^{\mathrm{T}} \in \mathbb{R}^n$ is a sufficiently differentiable function vector, $\boldsymbol{g}(\boldsymbol{x}_k) = [\boldsymbol{g}_1(\boldsymbol{x}_k) \quad \boldsymbol{g}_2(\boldsymbol{x}_k) \quad \cdots \quad \boldsymbol{g}_l(\boldsymbol{x}_k)] \in \mathbb{R}^{n\times l}$, $\boldsymbol{g}_i(\boldsymbol{x}_k) \in \mathbb{R}^n$ denotes the modeling error distribution matrix determining how the modeling error is introduced to the system dynamics, $\boldsymbol{h}(\boldsymbol{x}_k) = [h_1(\boldsymbol{x}_k) \quad h_2(\boldsymbol{x}_k) \quad \cdots \quad h_n(\boldsymbol{x}_k)]^{\mathrm{T}} \in \mathbb{R}^m$ is a sufficiently differentiable function, $\boldsymbol{d}_k \in \mathbb{R}^l$ denotes the modeling error, and $\boldsymbol{v}_k \in \mathbb{R}^m$ represents the measurement noise, which is assumed to be zero-mean Gaussian white-noise distributed process with

$$\mathbb{E}(\boldsymbol{v}_k) = \boldsymbol{0} \tag{2.34}$$

$$\mathbb{E}(\boldsymbol{v}_k \boldsymbol{v}_l^{\mathrm{T}}) = \boldsymbol{R}_k \delta_{kl} \tag{2.35}$$

where $\boldsymbol{R}_k \in \mathbb{R}^{m\times m}$ is a positive-definite measurement covariance matrix.

2.5.1 Derivation of predictive filter

For the nonlinear stochastic discrete system (2.32)–(2.33), it is assumed that the estimation of its states and output are determined by a preliminary model given by

$$\hat{\boldsymbol{x}}_{k+1} = \boldsymbol{f}(\hat{\boldsymbol{x}}_k) + \boldsymbol{g}(\hat{\boldsymbol{x}}_k)\hat{\boldsymbol{d}}_k \tag{2.36}$$

$$\hat{\boldsymbol{y}}_k = \boldsymbol{h}(\hat{\boldsymbol{x}}_k) \tag{2.37}$$

where $\hat{\boldsymbol{x}}_k$ and $\hat{\boldsymbol{y}}_k$ are the estimation of the state $\boldsymbol{x}_k$ and the output $\boldsymbol{y}_k$, respectively. $\hat{\boldsymbol{d}}_k$ is the to-be-determined estimation of the modeling error $\boldsymbol{d}_k$.

Based on (2.32)–(2.33), expanding the measurement output $\boldsymbol{y}_k$ about $\boldsymbol{x}_k$ in a Taylor series from one sampling instant to the next yields

$$\boldsymbol{y}_{k+1} = \boldsymbol{y}_k + \boldsymbol{Z}(\boldsymbol{x}_k, \Delta t) + \boldsymbol{\Lambda}(\Delta t)\boldsymbol{U}(\boldsymbol{x}_k)\boldsymbol{d}_k + \boldsymbol{\varepsilon}(\boldsymbol{x}_k, \boldsymbol{d}_k) + \boldsymbol{v}_k \tag{2.38}$$

with the sampling interval $\Delta t = t_{k+1} - t_k$ and the high-order discretization error $\boldsymbol{\varepsilon}(\boldsymbol{x}_k, \boldsymbol{d}_k) \in \mathbb{R}^m$. Moreover, expanding (2.37) about the state estimation $\hat{\boldsymbol{x}}_k$ in a

Taylor series and ignoring the high-order term $\boldsymbol{\varepsilon}(\hat{\boldsymbol{x}}_k, \hat{\boldsymbol{d}}_k)$ lead to

$$\hat{\boldsymbol{y}}_{k+1} = \hat{\boldsymbol{y}}_k + \boldsymbol{Z}(\hat{\boldsymbol{x}}_k, \Delta t) + \boldsymbol{\Lambda}(\Delta t)\boldsymbol{U}(\hat{\boldsymbol{x}}_k)\hat{\boldsymbol{d}}_k \tag{2.39}$$

where $\boldsymbol{\Lambda}(\Delta t) \in \mathbb{R}^{m\times m}$ is a diagonal matrix given by

$$\boldsymbol{\Lambda}(\Delta t) = \begin{bmatrix} \frac{(\Delta t)^{p_1}}{p_1!} & 0 & \cdots & 0 \\ 0 & \frac{(\Delta t)^{p_2}}{p_2!} & \cdots & 0 \\ \vdots & \vdots & \ddots & \vdots \\ 0 & 0 & \cdots & \frac{(\Delta t)^{p_m}}{p_m!} \end{bmatrix} \tag{2.40}$$

Here, $p_i \in \mathbb{R}_+$ is the lowest order of the derivative of $h_i(\hat{\boldsymbol{x}}_k)$ in which any component of the modeling error $\boldsymbol{d}_k$ first appears. $\boldsymbol{Z}(\hat{\boldsymbol{x}}_k, \Delta t) \in \mathbb{R}^m$ is a vector given by $\boldsymbol{Z}(\hat{\boldsymbol{x}}_k, \Delta t) = [Z_1(\hat{\boldsymbol{x}}_k, \Delta t) \quad Z_2(\hat{\boldsymbol{x}}_k, \Delta t) \quad \cdots \quad Z_m(\hat{\boldsymbol{x}}_k, \Delta t)]^{\mathrm{T}}$ with

$$Z_i(\hat{\boldsymbol{x}}_k, \Delta t) = \sum_{j=1}^{p_i} \frac{\Delta t^j}{j!} L_f^j(h_i(\hat{\boldsymbol{x}}_k)),\ i = 1, 2, \cdots, m \tag{2.41}$$

and $L_f^j(h_i(\hat{\boldsymbol{x}}_k))$, $i = 1, 2, \cdots, m$, being the kth-order Lie-derivative defined by

$$L_f^j(h_i(\hat{\boldsymbol{x}}_k)) = \begin{cases} h_i(\hat{\boldsymbol{x}}_k),\ j = 0 \\ \dfrac{\partial L_f^{j-1} h_i(\hat{\boldsymbol{x}}_k)}{\partial \hat{\boldsymbol{x}}} \boldsymbol{f},\ j = 1, 2, \cdots, p_i \end{cases} \tag{2.42}$$

The matrix $\boldsymbol{U}(\hat{\boldsymbol{x}}_k) = [\boldsymbol{U}_1(\hat{\boldsymbol{x}}_k) \quad \boldsymbol{U}_2(\hat{\boldsymbol{x}}_k) \quad \cdots \quad \boldsymbol{U}_m(\hat{\boldsymbol{x}}_k)]^{\mathrm{T}} \in \mathbb{R}^{m\times l}$ is a matrix with $\boldsymbol{U}_i(\hat{\boldsymbol{x}}_k) = [U_{i1}(\hat{\boldsymbol{x}}_k) \quad U_{i2}(\hat{\boldsymbol{x}}_k) \quad \cdots \quad U_{il}(\hat{\boldsymbol{x}}_k)]^{\mathrm{T}} \in \mathbb{R}^{m\times l}$, $i = 1, 2, \cdots, m$, where

$$U_{ij}(\hat{\boldsymbol{x}}_k) = L_{g_j}(L_f^{p_i-1}(h_i(\hat{\boldsymbol{x}}_k))) = \frac{\partial L_f^{p_i-1}(h_i(\hat{\boldsymbol{x}}_k))}{\partial \bar{\boldsymbol{x}}} \boldsymbol{g}_j,\ j = 1, 2, \cdots, l \tag{2.43}$$

Based on the Minimum Modeling Error (MME) estimation principle, a cost function consisting of the weighted sum square of the measurement-minus-estimate residuals plus the weighted sum square of the model correction term to be minimized, is given by

$$J(\boldsymbol{d}_k) = \frac{1}{2}(\boldsymbol{y}_{k+1} - \hat{\boldsymbol{y}}_{k+1})^{\mathrm{T}} \boldsymbol{R}_k^{-1} (\boldsymbol{y}_{k+1} - \hat{\boldsymbol{y}}_{k+1}) + \frac{1}{2}\boldsymbol{d}_k^{\mathrm{T}} \boldsymbol{W}_E \boldsymbol{d}_k \tag{2.44}$$

where $\boldsymbol{W}_E \in \mathbb{R}^{l\times l}$ is a semi-positive definite weighting matrix.

Substituting (2.39) into (2.44) and solving $\frac{\partial J(\boldsymbol{d}_k)}{\partial \boldsymbol{d}_k} = \mathbf{0}$, we can obtain the minimum modeling error estimation as

$$\begin{aligned}\hat{\boldsymbol{d}}_k = &-\left((\boldsymbol{\Lambda}(\Delta t)\boldsymbol{U}(\hat{\boldsymbol{x}}_k))^{\mathrm{T}}\boldsymbol{R}_k^{-1}\boldsymbol{\Lambda}(\Delta t)\boldsymbol{U}(\hat{\boldsymbol{x}}_k) + \boldsymbol{W}_E\right)^{-1} \\ &\times (\boldsymbol{\Lambda}(\Delta t)\boldsymbol{U}(\hat{\boldsymbol{x}}_k))^{\mathrm{T}}\boldsymbol{R}_k^{-1}(\boldsymbol{Z}(\hat{\boldsymbol{x}}_k, \Delta t) + \hat{\boldsymbol{y}}_k - \boldsymbol{y}_{k+1})\end{aligned} \tag{2.45}$$

which also satisfies the covariance constraint.

Based on the preceding calculations, the classical predictive filter (PF) design can be summarized in Algorithm 2.1. Moreover, the numerical integral method is adopted to update the state estimation in Algorithm 2.1.

Algorithm 2.1: The classical predictive filter.

Input Data: Initial estimation $\hat{\boldsymbol{x}}_0$, $\hat{\boldsymbol{y}}_0$, and measurement $\boldsymbol{y}_k$
Result: State estimation $\hat{\boldsymbol{x}}_k$
begin
  for $k = 0, 1, 2 \cdots$ **do**
    Calculating $\hat{\boldsymbol{d}}_k$ by using
    $$\begin{aligned}\hat{\boldsymbol{d}}_k = &-\left((\boldsymbol{\Lambda}(\Delta t)\boldsymbol{U}(\hat{\boldsymbol{x}}_k))^{\mathrm{T}}\boldsymbol{R}_k^{-1}\boldsymbol{\Lambda}(\Delta t)\boldsymbol{U}(\hat{\boldsymbol{x}}_k) + \boldsymbol{W}_E\right)^{-1} \\ &\times(\boldsymbol{\Lambda}(\Delta t)\boldsymbol{U}(\hat{\boldsymbol{x}}_k))^{\mathrm{T}}\boldsymbol{R}_k^{-1}(\boldsymbol{Z}(\hat{\boldsymbol{x}}_k, \Delta t) + \hat{\boldsymbol{y}}_k - \boldsymbol{y}_{k+1})\end{aligned}$$
    Updating the state and the output estimation using
    $\hat{\boldsymbol{x}}_{k+1} = \boldsymbol{f}(\hat{\boldsymbol{x}}_k) + \boldsymbol{g}(\hat{\boldsymbol{x}}_k)\hat{\boldsymbol{d}}_k$
    $\hat{\boldsymbol{y}}_k = \boldsymbol{h}(\hat{\boldsymbol{x}}_k)$
  end
end

2.5.2 Estimation performance of predictive filter

2.5.2.1 *Estimation performance of modeling error*

To quantitatively analyze the state estimation performance of the predictive filtering Algorithm 2.1, the estimated modeling error (2.45) is rewritten as

$$\hat{\boldsymbol{d}}_k = -(\underline{\boldsymbol{Z}}(\hat{\boldsymbol{x}}_k, \Delta t) + \underline{\boldsymbol{h}}(\hat{\boldsymbol{x}}_k)) \tag{2.46}$$

where $\underline{\boldsymbol{Z}}(\hat{\boldsymbol{x}}_k, \Delta t) = \boldsymbol{M}(\hat{\boldsymbol{x}}_k)\boldsymbol{Z}(\hat{\boldsymbol{x}}_k, \Delta t)$, $\underline{\boldsymbol{h}}(\hat{\boldsymbol{x}}_k) = \boldsymbol{M}(\hat{\boldsymbol{x}}_k)(\boldsymbol{h}(\hat{\boldsymbol{x}}_k) - \boldsymbol{y}_{k+1})$, and

$$\begin{aligned}\boldsymbol{M}(\hat{\boldsymbol{x}}_k) = &\Big((\boldsymbol{\Lambda}(\Delta t)\boldsymbol{U}(\hat{\boldsymbol{x}}_k))^{\mathrm{T}}\boldsymbol{R}_k^{-1}\boldsymbol{\Lambda}(\Delta t)\boldsymbol{U}(\hat{\boldsymbol{x}}_k) \\ &+ \boldsymbol{W}_E\Big)^{-1}(\boldsymbol{\Lambda}(\Delta t)\boldsymbol{U}(\hat{\boldsymbol{x}}_k))^{\mathrm{T}}\boldsymbol{R}_k^{-1}\end{aligned} \tag{2.47}$$

Similarly, using (2.38), the real modeling error $\boldsymbol{d}_k$ can be represented by

$$\boldsymbol{d}_k = -(\underline{\boldsymbol{Z}}(\boldsymbol{x}_k, \Delta t) + \underline{\boldsymbol{h}}(\boldsymbol{x}_k)) \tag{2.48}$$

where $\underline{Z}(\boldsymbol{x}_k, \Delta t) = \boldsymbol{M}(\boldsymbol{x}_k)\boldsymbol{Z}(\boldsymbol{x}_k, \Delta t)$, $\underline{\boldsymbol{h}}(\boldsymbol{x}_k) = \boldsymbol{M}(\boldsymbol{x}_k)(\boldsymbol{h}(\boldsymbol{x}_k) + \boldsymbol{v}_k - \boldsymbol{y}_{k+1})$, and

$$\begin{aligned} \boldsymbol{M}(\boldsymbol{x}_k) = \Big(& (\boldsymbol{\Lambda}(\Delta t)\boldsymbol{U}(\boldsymbol{x}_k))^{\mathrm{T}} \boldsymbol{R}_k^{-1} \boldsymbol{\Lambda}(\Delta t)\boldsymbol{U}(\boldsymbol{x}_k) \\ & + \boldsymbol{W}_E \Big)^{-1} (\boldsymbol{\Lambda}(\Delta t)\boldsymbol{U}(\boldsymbol{x}_k))^{\mathrm{T}} \boldsymbol{R}_k^{-1} \end{aligned} \tag{2.49}$$

We can apply the Taylor series expansion to (2.48). Let $\boldsymbol{x}_k = \hat{\boldsymbol{x}}_k + \Delta \boldsymbol{x}$, where $\Delta \boldsymbol{x} \in \mathbb{R}^n$ is a zero-mean random variable with covariance $\boldsymbol{P}_{x,k}$, then the Taylor series expansion of $\boldsymbol{d}_k$ about $\hat{\boldsymbol{x}}_k$ can be expressed as

$$\boldsymbol{d}_k = -\underline{\boldsymbol{Z}}(\hat{\boldsymbol{x}}_k, \Delta t) - \underline{\boldsymbol{h}}(\hat{\boldsymbol{x}}_k) - \sum_{i=1}^{\infty} \frac{\boldsymbol{D}_{\Delta x}^i \underline{\boldsymbol{Z}}}{i!} - \sum_{i=1}^{\infty} \frac{\boldsymbol{D}_{\Delta x}^i \underline{\boldsymbol{h}}}{i!} \tag{2.50}$$

where $\boldsymbol{D}_{\Delta x}^i \boldsymbol{Z} \in \mathbb{R}^l$ and $\boldsymbol{D}_{\Delta x}^i \boldsymbol{h} \in \mathbb{R}^l$ are the ith order term in the multidimensional Taylor Series, and

$$\boldsymbol{D}_{\Delta x} \underline{\boldsymbol{Z}} = \boldsymbol{G}_{\underline{Z}} \Delta \boldsymbol{x} \tag{2.51}$$

$$\boldsymbol{D}_{\Delta x} \underline{\boldsymbol{h}} = \boldsymbol{G}_{\underline{h}} \Delta \boldsymbol{x} \tag{2.52}$$

Here, $\boldsymbol{G}_{\underline{h}} \in \mathbb{R}^{l \times n}$ and $\boldsymbol{G}_{\underline{Z}} \in \mathbb{R}^{l \times n}$ are the Jacobian matrices. Moreover, $\boldsymbol{D}_{\Delta x}$ can be rewritten and interpreted as the scalar operator

$$\boldsymbol{D}_{\Delta x} = \sum_{j=1}^{n} \Delta x_j \frac{\partial}{\partial x_j} \tag{2.53}$$

which will act on $\underline{\boldsymbol{h}}(\cdot)$ and $\underline{\boldsymbol{Z}}$ on a component-by-component basis. In (2.53), Δx_j is the jth component of $\Delta \boldsymbol{x}$, and $\frac{\partial}{\partial x_j}$ is the normal partial derivative operator with respect to x_j (*i.e.*, the jth component of $\boldsymbol{x}_k$).

In accordance with (2.53), $\boldsymbol{D}_{\Delta x}^i \underline{\boldsymbol{Z}} \in \mathbb{R}^l$ and $\boldsymbol{D}_{\Delta x}^i \underline{\boldsymbol{h}} \in \mathbb{R}^l$ have the form of

$$\frac{\boldsymbol{D}_{\Delta x}^i \underline{\boldsymbol{h}}}{i!} = \frac{1}{i!} \left(\sum_{j=1}^{n} \Delta x_j \frac{\partial}{\partial x_j} \right)^i \underline{\boldsymbol{h}}(\boldsymbol{x})\big|_{\boldsymbol{x}=\hat{\boldsymbol{x}}_k}, \quad i \in \mathbb{N} \tag{2.54}$$

$$\frac{\boldsymbol{D}_{\Delta x}^i \underline{\boldsymbol{Z}}}{i!} = \frac{1}{i!} \left(\sum_{j=1}^{n} \Delta x_j \frac{\partial}{\partial x_j} \right)^i \underline{\boldsymbol{Z}}(\boldsymbol{x})\big|_{\boldsymbol{x}=\hat{\boldsymbol{x}}_k}, \quad i \in \mathbb{N} \tag{2.55}$$

Because $\boldsymbol{x}_k$ is a Gaussian random variable with zero mean and covariance $\boldsymbol{P}_{x,k}$, the distribution of $\Delta \boldsymbol{x}$ is symmetrical, such that all odd-order moments are zero, *i.e.*,

$$\mathbb{E}\left(\frac{D_{\Delta x}^{2i+1}\underline{h}}{i!}\right)=\mathbb{E}\left(\frac{D_{\Delta x}^{2i+1}\underline{Z}}{i!}\right)=\mathbf{0},\ i\in\mathbb{N}_0 \tag{2.56}$$

Invoking $\mathbb{E}(\Delta\boldsymbol{x}\Delta\boldsymbol{x}^{\mathrm{T}})=\boldsymbol{P}_{x,k}$, we can compute the expectation value of the 2nd order term as

$$\begin{aligned}\mathbb{E}\left(\frac{D_{\Delta x}^{2}\underline{h}}{2!}\right)&=\mathbb{E}\left(\frac{D_{\Delta x}(D_{\Delta x}\underline{h})}{2!}\right)=\mathbb{E}\left(\frac{((\Delta\boldsymbol{x})^{\mathrm{T}}\nabla(\Delta\boldsymbol{x})^{\mathrm{T}}\nabla)\underline{h}(\boldsymbol{x})\big|_{\boldsymbol{x}=\hat{\boldsymbol{x}}_k}}{2!}\right)\\&=\frac{(\nabla^{\mathrm{T}}\boldsymbol{P}_{x,k}\nabla)\underline{h}(\boldsymbol{x})|_{\boldsymbol{x}=\hat{\boldsymbol{x}}_k}}{2!}\end{aligned} \tag{2.57}$$

$$\mathbb{E}\left(\frac{D_{\Delta x}^{2}\underline{Z}}{2!}\right)=\frac{(\nabla^{\mathrm{T}}\boldsymbol{P}_{x,k}\nabla)\underline{Z}(\boldsymbol{x})|_{\boldsymbol{x}=\hat{\boldsymbol{x}}_k}}{2!} \tag{2.58}$$

$$\mathbb{E}((D_{\Delta x}\underline{h})(D_{\Delta x}\underline{h})^{\mathrm{T}})=\boldsymbol{G}_{\underline{h}}\mathbb{E}(\Delta\boldsymbol{x}(\Delta\boldsymbol{x})^{\mathrm{T}})\boldsymbol{G}_{\underline{h}}^{\mathrm{T}}=\boldsymbol{G}_{\underline{h}}\boldsymbol{P}_{x,k}\boldsymbol{G}_{\underline{h}}^{\mathrm{T}} \tag{2.59}$$

$$\mathbb{E}((D_{\Delta x}\underline{Z})(D_{\Delta x}\underline{Z})^{\mathrm{T}})=\boldsymbol{G}_{\underline{Z}}\mathbb{E}(\Delta\boldsymbol{x}(\Delta\boldsymbol{x})^{\mathrm{T}})\boldsymbol{G}_{\underline{Z}}^{\mathrm{T}}=\boldsymbol{G}_{\underline{Z}}\boldsymbol{P}_{x,k}\boldsymbol{G}_{\underline{Z}}^{\mathrm{T}} \tag{2.60}$$

$$\mathbb{E}((D_{\Delta x}\underline{h})(D_{\Delta x}\underline{Z})^{\mathrm{T}})=\boldsymbol{G}_{\underline{h}}\mathbb{E}(\Delta\boldsymbol{x}(\Delta\boldsymbol{x})^{\mathrm{T}})\boldsymbol{G}_{\underline{Z}}^{\mathrm{T}}=\boldsymbol{G}_{\underline{h}}\boldsymbol{P}_{x,k}\boldsymbol{G}_{\underline{Z}}^{\mathrm{T}} \tag{2.61}$$

$$\mathbb{E}((D_{\Delta x}\underline{Z})(D_{\Delta x}\underline{h})^{\mathrm{T}})=\boldsymbol{G}_{\underline{Z}}\mathbb{E}(\Delta\boldsymbol{x}(\Delta\boldsymbol{x})^{\mathrm{T}})\boldsymbol{G}_{\underline{h}}^{\mathrm{T}}=\boldsymbol{G}_{\underline{Z}}\boldsymbol{P}_{x,k}\boldsymbol{G}_{\underline{h}}^{\mathrm{T}} \tag{2.62}$$

Applying (2.57)–(2.62), it can be obtained that the posterior mean of the modeling error is given by the mean value of (2.50), which is determined as

$$\begin{aligned}\bar{\boldsymbol{d}}_k&=\mathbb{E}\left(\boldsymbol{d}_k\right)=\hat{\boldsymbol{d}}_k-\mathbb{E}\left(D_{\Delta x}\underline{Z}+\frac{D_{\Delta x}^{2}\underline{Z}}{2}+\cdots+D_{\Delta x}\underline{h}+\frac{D_{\Delta x}^{2}\underline{h}}{2}+\cdots\right)\\&=\hat{\boldsymbol{d}}_k-\frac{(\nabla^{\mathrm{T}}\boldsymbol{P}_{x,k}\nabla)\underline{h}(\boldsymbol{x})|_{\boldsymbol{x}=\hat{\boldsymbol{x}}_k}}{2!}-\frac{(\nabla^{\mathrm{T}}\boldsymbol{P}_{x,k}\nabla)\underline{Z}(\boldsymbol{x})|_{\boldsymbol{x}=\hat{\boldsymbol{x}}_k}}{2!}\\&\quad-\mathbb{E}\left(\frac{D_{\Delta x}^{2}\underline{h}}{4!}+\frac{D_{\Delta x}^{2}\underline{Z}}{4!}+\cdots\right)\end{aligned} \tag{2.63}$$

which can also be used to compute the covariance matrix of the modeling error as follows.

$$
\begin{aligned}
\boldsymbol{P}_{d,k} &\triangleq \mathbb{E}((\boldsymbol{d}_k - \bar{\boldsymbol{d}}_k)(\boldsymbol{d}_k - \bar{\boldsymbol{d}}_k)^{\mathrm{T}}) \\
&= \mathbb{E}(\boldsymbol{d}_k \boldsymbol{d}_k^{\mathrm{T}}) - \bar{\boldsymbol{d}}_k \bar{\boldsymbol{d}}_k^{\mathrm{T}} \\
&= \mathbb{E}\left(\sum_{i=1}^{\infty}\sum_{j=1}^{\infty} \frac{1}{i!j!} (\boldsymbol{D}_{\Delta x}^{i}(\underline{\boldsymbol{Z}}+\underline{\boldsymbol{h}}))(\boldsymbol{D}_{\Delta x}^{i}(\underline{\boldsymbol{Z}}+\underline{\boldsymbol{h}}))^{\mathrm{T}}\right) \\
&\quad - \sum_{i=1}^{\infty}\sum_{j=1}^{\infty} \frac{1}{i!j!} \mathbb{E}(\boldsymbol{D}_{\Delta x}^{i}(\underline{\boldsymbol{Z}}+\underline{\boldsymbol{h}}))\mathbb{E}((\boldsymbol{D}_{\Delta x}^{i}(\underline{\boldsymbol{Z}}+\underline{\boldsymbol{h}}))^{\mathrm{T}}) \\
&= \boldsymbol{G}_{\underline{h}} \boldsymbol{P}_{x,k} \boldsymbol{G}_{\underline{h}}^{\mathrm{T}} + \boldsymbol{G}_{\underline{Z}} \boldsymbol{P}_{x,k} \boldsymbol{G}_{\underline{h}}^{\mathrm{T}} + \boldsymbol{G}_{\underline{h}} \boldsymbol{P}_{x,k} \boldsymbol{G}_{\underline{Z}}^{\mathrm{T}} \\
&\quad - \mathbb{E}\left(\frac{\boldsymbol{D}_{\Delta x}^{2!}(\underline{\boldsymbol{Z}}+\underline{\boldsymbol{h}})}{2!}\right)\mathbb{E}\left(\left(\frac{\boldsymbol{D}_{\Delta x}^{2!}(\underline{\boldsymbol{Z}}+\underline{\boldsymbol{h}})}{2!}\right)^{\mathrm{T}}\right) \\
&\quad \underbrace{+\mathbb{E}\left(\sum_{i=1}^{\infty}\sum_{j=1}^{\infty} \frac{1}{i!j!} (\boldsymbol{D}_{\Delta x}^{i}(\underline{\boldsymbol{Z}}+\underline{\boldsymbol{h}}))(\boldsymbol{D}_{\Delta x}^{j}(\underline{\boldsymbol{Z}}+\underline{\boldsymbol{h}}))^{\mathrm{T}}\right)}_{\text{condition\_1}} \\
&\quad \underbrace{- \sum_{i=1}^{\infty}\sum_{j=1}^{\infty} \frac{1}{(2i)!(2j)!} \mathbb{E}(\boldsymbol{D}_{\Delta x}^{2i}(\underline{\boldsymbol{Z}}+\underline{\boldsymbol{h}}))\mathbb{E}((\boldsymbol{D}_{\Delta x}^{2j}(\underline{\boldsymbol{Z}}+\underline{\boldsymbol{h}}))^{\mathrm{T}})}_{\text{condition\_2}}
\end{aligned}
\tag{2.64}
$$

where "condition_1" denotes $ij > 1$ and $i + j$ are odd number, and "condition_2" denotes $ij > 1$.

Theorem 2.1. *The estimated modeling error $\hat{\boldsymbol{d}}_k$ achieved by the predictive filtering Algorithm 2.1 is accurate to the first-order accuracy of the real modeling error $\boldsymbol{d}_k$ and the covariance of the estimated modeling error $\hat{\boldsymbol{d}}_k$ is accurate to the second-order accuracy of the real covariance of $\boldsymbol{d}_k$.*

Proof. This can be directly proved from (2.63) and (2.64). □

2.5.2.2 *State estimation performance of predictive filter*

Regardless of any filtering approach, the following new variables are introduced first for the nonlinear discrete system (2.32)–(2.33).

$$
\eta(\boldsymbol{x}_k) = \boldsymbol{g}(\boldsymbol{x}_k)\boldsymbol{d}_k - \boldsymbol{g}(\boldsymbol{x}_k)\boldsymbol{M}(\boldsymbol{x}_k)\boldsymbol{v}_k \tag{2.65}
$$

$$
\eta(\hat{\boldsymbol{x}}_k) = -\boldsymbol{g}(\hat{\boldsymbol{x}}_k)\hat{\boldsymbol{d}}_k \tag{2.66}
$$

Then, it is got from (2.32)–(2.33), (2.36)–(2.37), and (2.46)–(2.49) that

$$\boldsymbol{x}_{k+1} = \boldsymbol{x}_k + \Delta t \boldsymbol{f}(\boldsymbol{x}_k) + \Delta t \boldsymbol{\eta}(\boldsymbol{x}_k) - \Delta t \boldsymbol{g}(\boldsymbol{x}_k)\boldsymbol{M}(\boldsymbol{x}_k)\boldsymbol{v}_k + \boldsymbol{\mu}(\boldsymbol{x}_k, \boldsymbol{d}_k) \tag{2.67}$$

and

$$\hat{\boldsymbol{x}}_{k+1} = \hat{\boldsymbol{x}}_k + \Delta t \boldsymbol{f}(\hat{\boldsymbol{x}}_k) + \Delta t \boldsymbol{\eta}(\hat{\boldsymbol{x}}_k) + \boldsymbol{\mu}(\hat{\boldsymbol{x}}_k, \hat{\boldsymbol{d}}_k) \tag{2.68}$$

where $\boldsymbol{\mu}(\boldsymbol{x}_k, \boldsymbol{d}_k) \in \mathbb{R}^n$ and $\boldsymbol{\mu}(\hat{\boldsymbol{x}}_k, \hat{\boldsymbol{d}}_k) \in \mathbb{R}^n$ are the high-order linearization and the high-order discretization errors, respectively.

Let $\tilde{\boldsymbol{x}}_{k+1} = \boldsymbol{x}_{k+1} - \hat{\boldsymbol{x}}_{k+1}$ be the state estimation error, it can be obtained from (2.67)–(2.68) that

$$\begin{aligned} \tilde{\boldsymbol{x}}_{k+1} = {} & \tilde{\boldsymbol{x}}_k + \Delta t(\boldsymbol{f}(\boldsymbol{x}_k) - \boldsymbol{f}(\hat{\boldsymbol{x}}_k)) + \Delta t(\boldsymbol{\eta}(\boldsymbol{x}_k) - \boldsymbol{\eta}(\hat{\boldsymbol{x}}_k)) \\ & - \Delta t \bar{\boldsymbol{B}}(\hat{\boldsymbol{x}}_k)\boldsymbol{v}_k + \boldsymbol{\gamma}(\boldsymbol{x}_k, \boldsymbol{d}_k, \hat{\boldsymbol{x}}_k, \hat{\boldsymbol{d}}_k) \end{aligned} \tag{2.69}$$

with

$$\bar{\boldsymbol{B}}(\hat{\boldsymbol{x}}_k) = \boldsymbol{g}(\hat{\boldsymbol{x}}_k)\boldsymbol{M}(\hat{\boldsymbol{x}}_k) \tag{2.70}$$

and

$$\boldsymbol{\gamma}(\boldsymbol{x}_k, \boldsymbol{d}_k, \hat{\boldsymbol{x}}_k, \hat{\boldsymbol{d}}_k) = \boldsymbol{\mu}(\boldsymbol{x}_k, \boldsymbol{d}_k) - \boldsymbol{\mu}(\hat{\boldsymbol{x}}_k, \hat{\boldsymbol{d}}_k) - \boldsymbol{\varepsilon}(\bar{\boldsymbol{B}}(\hat{\boldsymbol{x}}_k))\boldsymbol{v}_k \tag{2.71}$$

where $\boldsymbol{\varepsilon}(\bar{\boldsymbol{B}}(\hat{\boldsymbol{x}}_k))$ is the high-order discretization error given by

$$\boldsymbol{\varepsilon}(\bar{\boldsymbol{B}}(\hat{\boldsymbol{x}}_k)) = \Delta t \boldsymbol{g}(\boldsymbol{x}_k)\boldsymbol{M}(\boldsymbol{x}_k) - \Delta t \boldsymbol{g}(\hat{\boldsymbol{x}}_k)\boldsymbol{M}(\hat{\boldsymbol{x}}_k) \tag{2.72}$$

Since the state estimation error covariance is mathematically defined as $\boldsymbol{P}_{\tilde{x},k+1} = \mathbb{E}(\tilde{\boldsymbol{x}}_{k+1}\tilde{\boldsymbol{x}}_{k+1}^{\mathrm{T}})$, inserting (2.69) in it yields

$$\begin{aligned} \boldsymbol{P}_{\tilde{x},k+1} = {} & \boldsymbol{P}_{\tilde{x},k} + \Delta t \mathbb{E}\Big(\tilde{\boldsymbol{x}}_k \left(\boldsymbol{f}(\boldsymbol{x}_k) - \boldsymbol{f}(\hat{\boldsymbol{x}}_k)\right)^{\mathrm{T}} + \tilde{\boldsymbol{x}}_k \left(\boldsymbol{\eta}(\boldsymbol{x}_k) - \boldsymbol{\eta}(\hat{\boldsymbol{x}}_k)\right)^{\mathrm{T}} \\ & + \left(\boldsymbol{f}(\boldsymbol{x}_k) - \boldsymbol{f}(\hat{\boldsymbol{x}}_k)\right)\tilde{\boldsymbol{x}}_k^{\mathrm{T}} + \left(\boldsymbol{\eta}(\boldsymbol{x}_k) - \boldsymbol{\eta}(\hat{\boldsymbol{x}}_k)\right)\tilde{\boldsymbol{x}}_k^{\mathrm{T}}\Big) \\ & + (\Delta t)^2 \mathbb{E}\Big(\left(\boldsymbol{f}(\boldsymbol{x}_k) - \boldsymbol{f}(\hat{\boldsymbol{x}}_k)\right)\left(\boldsymbol{f}(\boldsymbol{x}_k) - \boldsymbol{f}(\hat{\boldsymbol{x}}_k)\right)^{\mathrm{T}} \\ & + \left(\boldsymbol{\eta}(\boldsymbol{x}_k) - \boldsymbol{\eta}(\hat{\boldsymbol{x}}_k)\right)\left(\boldsymbol{\eta}(\boldsymbol{x}_k) - \boldsymbol{\eta}(\hat{\boldsymbol{x}}_k)\right)^{\mathrm{T}} \\ & + \left(\boldsymbol{f}(\boldsymbol{x}_k) - \boldsymbol{f}(\hat{\boldsymbol{x}}_k)\right)\left(\boldsymbol{\eta}(\boldsymbol{x}_k) - \boldsymbol{\eta}(\hat{\boldsymbol{x}}_k)\right)^{\mathrm{T}} \\ & + \left(\boldsymbol{\eta}(\boldsymbol{x}_k) - \boldsymbol{\eta}(\hat{\boldsymbol{x}}_k)\right)\left(\boldsymbol{f}(\boldsymbol{x}_k) - \boldsymbol{f}(\hat{\boldsymbol{x}}_k)\right)^{\mathrm{T}}\Big) \\ & + (\Delta t)^2 \bar{\boldsymbol{B}}(\hat{\boldsymbol{x}}_k)\boldsymbol{R}_k\bar{\boldsymbol{B}}(\hat{\boldsymbol{x}}_k)^{\mathrm{T}} + \boldsymbol{o}_1(\boldsymbol{x}_k, \boldsymbol{d}_k, \hat{\boldsymbol{x}}_k, \hat{\boldsymbol{d}}_k) + \boldsymbol{o}_2(\boldsymbol{v}_k) \end{aligned} \tag{2.73}$$

where $\boldsymbol{o}_1(\boldsymbol{x}_k, \boldsymbol{d}_k, \hat{\boldsymbol{x}}_k, \hat{\boldsymbol{d}}_k)$ is the polynomial sum of $\boldsymbol{\gamma}(\boldsymbol{x}_k, \boldsymbol{d}_k, \hat{\boldsymbol{x}}_k, \hat{\boldsymbol{d}}_k)$, and $\boldsymbol{o}_2(\boldsymbol{v}_k)$ is the polynomial sum of $\boldsymbol{v}_k$.

Ignoring the higher-order terms in (2.73) and denoting the covariance of the state estimation ensured by the predictive filter as $(\boldsymbol{P}_{\tilde{x},k+1})_{\mathrm{PF}}$, then the following theorem is valid.

Theorem 2.2. *When applying the predictive filter in Algorithm 2.1 to the nonlinear discrete system* (2.32)–(2.33), *the resulted covariance of the state estimation satisfies*

$$(\boldsymbol{P}_{\tilde{x},k+1})_{PF} = \boldsymbol{A}_k(\boldsymbol{P}_{\tilde{x},k})_{PF}\boldsymbol{A}_k^{\mathrm{T}} + (\Delta t)^2\bar{\boldsymbol{B}}_k\boldsymbol{R}_k\bar{\boldsymbol{B}}_k^{\mathrm{T}} + \tilde{\boldsymbol{P}}_k \tag{2.74}$$

where

$$\boldsymbol{A}_k = \boldsymbol{I}_n + \Delta t\boldsymbol{G}_f + \Delta t\boldsymbol{G}_\eta \tag{2.75}$$

$$\tilde{\boldsymbol{P}}_k = \Delta t\bar{\boldsymbol{\Sigma}}_1 + (\Delta t)^2\bar{\boldsymbol{\Sigma}}_2 + \boldsymbol{o}_1(\boldsymbol{x}_k, \boldsymbol{d}_k, \hat{\boldsymbol{x}}_k, \hat{\boldsymbol{d}}_k) + \boldsymbol{o}_2(\boldsymbol{v}_k) \tag{2.76}$$

$$\bar{\boldsymbol{\Sigma}}_1 = \sum_{i=1}^{\infty}\frac{1}{(2i+1)!}\mathbb{E}\left(\tilde{\boldsymbol{x}}_k(\boldsymbol{D}_{\tilde{\boldsymbol{x}}_k}^{2i+1}(\boldsymbol{f}+\boldsymbol{\eta}))^{\mathrm{T}} + \boldsymbol{D}_{\tilde{\boldsymbol{x}}_k}^{2i+1}(\boldsymbol{f}+\boldsymbol{\eta})\tilde{\boldsymbol{x}}_k^{\mathrm{T}}\right) \tag{2.77}$$

$$\bar{\boldsymbol{\Sigma}}_2 = \mathbb{E}\left(\underbrace{\sum_{i=1}^{\infty}\sum_{j=1}^{\infty}\frac{1}{i!j!}\boldsymbol{D}_{\tilde{\boldsymbol{x}}_k}^{i}(\boldsymbol{f}+\boldsymbol{\eta})(\boldsymbol{D}_{\tilde{\boldsymbol{x}}_k}^{j}(\boldsymbol{f}+\boldsymbol{\eta}))^{\mathrm{T}}}_{condition\_1}\right) \tag{2.78}$$

Proof. This theorem can be directly proved by using (2.73) with its high-order terms omitted. □

2.6 Summary

This is a preliminary chapter. Some basic definitions and mathematical lemmas related to the probability theory was presented, which are useful to the development and theoretical analysis of filters to be developed in the rest of this book. The desired properties or performance that any filter should have were introduced. At last, the classical predictive filter including its state estimation procedures and state estimation performance was introduced in detail. It was proved that the predictive filter can achieve a first-order estimation accuracy of the real modeling error. The covariance of the estimated modeling error achieved by the predictive filter is accurate to the second-order accuracy of the modeling error's real covariance.

Chapter 3

Modeling of microsatellite control system

3.1 Introduction

When investigating the attitude and the orbit motion of a single microsatellite, the problem of establishing its dynamic model has been well studied. Note that multiple microsatellites flying in formation have attracted significant attention in the past decade. For microsatellite formation flying (SFF), the relative motion control system should be modeled, which is important for the controller design and state estimation. The Clohessy-Wiltshire (CW) equation is the well-known linear model to describe the relative motion dynamics of SFF in the early years (Dang and Zhang, 2012; Clohessy and Wiltshire, 1960). The Tschauner-Hempel model (T-H) is another solution to this problem (Tschauner and Hempel, 1965). However, both the CW equation and the T-H model does not consider any perturbation force. In fact, the motion of the formation flying microsatellites is fully perturbed by external disturbance forces such as the J_2 perturbation, the atmospheric drag force, and the solar radiation pressure force. Particularly, the J_2 perturbation and the atmospheric drag perturbation are two primary external forces acting on SFF. Hence, the CW equation and the T-H model are not capable of accurately describing the relative motion of SFF.

In the past two decades, the accurately modeling problem of the relative motion of SFF has been extensively studied (Mishne, 2004; Morgan and Chung, 2012). The CW equation was modified in Carter and Humi (2002b) and Carter and Humi (2002a) to describe the relation motion of SFF in the presence of the atmospheric drag. Because the J_2 perturbation was not considered, the model presented in Carter and Humi (2002b) and Carter and Humi (2002a) is not accurate enough. In Cho and Park (2009), the T-H model was modified with the J_2 perturbation for an elliptical reference orbit accounted. However, the modeling error increases along with the increase of the reference orbit's eccentricity. An analytical solution for the radius of a reference orbit affected by the J_2 perturbation was presented in Wei et al. (2013) to approximate the actual reference orbit, while the unperturbed standard orbit was not used. A state transition matrix was developed in Gim and Alfriend (2003) to describe the effects of the reference orbit eccentricity and the gravitational perturbation. In Schaub and Alfriend (2001), the J_2 invariant relative orbits were designed for SFF. The effect of the eccentricity on the relative motion of SFF was analyzed in Sen-

Predictive Filtering for Microsatellite Control System. https://doi.org/10.1016/B978-0-12-821865-5.00016-4

gupta and Vadali (2007). Based on the mathematical model presented in Schaub and Alfriend (2001), another analytical state transition matrix was reported in Hamel and Lafontaine (2007) to describe the relative motion of two formation flying satellites in an elliptical orbit affected by the J_2 perturbation force. In Schweigart and Sedwick (2002), an S-S dynamic model that is linear was presented with the J_2 perturbation considered only. The S-S model was improved in Roberts and Roberts (2004) to establish a set of time-varying linear dynamic equations by utilizing the actual gradient of the J_2 perturbation directly. Based on the S-S model, the relative motion equation with the atmospheric drag perturbation force considered was established in Reid and Misra (2011).

This chapter presents another approach to establish a linear dynamic model of the relative motion of two formation flying microsatellites in an elliptical low-orbit with small eccentricity. The J_2 and the atmospheric perturbation forces are considered. When investigating the attitude and the orbit motion of a single microsatellite, the problem of establishing its dynamic model has been well studied. Hence, the control system model of a single microsatellite is presented in this chapter directly based on the standard textbook like Sidi (1997).

3.2 Definition of coordinate frames

Consider $N_0 \in \mathbb{N}$ microsatellites flying in space, the coordinate frames used to describe their kinematics and dynamics are illustrated in Fig. 3.1. The Earth-centered inertial coordinate frame $\mathfrak{F}_I(O_I X_I Y_I Z_I)$ with its origin O_I at the center of the Earth is used to determine the orbital position of the microsatellite. The X_I axis is aligned with the mean equinox. The Z_I axis is aligned with the Earth's spin axis or the celestial North Pole. The Y_I axis is rotated by 90 degrees East about the celestial equator. One commonly used $\mathfrak{F}_I$ is defined with the Earth's Mean Equator and Equinox at 12 : 00 Terrestrial Time on 1 January 2000. The coordinate system $\mathfrak{F}_{Bi}(O_{Bi} X_{Bi} Y_{Bi} Z_{Bi})$ is the body-fixed reference frame of the ith microsatellite, $i = 1, 2, \cdots, N_0$. The Local-Vertical-Local-Horizontal (LVLH) coordinate frame $\mathfrak{F}_r(O_r X_r Y_r Z_r)$ is used to describe the relative motion dynamics of a microsatellite with respect to a reference orbit. Its origin O_r is on the reference orbit. The X_r axis points to the radial direction, the Z_r axis is perpendicular to the reference orbital plane and points to the direction of reference orbital momentum vector, and the Y_r axis is normal to the X_r and Z_r axes. It defines a right-handed orthogonal reference coordinate frame attached to the reference orbit.

3.3 Modeling of single microsatellite control system

When considering the use of a single microsatellite only to carry out some aerospace missions, its control system includes the attitude and the orbit control subsystems. These two control systems are modeled in this section.

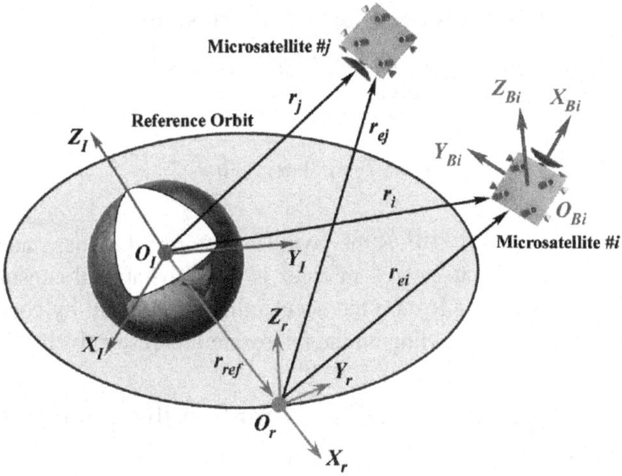

FIGURE 3.1 Coordinate frames defined for dynamics modeling.

3.3.1 Mathematical model of attitude control system

For a single microsatellite, *i.e.*, $N_0 = 1$, the coordinate frame $\mathcal{F}_{Bi}(O_{Bi}X_{Bi}Y_{Bi}Z_{Bi})$ in Fig. 3.1 is denoted as $\mathcal{F}_B(O_BX_BY_BZ_B)$. The unit quaternion representation is adopted in this book to represent the attitude of this microsatellite. Then, the microsatellite attitude control system can be mathematically modeled as

$$\dot{\boldsymbol{q}} = \boldsymbol{\Psi}(\boldsymbol{\omega}_{BI})\boldsymbol{q} = \boldsymbol{E}(\boldsymbol{q})\boldsymbol{\omega}_{BI} \tag{3.1}$$

$$\boldsymbol{J}\dot{\boldsymbol{\omega}}_{BI} = -\boldsymbol{\omega}_{BI}^{\times}\boldsymbol{J}\boldsymbol{\omega}_{BI} + \boldsymbol{u}_c + \boldsymbol{u}_d \tag{3.2}$$

where the unit quaternion $\boldsymbol{q} = [q_1 \quad \boldsymbol{q}_{24}^{\mathrm{T}}]^{\mathrm{T}} \in \mathbb{R}^4, q_1 \in \mathbb{R}, \boldsymbol{q}_{24} = [q_2 \quad q_3 \quad q_4]^{\mathrm{T}} \in \mathbb{R}^3$, denotes the attitude orientation of the microsatellite in $\mathcal{F}_B$ with respect to $\mathcal{F}_I$. $\boldsymbol{\omega}_{BI} = [\omega_{BI}^1 \quad \omega_{BI}^2 \quad \omega_{BI}^3]^{\mathrm{T}} \in \mathbb{R}^3$ is the angular velocity of the microsatellite with respect to $\mathcal{F}_I$ and expressed in $\mathcal{F}_B$. $\boldsymbol{J} \in \mathbb{R}^{3\times 3}$ is the total inertia matrix of the microsatellite. $\boldsymbol{u}_c \in \mathbb{R}^3$ is the total control torque. $\boldsymbol{u}_d \in \mathbb{R}^3$ denotes the modeling error induced by external disturbance torque, uncertain parameters, and other uncertainties. The matrices $\boldsymbol{\Psi}(\boldsymbol{\omega}_{BI}) \in \mathbb{R}^{4\times 4}$ and $\boldsymbol{E}(\boldsymbol{q}) \in \mathbb{R}^{4\times 3}$ are defined as

$$\boldsymbol{\Psi}(\boldsymbol{\omega}_{BI}) = \frac{1}{2}\begin{bmatrix} 0 & -\boldsymbol{\omega}_{BI}^{\mathrm{T}} \\ \boldsymbol{\omega}_{BI} & -\boldsymbol{\omega}_{BI}^{\times} \end{bmatrix} \tag{3.3}$$

$$\boldsymbol{E}(\boldsymbol{q}) = \frac{1}{2}\begin{bmatrix} -\boldsymbol{q}_{24}^{\mathrm{T}} \\ q_1\boldsymbol{I}_3 + \boldsymbol{q}_{24}^{\times} \end{bmatrix} \tag{3.4}$$

3.3.2 Mathematical model of orbit control system

For any orbital microsatellite, its orbit control system can be mathematically modeled as (Cao and Misra, 2015)

$$\ddot{\boldsymbol{r}} = -\frac{\mu_0}{r^3}\boldsymbol{r} + \boldsymbol{u}_T + \boldsymbol{u}_P \tag{3.5}$$

where $\boldsymbol{r} \in \mathbb{R}^3$, $\boldsymbol{r} = [r \quad 0 \quad 0]^{\mathrm{T}}$ is the position vector of the microsatellite with respect to $\mathfrak{F}_I$. $\mu_0 = 3.986 \times 10^{14}$ m^3/sec^2 is the gravitational constant of the Earth. $\boldsymbol{u}_T \in \mathbb{R}^3$ and $\boldsymbol{u}_P \in \mathbb{R}^3$ are the accelerations generated by control force and the perturbation force acting on the microsatellite, respectively.

3.4 Modeling of microsatellite formation flying control system

Because the functionality of a single complex satellite can be distributed between a cluster of closely formation flying microsatellites, formation flying of two microsatellites in low orbit is extensively considered in this book, *i.e.*, $N_0 = 2$. This formation flying consists of a chief microsatellite and a deputy microsatellite. Meanwhile, the LVLH coordinate frame $\mathfrak{F}_r$ in Fig. 3.1 is usually defined with its origin O_r located in the center of the mass of the chief microsatellite. The X_r axis points to the radial direction of the chief microsatellite, the Z_r axis is perpendicular to the orbital plane of the chief microsatellite and points to the direction of the chief microsatellite's orbital momentum vector, and the Y_r axis is normal to the X_r and Y_r axes.

For any low-orbit microsatellite, the J_2 perturbation and the atmospheric drag are two main perturbation forces. Taking these two perturbation forces into account, when there is no control force acting on the chief microsatellite, its motion with respect to the Earth is governed by the two-body equation (Vallado, 2005)

$$\ddot{\boldsymbol{r}}_c = -\frac{\mu_0}{r_c^3}\boldsymbol{r}_c + \boldsymbol{f}_{J_2,c} + \boldsymbol{f}_{drag,c} \tag{3.6}$$

where $\boldsymbol{r}_c \in \mathbb{R}^3$, $\boldsymbol{r}_c = [r_c \quad 0 \quad 0]^{\mathrm{T}}$ is the position vector of the chief microsatellite with respect to $\mathfrak{F}_I$. $\boldsymbol{f}_{J_2,c} = [f_{J_2,c,x} \quad f_{J_2,c,y} \quad f_{J_2,c,z}]^{\mathrm{T}} \in \mathbb{R}^3$ and $\boldsymbol{f}_{drag,c} = [f_{drag,c,x} \quad f_{drag,c,y} \quad f_{drag,c,z}]^{\mathrm{T}} \in \mathbb{R}^3$ are the accelerations due to the J_2 perturbation and the atmospheric drag, respectively.

Let the relative position of the deputy microsatellite from the origin of the LVLH be defined by $\boldsymbol{r}_e = [x \quad y \quad z]^{\mathrm{T}} \in \mathbb{R}^3$, then the motion of the deputy microsatellite with the J_2 perturbation and the atmospheric drag is described as

$$\ddot{\boldsymbol{r}}_d = -\frac{\mu_0}{r_d^3}\boldsymbol{r}_d + \boldsymbol{f}_{J_2,d} + \boldsymbol{f}_{drag,d} \tag{3.7}$$

where $r_d = \sqrt{(r_c + x)^2 + y^2 + y^2}$ is the distance from the origin of $\mathcal{F}_I$ to the deputy microsatellite. $\boldsymbol{f}_{J_2,d} = [f_{J_2,d,x} \quad f_{J_2,d,y} \quad f_{J_2,d,z}]^{\mathrm{T}} \in \mathbb{R}^3$ and $\boldsymbol{f}_{drag,d} = [f_{drag,d,x} \quad f_{drag,d,y} \quad f_{drag,d,z}]^{\mathrm{T}} \in \mathbb{R}^3$ denote the accelerations due to the J_2 perturbation and the atmospheric drag, respectively.

From (3.6) and (3.7), the motion of the deputy microsatellite relative to the chief microsatellite can be established as

$$\ddot{\boldsymbol{r}}_e = \ddot{\boldsymbol{r}}_d - \ddot{\boldsymbol{r}}_c = -\frac{\mu_0}{r_d^3}\boldsymbol{r}_d + \frac{\mu_0}{r_c^3}\boldsymbol{r}_c + \boldsymbol{f}_{J_2,d} - \boldsymbol{f}_{J_2,c} + \boldsymbol{f}_{drag,d} - \boldsymbol{f}_{drag,c} \tag{3.8}$$

Because the coordinate frame $\mathcal{F}_r$ attached to the chief microsatellite is rotating, the rotational terms should be considered when calculating the relative acceleration and velocity between the chief and the deputy microsatellites. Therefore, the relative motion (3.8) can be rewritten as

$$\begin{aligned}&\ddot{\boldsymbol{r}}_e + 2(\boldsymbol{\omega}_0)^\times \boldsymbol{r}_e + (\boldsymbol{\omega}_0)^\times((\boldsymbol{\omega}_0)^\times \boldsymbol{r}_e) + (\dot{\boldsymbol{\omega}}_0)^\times \boldsymbol{r}_e \\ &= -\frac{\mu_0}{r_d^3}\boldsymbol{r}_d + \frac{\mu_0}{r_c^3}\boldsymbol{r}_c + \boldsymbol{f}_{J_2,d} - \boldsymbol{f}_{J_2,c} + \boldsymbol{f}_{drag,d} - \boldsymbol{f}_{drag,c}\end{aligned} \tag{3.9}$$

where $\boldsymbol{\omega}_0 = [\omega_{0x} \quad 0 \quad \omega_{0z}]^{\mathrm{T}}$ is the chief microsatellite's orbital angular velocity. $\dot{\boldsymbol{\omega}}_0 = [\dot{\omega}_{0x} \quad 0 \quad \dot{\omega}_{0z}]^{\mathrm{T}}$ is the angular acceleration of the chief microsatellite's orbit.

3.4.1 Definition of dimensionless expressions

To reduce the computation time needed for the filtering algorithms in the following chapters, dimensionless expressions are the required tool to fairly compare the state estimation performance achieved by different filters. This recommends that all the states of the relative motion (3.9) should be given in a dimensionless form. Hence, all the states of the relative motion (3.9) are non-dimensionalized by preliminarily defining a dimensionless time parameter as

$$\tau = \upsilon_{o0} t \tag{3.10}$$

where $\upsilon_{o0} = \sqrt{\frac{\mu_0}{a_{c0}^3}} \in \mathbb{R}_+$ is the initial mean orbital rate of the chief microsatellite, $a_{c0} \in \mathbb{R}_+$ is the initial semi-major axis of the chief microsatellite's orbit. The reason of using the initial orbital rate υ_{o0} but not the instantaneous mean orbital rate $\upsilon_o = \sqrt{\frac{\mu_0}{a_c^3}} \in \mathbb{R}_+$ is that the scaling between τ and the time t should remain as a constant. $a_c \in \mathbb{R}_+$ is the instantaneous semi-major axis of the chief microsatellite's orbit. This can not be achieved by using υ_o, because υ_o is time-varying due to the drag forces.

In addition to introduce the dimensionless time parameter τ, the terms x, y, and z in the relative position vector $\boldsymbol{r}_e$ are transformed into dimensionless distances with respect to a reference length parameter $L_r \in \mathbb{R}_+$. More specifically, the dimensionless expression of x, y, and z are defined as

$$\hat{x} = \frac{x}{L_r} \tag{3.11}$$

$$\hat{y} = \frac{y}{L_r} \tag{3.12}$$

$$\hat{z} = \frac{z}{L_r} \tag{3.13}$$

Based on which, the velocities and the accelerations can be obtained as

$$\dot{x} = \frac{dx}{dt} = \frac{d(\hat{x}L_r)}{d\left(\frac{\tau}{v_{o0}}\right)} = v_{o0}L_r\hat{x}' \tag{3.14}$$

$$\dot{y} = v_{o0}L_r\hat{y}' \tag{3.15}$$

$$\dot{z} = v_{o0}L_r\hat{z}' \tag{3.16}$$

$$\dot{\boldsymbol{r}}_e = v_{o0}L_r\hat{\boldsymbol{r}}_e' \tag{3.17}$$

$$\ddot{x} = \frac{d^2x}{dt^2} = \frac{d^2(\hat{x}L_r)}{d\left(\frac{\tau}{v_{o0}}\right)^2} = v_{o0}^2L_r\hat{x}'' \tag{3.18}$$

$$\ddot{y} = v_{o0}^2L_r\hat{y}'' \tag{3.19}$$

$$\ddot{z} = v_{o0}^2L_r\hat{z}'' \tag{3.20}$$

$$\ddot{\boldsymbol{r}}_e = v_{o0}^2L_r\hat{\boldsymbol{r}}_e'' \tag{3.21}$$

where $\hat{x}' = \frac{d\hat{x}}{d\tau}$, $\hat{y}' = \frac{d\hat{y}}{d\tau}$, $\hat{z}' = \frac{d\hat{z}}{d\tau}$, $\hat{\boldsymbol{r}}_e' = \frac{d\boldsymbol{r}_e}{d\tau}$, $\hat{x}'' = \frac{d^2x}{d\tau^2}$, $\hat{y}'' = \frac{d^2y}{d\tau^2}$, $\hat{z}'' = \frac{d^2z}{d\tau^2}$, and $\hat{\boldsymbol{r}}_e'' = \frac{d^2\boldsymbol{r}_e}{d\tau^2}$.

For (3.14)–(3.21), $L_r = d_0$ can be set, where $d_0 \in \mathbb{R}_+$ is a value associated with the geometry of the desired formation motion geometry. For example, d_0 can be determined as the desired radius of a projected circular formation or the desired distance between the microsatellite in an in-track, leader-follower formation. As a result, $\hat{x}$, $\hat{y}$, and $\hat{z}$ are ensured to have the order of one.

3.4.2 Assumption of small eccentricity

To simplify analysis, the chief microsatellite orbit is assumed to have a small eccentricity. The eccentricity terms with the order higher than one is neglected. Based on this, some important orbital parameters can be simplified (Francis, 2012). The semilatus rectum $p \in \mathbb{R}_+$ can be defined as

$$p = a_c(1 - e^2) \approx a_c \tag{3.22}$$

where $e \in \mathbb{R}_+$ is the small eccentricity of the chief microsatellite's orbit.

Based on the assumption of small eccentricity and in conjunction with a binomial expansion, the powers of r_c can be simplified as

$$r_c^k = \left(\frac{a_c(1-e^2)}{1+e\cos f}\right)^k \approx a_c^k(1+e\cos f)^{-k} \approx a_c^k(1-ke\cos f),\ k \in \mathbb{N} \quad (3.23)$$

where $f \in \mathbb{R}$ is the true anomaly of the chief microsatellite's orbit. Moreover, the following term in (3.9) can be simplified as (Francis, 2012)

$$\frac{\mu_0}{r_c^3}\boldsymbol{r}_c - \frac{\mu_0}{r_d^3}\boldsymbol{r}_d = \begin{bmatrix} 2 & 0 & 0 \\ 0 & -1 & 0 \\ 0 & 0 & -1 \end{bmatrix} v_o^2(1+3e\cos f)\boldsymbol{r}_e \quad (3.24)$$

Applying (3.11)–(3.21) and (3.23), the non-dimensionalized version of (3.9) can be established as

$$\begin{aligned} \hat{x}'' = &\ 2\hat{\omega}_{0z}\hat{y}' - \hat{\omega}_{0x}\hat{\omega}_{0z}\hat{z} + \hat{\omega}'_{0z}\hat{y} + \hat{\omega}_{0z}^2\hat{x} + 2\hat{v}_o^2(1+3e\cos f)\hat{x} \\ &+ \Delta\hat{f}_{J_2,x} + \Delta\hat{f}_{drag,x} \end{aligned} \quad (3.25)$$

$$\begin{aligned} \hat{y}'' = &-2\hat{\omega}_{0z}\hat{x}' + 2\hat{\omega}_{0x}\hat{z}' + \hat{\omega}'_{0x}\hat{z} - \hat{\omega}'_{0z}\hat{x} + \hat{\omega}_{0x}^2\hat{y} + \hat{\omega}_z^2\hat{y} \\ &- \hat{v}_o^2(1+3e\cos f)\,\hat{y} + \Delta\hat{f}_{J_2,y} + \Delta\hat{f}_{drag,y} \end{aligned} \quad (3.26)$$

$$\begin{aligned} \hat{z}'' = &-2\hat{\omega}_{0x}\hat{y}' - \hat{\omega}_{0x}\hat{\omega}_{0z}\hat{x} - \hat{\omega}'_{0x}\hat{y} + \hat{\omega}_{0x}^2\hat{z} - \hat{v}_o^2(1+3e\cos f)\,\hat{z} \\ &+ \Delta\hat{f}_{J_2,z} + \Delta\hat{f}_{drag,z} \end{aligned} \quad (3.27)$$

where $\hat{\omega}_{0x} = \frac{\omega_{0x}}{v_{o0}}$, $\hat{\omega}_{0z} = \frac{\omega_{0z}}{v_{o0}}$, $\hat{\omega}'_{0x} = \frac{\dot{\omega}_{0x}}{v_{o0}^2}$, $\hat{\omega}'_{0z} = \frac{\dot{\omega}_{0z}}{v_{o0}^2}$, $\hat{v}_o = \frac{v_o}{v_{o0}}$, and

$$\Delta\hat{f}_{J_2,i} = \frac{f_{J_2,d,i} - f_{J_2,c,i}}{v_{o0}^2 d_0},\ i = x, y, z \quad (3.28)$$

$$\Delta\hat{f}_{drag,i} = \frac{f_{drag,d,i} - f_{drag,c,i}}{v_{o0}^2 d_0},\ i = x, y, z \quad (3.29)$$

3.4.3 J_2 and atmospheric drag perturbations

The acceleration due to the J_2 perturbations needed to propagate the chief microsatellite's orbit is given as follows (Schweigart and Sedwick, 2002).

$$\boldsymbol{f}_{J_2,c} = \begin{bmatrix} f_{J_2,c,x} \\ f_{J_2,c,y} \\ f_{J_2,c,z} \end{bmatrix} = -\frac{3J_2\mu_0 R_e^2}{2r_c^4}\begin{bmatrix} 1-3\sin^2 i\sin^2\theta \\ 2\sin^2 i\sin\theta\cos\theta \\ 2\sin i\cos i\sin\theta \end{bmatrix} \quad (3.30)$$

where $J_2 \in \mathbb{R}$ is the dominant harmonics in the oblateness perturbation. $i \in \mathbb{R}$ is the inclination angle of the orbit of the chief microsatellite. $\theta \in \mathbb{R}$ is the true latitude satisfying $\theta = f + \omega$ with the argument of perigee $\omega \in \mathbb{R}$. $R_e \in \mathbb{R}$ is the equatorial radius of the Earth.

Based on the small eccentricity assumption, applying (3.23) can simplify (3.30) as

$$\boldsymbol{f}_{J_2,c} = -\frac{3J_2\mu_0 R_e^2}{2a_c^4}(1+4e\cos f)\begin{bmatrix} 1-3\sin^2 i\sin^2\theta \\ 2\sin^2 i\sin\theta\cos\theta \\ 2\sin i\cos i\sin\theta \end{bmatrix} \tag{3.31}$$

In accordance, its dimensionless expression is given by

$$\hat{\boldsymbol{f}}_{J_2,c} = -\frac{3J_2\hat{v}_0^2\hat{R}_e^2}{2\hat{a}_c}(1+4e\cos f)\begin{bmatrix} 1-3\sin^2 i\sin^2\theta \\ 2\sin^2 i\sin\theta\cos\theta \\ 2\sin i\cos i\sin\theta \end{bmatrix} \tag{3.32}$$

where $\hat{a}_c = \frac{a_c}{d_0}$ and $\hat{R}_e = \frac{R_e}{d_0}$.

For low-altitude microsatellites, the atmospheric drag is another primary perturbation. This severely deteriorates the dynamic performance of microsatellite formation flying. The acceleration of the atmospheric drag on the chief microsatellite is given by

$$\boldsymbol{f}_{drag,c} = -\frac{c_d^c A_c}{2m_c}\rho_c||\boldsymbol{V}_r||\boldsymbol{V}_r \tag{3.33}$$

where $c_d^c \in \mathbb{R}$ is the chief microsatellite's drag coefficient, $A_c \in \mathbb{R}$ is the cross sectional area, $m_c \in \mathbb{R}$ is the chief microsatellite's mass, $\rho_c \in \mathbb{R}$ is the local atmospheric density, and $\boldsymbol{V}_r \in \mathbb{R}^3$ is the chief microsatellite's velocity relative to the rotating atmosphere. Moreover, the atmosphere is assumed to rotate at the same angular velocity as the Earth, which is given as

$$\boldsymbol{V}_r = \boldsymbol{V} - (\boldsymbol{w}_E)^{\times}\boldsymbol{r}_c \tag{3.34}$$

where $\boldsymbol{V} \in \mathbb{R}^3$ is the chief microsatellite's absolute velocity and $\boldsymbol{w}_E \in \mathbb{R}^3$ is the angular velocity of the Earth.

The atmospheric drag effects are expected to be expressed in the Hill frame. Then, $\boldsymbol{V}_r$ should be expressed first in the Hill coordinates. A detailed explanation of the coordinate transformations leading to the expression of $\boldsymbol{V}_r$ in the Hill frame is provided in Vallado (2005). In general, the expression for the chief microsatellite's velocity relative to the rotating atmosphere is determined as

$$\boldsymbol{V}_r = \begin{bmatrix} \dot{x}+\dot{r}_c - y(\dot{f}-w_E\cos i) - zw_E\cos\theta\sin i \\ \dot{y}+(r_c+x)(\dot{f}-w_E\cos i)+zw_E\sin\theta\sin i \\ \dot{z}+(r_c+x)w_E\cos\theta\sin i - yw_E\sin\theta\sin i \end{bmatrix} \tag{3.35}$$

where $w_E = ||\boldsymbol{w}_E||$.

Define a dimensionless term for the drag coefficient as

$$\hat{\beta} = \frac{c_d^c A_c}{m_c} \rho_c d_0 \tag{3.36}$$

Then, the dimensionless drag expression of (3.33) can be obtained as

$$\hat{\boldsymbol{f}}_{drag,c} = \frac{1}{v_{o0}^2 d_0} \boldsymbol{f}_{drag,c} = -\frac{1}{2} \hat{\beta} ||\hat{\boldsymbol{V}}_r|| \hat{\boldsymbol{V}}_r \tag{3.37}$$

where the dimensionless velocity $\hat{\boldsymbol{V}}_r$ is specified as

$$\hat{\boldsymbol{V}}_r = \frac{1}{v_{o0} d_0} \boldsymbol{V}_r = \begin{bmatrix} \hat{x}' + \hat{r}_c' - \hat{y}(f' - \hat{w}_E \cos i) - \hat{z} \hat{w}_E \cos\theta \sin i \\ \hat{y}' + (\hat{r}_c + \hat{x})(f' - \hat{w}_E \cos i) + \hat{z} \hat{w}_E \sin\theta \sin i \\ \hat{z}' + (\hat{r}_c + \hat{x}) \hat{w}_E \cos\theta \sin i - \hat{y} \hat{w}_E \sin\theta \sin i \end{bmatrix} \tag{3.38}$$

with $\hat{r}_c' = \frac{\dot{r}_c}{v_{o0} d_0}$, $\hat{r}_c = \frac{r_c}{d_0}$, $f' = \frac{\dot{f}}{v_{o0}}$, and $\hat{w}_E = \frac{\omega_E}{v_{o0}}$. Moreover, based on the assumption of small eccentricity and the chain rule, it follows that

$$\hat{r}_c' = \frac{1}{v_{o0} d_0} \frac{dr_c}{df} \frac{df}{dt} = \hat{a}_c e \sin f (1 + 2e \cos f) f' \approx \hat{a}_c e f' \sin f \tag{3.39}$$

Because we consider the perturbations on the chief microsatellite placed at the origin of the Hill frame, x, y, and z and their derivative are all set to zero to simplify the above expression. Hence, we have

$$\begin{aligned} \hat{\boldsymbol{V}}_r = \begin{bmatrix} v_x \\ v_y \\ v_z \end{bmatrix} &= \begin{bmatrix} \hat{r}_c' \\ \hat{r}_c (f' - \hat{w}_E \cos i) \\ \hat{r}_c \hat{w}_E \cos\theta \sin i \end{bmatrix} \\ &= \hat{a}_c \begin{bmatrix} e f' \sin f \\ (1 - e\cos f)(f' - \hat{w}_E \cos i) \\ (1 - e \cos f) \hat{w}_E \cos\theta \sin i \end{bmatrix} \end{aligned} \tag{3.40}$$

and $||\hat{\boldsymbol{V}}_r|| = \sqrt{v_x^2 + v_y^2 + v_z^2}$. Since v_x^2 contains high-order terms of e, v_x^2 is negligible. Then, we can simplify $||\hat{\boldsymbol{V}}_r||$ as

$$||\hat{\boldsymbol{V}}_r|| \approx \hat{a}_c (1 - e \cos f) \bar{V} \tag{3.41}$$

where $\bar{V} = \sqrt{(f' - \hat{w}_E \cos i)^2 + \hat{w}_E^2 \cos^2\theta \sin^2 i}$.

Due to the fact that $\hat{w}_E \ll f'$ holds, $\bar{V}$ can be simplified as follows by omitting the high-order terms $o\left(\frac{\hat{w}_E}{f'}\right)^3$.

$$\begin{aligned}\bar{V} &= f'\sqrt{1-2\frac{\hat{w}_E}{f'}\cos i+\left(\frac{\hat{w}_E}{f'}\right)^2(1-\sin^2 i\sin^2\theta)}\\ &= f'\left(1-\frac{\hat{w}_E}{f'}\cos i+\frac{1}{2}\left(\frac{\hat{w}_E}{f'}\right)^2(1-\sin^2 i\sin^2\theta-\cos^2 i)+o\left(\left(\frac{\hat{w}_E}{f'}\right)^3\right)\right)\\ &\approx f'-\hat{w}_E\cos i+\frac{\hat{w}_E^2}{2f'}\sin^2 i\cos^2\theta\end{aligned} \tag{3.42}$$

Inserting (3.42) into (3.41) yields

$$||\hat{V}_r|| \approx \hat{a}_c(1-e\cos f)\left(f'-\hat{w}_E\cos i+\frac{\hat{w}_E^2}{2f'}\sin^2 i\cos^2\theta\right) \tag{3.43}$$

Then, the dimensionless acceleration due to the atmospheric drag can be finally expressed as

$$\hat{f}_{drag,c} = -\begin{bmatrix} \frac{\hat{\beta}\hat{a}_c^2 e}{2}\sin f\left(f'^2-\hat{w}_E f'\cos i+\frac{1}{2}\hat{w}_E^2\cos^2\theta\sin^2 i\right)\\ \frac{\hat{\beta}\hat{a}_c^2 e(1-2e\cos f)}{2}\left(f'^2-2\hat{w}_E f'\cos i+\frac{\hat{w}_E^2\cos^2\theta\sin^2 i}{2}+\hat{w}_E^2\cos^2 i\right)\\ \frac{\hat{\beta}\hat{a}_c^2 e(1-2e\cos f)}{2}\left(\hat{w}_E f'\cos\theta\sin i-\hat{w}_E^2\cos\theta\sin i\cos i\right)\end{bmatrix} \tag{3.44}$$

where all of the terms containing the high-order terms $o\left(\frac{\hat{\omega}_e}{f'}\right)^3$ is eliminated.

3.4.4 Orbit propagation of the chief microsatellite

The final differential equations that govern the propagation of the orbital elements describing the chief microsatellite's orbit are presented in this section. To finish this work, the Gauss's planetary equations (Junkins and Schaub, 2009) are modified to express them in terms of derivatives with respect to τ and then used in conjunction with the perturbative accelerations calculated in (3.32) and (3.44), and it follows that

$$\frac{d\hat{a}_c}{d\tau} = \frac{2\hat{a}_c^2}{\hat{h}}(e \sin f \hat{a}_x + (1 + e \cos f)\hat{a}_y) \tag{3.45}$$

$$\frac{di}{d\tau} = \frac{\hat{a}_c(1-e^2)\cos\theta}{\hat{h}(1+e\cos f)}\hat{a}_z \tag{3.46}$$

$$\frac{de}{d\tau} = \frac{1}{\hat{h}}(1-e^2)\left(\hat{a}_c \sin f \hat{a}_x + \hat{a}_c\hat{a}_y \cos f + \frac{\hat{a}_c(\cos f + e)}{1 + e\cos f}\hat{a}_y\right) \tag{3.47}$$

$$\frac{d\Omega}{d\tau} = \frac{\hat{a}_c(1-e^2)\sin\theta}{\hat{h}(1+e\cos f)\sin i}\hat{a}_z \tag{3.48}$$

$$\begin{aligned}\frac{d\omega}{d\tau} &= \frac{1}{\hat{h}e}(-\hat{a}_c(1-e^2)\cos f \hat{a}_x + \hat{a}_c(1-e^2)\sin f \hat{a}_y) \\ &+ \frac{1}{\hat{h}e}\frac{\hat{a}_c(1-e^2)}{1+e\cos f}\sin f \hat{a}_y - \frac{\hat{a}_c(1-e^2)\sin\theta\cos i}{h \sin i(1+e\cos f)}\hat{a}_z\end{aligned} \tag{3.49}$$

$$\begin{aligned}\frac{dM}{d\tau} &= \hat{v}_o + \frac{\sqrt{1-e^2}}{\hat{h}e}\left(\hat{a}_c(1-e^2)\cos f - \frac{\hat{a}_c(1-e^2)}{1+e\cos f}e\right)\hat{a}_x \\ &- \frac{\sqrt{1-e^2}}{\hat{h}e}\left(\hat{a}_c(1-e^2)\cos f - \frac{\hat{a}_c(1-e^2)}{1+e\cos f}e\right)\sin f \hat{a}_y\end{aligned} \tag{3.50}$$

where $\hat{h} = \frac{h}{v_{o0}d_0^2} \in \mathbb{R}$ is the non-dimensional angular momentum, $h \in \mathbb{R}$ is the angular momentum of the chief microsatellite, $\Omega \in \mathbb{R}$, $\omega \in \mathbb{R}$, and $M \in \mathbb{R}$ are the right ascension of the ascending node, the argument of perigee, and the mean anomaly for the chief microsatellite's orbit, respectively.

Invoking the assumption of small eccentricity, we can simplify (3.45)–(3.50) as

$$\frac{d\hat{a}_c}{d\tau} = \frac{2\hat{a}_c^2}{\hat{h}}(e \sin f \hat{a}_x + (1 + e \cos f)\hat{a}_y) \tag{3.51}$$

$$\frac{di}{d\tau} = \frac{\hat{a}_c(1-e^2)\cos\theta}{\hat{h}}\hat{a}_z \tag{3.52}$$

$$\frac{de}{d\tau} = \frac{\hat{a}_c}{\hat{h}}(\sin f \hat{a}_x + ((2 - e\cos f)\cos f + e)\hat{a}_y) \tag{3.53}$$

$$\frac{d\Omega}{d\tau} = \frac{\hat{a}_c(1-e\cos f)\sin\theta\cos i}{\hat{h}\sin i}\hat{a}_z \tag{3.54}$$

$$\frac{d\omega}{d\tau} = \frac{\hat{a}_c}{\hat{h}e}((2 - e\cos f)\sin f\hat{a}_y - \cos f\hat{a}_x) - \frac{\hat{a}_c(1 - e\cos f)\sin\theta}{\hat{h}\tan i}\hat{a}_z \quad (3.55)$$

$$\frac{dM}{d\tau} = \hat{v}_o + \frac{\hat{a}_c}{\hat{h}e}((\cos f - 2e)\hat{a}_x - (2 - e\cos f)\sin f\hat{a}_y) \quad (3.56)$$

To further simplify the analysis, we can orbit-average these equations. The averaged equations ignore the small periodic perturbations, but capture the larger, more significant, and secular drifts. To orbit-average these equations, the method given in Junkins and Schaub (2009) is applied:

$$\frac{d\bar{\xi}}{d\tau} = \frac{(1 - e^2)^{\frac{3}{2}}}{2\pi}\int_0^{2\pi}\frac{1}{(1 + e\cos f)^2}\frac{d\xi}{d\tau}df \quad (3.57)$$

$$\frac{d\bar{\xi}}{d\tau} \approx \frac{1}{2\pi}\int_0^{2\pi}(1 - 2e\cos f)\frac{d\xi}{d\tau}df \quad (3.58)$$

where $\xi \in \mathbb{R}$ represents any orbital element and $\bar{\xi} \in \mathbb{R}$ denotes the term that has been orbit-averaged. The orbit-averaged expressions for the J_2 perturbation can be found in Junkins and Schaub (2009). Based on the small eccentricity assumption, using the dimensionless time (3.9) leads to

$$\frac{d\bar{\hat{a}}}{d\tau} = 0 \quad (3.59)$$

$$\frac{d\bar{i}}{d\tau} = 0 \quad (3.60)$$

$$\frac{d\bar{e}}{d\tau} = 0 \quad (3.61)$$

$$\frac{d\bar{\Omega}}{d\tau} = -\frac{3\hat{R}_e^2}{2\hat{a}_c^2}J_2\hat{v}_o\cos i \quad (3.62)$$

$$\frac{d\bar{\omega}}{d\tau} = \frac{3\hat{R}_e^2}{4\hat{a}_c^2}J_2\hat{v}_o(5\cos^2 i - 1) \quad (3.63)$$

$$\frac{d\bar{M}}{d\tau} = \hat{v}_o + \frac{3\hat{R}_e^2}{4\hat{a}_c^2}J_2\hat{v}_o(3\cos^2 i - 1) \quad (3.64)$$

On the other hand, we can get similar expressions for the effects of the atmospheric drag. Substituting (3.44) into (3.51)–(3.56), one has

$$\begin{aligned}\frac{d\hat{a}_c}{d\tau} \approx -\frac{\hat{a}_c^4\hat{\beta}}{\hat{h}}(1 - 3e\cos f)\Big(f'^2 - 2\hat{w}_E f'\cos i \\ + \hat{w}_E^2\left(\frac{1}{2}\cos^2(f + \omega)\sin^2 i + \cos^2 i\right)\Big)\end{aligned} \quad (3.65)$$

$$\frac{di}{d\tau} \approx \frac{\hat{a}_c^4 \hat{\beta} \cos(f+\omega)}{2\hat{h}} 5e\cos f \hat{w}_E \cos(f+\omega) \sin i (f' - \hat{w}_E \cos i) - \frac{\hat{a}_c^4 \hat{\beta} \cos(f+\omega)}{2\hat{h}} \hat{w}_E \cos(f+\omega) \sin i (f' - \hat{w}_E \cos i) \tag{3.66}$$

$$\begin{aligned}\frac{de}{d\tau} \approx & -\frac{\hat{a}_c^3 \hat{\beta}}{2\hat{h}} e \sin^2 f \left(f'^2 - \hat{w}_E f' \cos i + \frac{\cos^2(f+\omega)\sin^2 i \hat{w}_E^2}{8} \right) \\ & - \frac{\hat{a}_c^3 \hat{\beta}}{2\hat{h}} (2\cos f + e - 9e\cos^2 f) \left(f'^2 - 2\hat{w}_E f' \cos i \right. \\ & \left. + \hat{w}_E^2 \left(\frac{1}{2} \cos^2(f+\omega) \sin^2 i + \cos^2 i \right) \right)\end{aligned} \tag{3.67}$$

$$\begin{aligned}\frac{d\Omega}{d\tau} \approx & -\frac{\hat{a}_c^3 \hat{\beta} \sin(f+\omega)}{2\hat{h}} (1 - 5e\cos f) \left(\hat{w}_E f' \cos(f+\omega) \right. \\ & \left. - \hat{w}_E^2 \cos(f+\omega) \cos i \right)\end{aligned} \tag{3.68}$$

$$\begin{aligned}\frac{d\omega}{d\tau} \approx & \frac{\hat{a}_c^3 \hat{\beta}}{2\hat{h}e} \left(e \sin f \cos f \left(f'^2 - \hat{w}_E f' \cos i + \frac{\hat{w}_E^2}{2} \cos^2(f+\omega) \sin^2 i \right) \right. \\ & - (2\sin f - 9e \sin f \cos f) \left(f'^2 - 2\hat{w}_E f' \cos i \right. \\ & \left. \left. + \hat{w}_E^2 \left(\frac{1}{2} \cos^2(f+\omega) \sin^2 i + \cos^2 i \right) \right) \right) \\ & + \frac{\hat{a}_c^3 \hat{\beta} \sin(f+\omega)}{2\hat{h}} \cos i \hat{w}_E \cos(f+\omega)(f' - \hat{w}_E \cos i) \\ & - \frac{5\hat{a}_c^3 \hat{\beta} \sin(f+\omega)}{2\hat{h}} \cos i e \cos f \hat{w}_E \cos(f+\omega)(f' - \hat{w}_E \cos i)\end{aligned} \tag{3.69}$$

$$\begin{aligned}\frac{M}{d\tau} \approx & \hat{v}_o - \frac{\hat{a}_c^3 \hat{\beta}}{2\hat{h}e} \left(e \sin f \cos f \left(f'^2 - \hat{w}_E f' \cos i \right. \right. \\ & \left. + \frac{1}{2} \hat{w}_E^2 \cos^2(f+\omega) \sin^2 i \right) - (2 - 9e\cos f) \sin f \left(f'^2 \right. \\ & \left. \left. - 2\hat{w}_E f' \cos i + \hat{w}_E^2 \left(\frac{1}{2} \cos^2(f+\omega) \sin^2 i + \cos^2 i \right) \right) \right)\end{aligned} \tag{3.70}$$

Then, applying the orbit-averaged method given in (3.57) and (3.58) results in

$$\frac{d\bar{\hat{a}}_c}{d\tau} \approx -\frac{\hat{a}_c^4 \hat{\beta}}{\hat{h}} \left(f'^2 - 2\hat{w}_E f' \cos i + \hat{w}_E^2 \left(\frac{1}{4} \sin^2 i + \cos^2 i \right) \right) \tag{3.71}$$

$$\frac{d\bar{i}}{d\tau} \approx -\frac{\hat{a}_c^3\hat{\beta}}{4\hat{h}}(\hat{w}_E f' \sin i - \hat{w}_E^2 \sin i \cos i) \tag{3.72}$$

$$\begin{aligned}\frac{d\bar{e}}{d\tau} \approx &-\frac{\hat{a}_c^4\hat{\beta}}{4\hat{h}} e f'^2 \left(1 - \frac{\hat{w}_E}{f'}\cos i + \left(\frac{\hat{w}_E}{f'}\right)^2 \sin^2 i \left(\frac{1}{8} + \frac{1}{4}\sin^2\omega\right)\right) \\ &- \frac{\hat{a}_c^3\hat{\beta}}{2\hat{h}} e f'^2 \left(1 - 2\frac{\hat{w}_E}{f'}\cos i + \left(\frac{\hat{w}_E}{f'}\right)^2 \left(\frac{1}{4}\sin^2 i + \cos^2 i\right)\right) \\ &- \frac{9\hat{a}_c^3\hat{\beta}e}{32\hat{h}}\Big(8f'^2 - 8\hat{w}_E f'(2-\cos i)\cos i \\ &+ \hat{w}_E^2 \sin^2 i(1 + 2\cos^2\omega)\Big)\end{aligned} \tag{3.73}$$

$$\frac{d\bar{\Omega}}{d\tau} \approx 0 \tag{3.74}$$

$$\frac{d\bar{\omega}}{d\tau} \approx -\frac{5\hat{a}_c^3\hat{\beta}}{8\hat{h}}\hat{w}_E^2 \sin^2 i \sin\omega\cos\omega \tag{3.75}$$

$$\frac{d\bar{M}}{d\tau} \approx \hat{v}_o + \frac{5\hat{a}_c^3\hat{\beta}}{8\hat{h}}\hat{w}_E^2 \sin^2 i \sin\omega\cos\omega \tag{3.76}$$

Combining (3.59)–(3.64) with (3.71)–(3.76), the final orbit-averaged expressions affected by the J_2 perturbation and the atmospheric drag can be established as follows. These equations will be used in the numerical simulations to describe the motion of the chief microsatellite.

$$\frac{d\bar{\hat{a}}_c}{d\tau} \approx -\frac{\hat{a}_c^4\hat{\beta}}{4\hat{h}}(4f'^2 - 8\hat{w}_E f'\cos i + \hat{w}_E^2(\sin^2 i + 4\cos^2 i)) \tag{3.77}$$

$$\frac{d\bar{i}}{d\tau} \approx -\frac{\hat{a}_c^3\hat{\beta}}{4\hat{h}} f'^2(\hat{w}_E f' \sin i - \hat{w}_E^2 \sin i \cos i) \tag{3.78}$$

$$\begin{aligned}\frac{d\bar{e}}{d\tau} \approx &-\frac{\hat{a}_c^4\hat{\beta}}{32\hat{h}} e(8f'^2 - 8\hat{w}_E f'\cos i + \hat{w}_E^2 \sin^2 i(1 + 2\sin^2\omega)) \\ &- \frac{\hat{a}_c^3\hat{\beta}}{8\hat{h}} e(4f'^2 - 8\hat{w}_E f'\cos i + \hat{w}_E^2(\sin^2 i + 4\cos^2 i)) \\ &- \frac{9\hat{a}_c^3\hat{\beta}}{32\hat{h}} e\Big(8f'^2 - 16\hat{w}_E f'\cos i + 8\hat{w}_E f'\cos^2 i \\ &+ \hat{w}_E^2 \sin^2 i(1 + 2\cos^2\omega)\Big)\end{aligned} \tag{3.79}$$

$$\frac{d\bar{\boldsymbol{\Omega}}}{d\tau} \approx -\frac{3\hat{R}_e^2}{2\hat{a}_c^2} J_2 \hat{v}_o \cos i \tag{3.80}$$

$$\frac{d\bar{\omega}}{d\tau} = \frac{3\hat{R}_e^2}{4\hat{a}_c^2} J_2 \hat{v}_o (5\cos^2 i - 1) - \frac{5\hat{a}_c^3 \hat{\beta}}{8\hat{h}} \hat{w}_E^2 \sin^2 i \sin\omega \cos\omega \tag{3.81}$$

$$\frac{d\bar{M}}{d\tau} \approx 2\hat{v}_o + \frac{3}{4} J_2 \hat{v}_o \frac{\hat{R}_e^2}{\hat{a}_c^2} (3\cos^2 i - 1) + \frac{5\hat{a}_c^3 \hat{\beta}}{8\hat{h}} \hat{w}_E^2 \sin^2 i \sin\omega \cos\omega \tag{3.82}$$

3.4.5 Effect of perturbation forces on relative motion

In the preceding subsections, the system equation of the chief microsatellite's motion under the effect of the atmospheric drag and the J_2 perturbations are mathematically modeled. In this subsection, the effect of those two perturbations acting on the relative motion between the deputy microsatellite and the chief microsatellite is quantitatively analyzed.

3.4.5.1 *Effect of the J_2 perturbation on relative motion*

Let $\nabla \boldsymbol{J}_2(\boldsymbol{r}_c) \in \mathbb{R}^{3\times 3}$ be the gradient of the J_2 perturbation field. To obtain the relative acceleration arising by this perturbation, the calculation method in Schweigart and Sedwick (2002) is used. It is obtained by multiplying the relative position vector by $\nabla \boldsymbol{J}_2(\boldsymbol{r}_c)$:

$$\Delta \boldsymbol{f}_{J_2} = \boldsymbol{f}_{J_2,d} - \boldsymbol{f}_{J_2,c} = \nabla \boldsymbol{J}_2(\boldsymbol{r}_c) \boldsymbol{r}_e \tag{3.83}$$

where $\nabla \boldsymbol{J}_2(\boldsymbol{r}_c)$ is specified as

$$\nabla \boldsymbol{J}_2(\boldsymbol{r}_c) = \\ \check{b} \begin{bmatrix} 4 - 12\sin^2 i \sin^2\theta & 4\sin^2 i \sin 2\theta & 4\sin 2i \sin\theta \\ 4\sin^2 i \sin 2\theta & \sin^2 i (7\sin^2\theta - 2) - 1 & -\sin 2i \cos\theta \\ 4\sin 2i \sin\theta & -\sin 2i \cos\theta & \sin^2 i (2 + 5\sin^2\theta) - 3 \end{bmatrix} \tag{3.84}$$

with $\check{b} = \frac{6\mu_0 J_2 R_e^2}{4 r_c^5}$.

The above gradient can be simplified by orbit-averaging it to capture only the secular effects and eliminate the periodic effects. To carry out this simplification, the following integral is introduced first.

$$\nabla \bar{\boldsymbol{J}}_2 = \frac{1}{2\pi} \int_0^{2\pi} \nabla \boldsymbol{J}_2 d\theta \tag{3.85}$$

Note that r_c is a function of θ through the previously defined relation

$$\frac{1}{r_c^5} \approx \frac{1}{a_c^5}(1 + 5e\cos f) = \frac{1}{a_c^5}(1 + 5e\cos(\theta - \boldsymbol{\omega})) \tag{3.86}$$

Based on this equation, it leaves (3.85) as

$$\nabla \bar{\boldsymbol{J}}_2 = \upsilon_o^2 \boldsymbol{Q} + \upsilon_o^2 e \tilde{\boldsymbol{Q}} \tag{3.87}$$

with $\boldsymbol{Q} \in \mathbb{R}^{3\times3}$ and $\tilde{\boldsymbol{Q}} \in \mathbb{R}^{3\times3}$ determined as

$$\boldsymbol{Q} = \begin{bmatrix} 4s & 0 & 0 \\ 0 & -s & 0 \\ 0 & 0 & -3s \end{bmatrix} \tag{3.88}$$

$$\tilde{\boldsymbol{Q}} = \begin{bmatrix} 0 & 0 & 4e\tilde{s}\sin\omega \\ 0 & 0 & -e\tilde{s}\cos\omega \\ 4e\tilde{s}\sin\omega & -e\tilde{s}\cos\omega & 0 \end{bmatrix} \tag{3.89}$$

where $s = \frac{3R_e^2 J_2}{8a_c^2}(1+3\cos 2i)$ and $\tilde{s} = \frac{15R_e^2 J_2}{4a_c^2}\sin 2i$.

To this end, the dimensionless relative acceleration due to the J_2 perturbation can be obtained from (3.11)–(3.13) and (3.87)–(3.89) as

$$\Delta\hat{\boldsymbol{f}}_{J_2} = \frac{1}{\upsilon_{o0}^2 d_0}\Delta\boldsymbol{f}_{J_2} = \hat{\upsilon}_o^2 \begin{bmatrix} 4s & 0 & 4e\tilde{s}\sin\omega \\ 0 & -s & -e\tilde{s}\cos\omega \\ 4e\tilde{s}\sin\omega & -e\tilde{s}\cos\omega & -3s \end{bmatrix}\begin{bmatrix} \hat{x} \\ \hat{y} \\ \hat{z} \end{bmatrix} \tag{3.90}$$

3.4.5.2 Effect of the drag perturbation on relative motion

The effects of the atmospheric drag on the motion of the deputy microsatellite relative to the chief microsatellite are analyzed in this part. Based on the atmospheric drag in (3.33) and its dimensionless version in (3.37), the dimensionless velocity term $\hat{\boldsymbol{V}}_r$ of (3.38) can be calculated as

$$\hat{\boldsymbol{V}}_r = \boldsymbol{V}_0 + \boldsymbol{V}_1 \tag{3.91}$$

where $\boldsymbol{V}_0 = [\upsilon_{0,x} \;\; \upsilon_{0,y} \;\; \upsilon_{0,z}] \in \mathbb{R}^3$ contains the zeroth order terms in $\hat{x}$, $\hat{y}$, $\hat{z}$, $\hat{x}'$, $\hat{y}'$, and $\hat{z}'$. $\boldsymbol{V}_0$ represents the absolute velocity of the chief microsatellite. $\boldsymbol{V}_1 = [\upsilon_{1,x} \;\; \upsilon_{1,y} \;\; \upsilon_{1,z}] \in \mathbb{R}^3$ contains the first-order terms in $\hat{x}$, $\hat{y}$, $\hat{z}$, $\hat{x}'$, $\hat{y}'$, and $\hat{z}'$, and it denotes the relative velocity of the deputy microsatellite relative to the chief microsatellite. Both are specified as

$$\boldsymbol{V}_0 = \begin{bmatrix} \frac{r_c'}{d_0} \\ \frac{r_c}{d_0}(f' - \hat{\omega}_e\cos i) \\ \frac{r_c}{d_0}\hat{\omega}_e\cos\theta\sin i \end{bmatrix} = \frac{\hat{a}_c}{d_0}\begin{bmatrix} e\sin f f' \\ (1-e\cos f)(f' - \hat{\omega}_e\cos i) \\ (1-e\cos f)\hat{\omega}_e\cos\theta\sin i \end{bmatrix} \tag{3.92}$$

$$\boldsymbol{V}_1 = \begin{bmatrix} \hat{x}' - \hat{y}(f' - \hat{\omega}_e \cos i) - \hat{z}\hat{\omega}_e \cos\theta \sin i \\ \hat{y}' + \hat{x}(f' - \hat{\omega}_e \cos i) + \hat{z}\hat{\omega}_e \sin\theta \sin i \\ \hat{z}' + \hat{x}\hat{\omega}_e \cos\theta \sin i - \hat{y}\hat{\omega}_e \sin\theta \sin i \end{bmatrix} \tag{3.93}$$

When the magnitude of $\boldsymbol{V}_0$ is of the order of 10^4, then the magnitude of $\boldsymbol{V}_1$ is of the order of one. This is quite larger. To linearize the dimensionless atmospheric drag, $||\hat{\boldsymbol{V}}_r||\hat{\boldsymbol{V}}_r$ should be linearized, and it can be linearized as

$$\begin{aligned} ||\hat{\boldsymbol{V}}_r||\hat{\boldsymbol{V}}_r &= \|\boldsymbol{V}_0 + \boldsymbol{V}_1\|\,(\boldsymbol{V}_0 + \boldsymbol{V}_1) \\ &= \sqrt{\|\boldsymbol{V}_0\|^2 + 2\boldsymbol{V}_0^{\mathrm{T}}\boldsymbol{V}_1 + \|\boldsymbol{V}_1\|^2}(\boldsymbol{V}_0 + \boldsymbol{V}_1) \\ &= \|\boldsymbol{V}_0\|\sqrt{1 + \frac{2\boldsymbol{V}_0^{\mathrm{T}}\boldsymbol{V}_1}{\|\boldsymbol{V}_0\|} + \frac{\|\boldsymbol{V}_1\|^2}{\|\boldsymbol{V}_0\|^2}}(\boldsymbol{V}_0 + \boldsymbol{V}_1) \end{aligned} \tag{3.94}$$

Applying a binomial series expansion to $\sqrt{||\boldsymbol{V}_0||^2 + 2\boldsymbol{V}_0^{\mathrm{T}}\boldsymbol{V}_1 + ||\boldsymbol{V}_1||^2}$ and ignoring the second or higher order terms yield

$$||\boldsymbol{V}_0 + \boldsymbol{V}_1||\,(\boldsymbol{V}_0 + \boldsymbol{V}_1) \approx ||\boldsymbol{V}_0||\boldsymbol{V}_0 + ||\boldsymbol{V}_0||\boldsymbol{V}_1 + \frac{\boldsymbol{V}_0^{\mathrm{T}}\boldsymbol{V}_1}{||\boldsymbol{V}_0||}\boldsymbol{V}_0 \tag{3.95}$$

The reason of ignoring higher-order terms is that there is significant difference of the magnitudes between $\boldsymbol{V}_0$ and $\boldsymbol{V}_1$. Actually, the orbit of the deputy microsatellite follows the chief microsatellite, the relative velocity $\boldsymbol{V}_1$ is significantly smaller in magnitude than the absolute velocity $\boldsymbol{V}_0$.

Invoking (3.95) leaves $||\boldsymbol{V}_0||$ and $||\boldsymbol{V}_0||\boldsymbol{V}_0$ as

$$\|\boldsymbol{V}_0\| \approx \frac{a_c}{d_0}(1 - e\cos f)\,f'\left(1 - \frac{\hat{w}_E}{f'}\cos i + \frac{1}{2}\left(\frac{\hat{w}_E}{f'}\right)^2 \sin^2 i \cos^2\theta\right) \tag{3.96}$$

$$\|\boldsymbol{V}_0\|\,\boldsymbol{V}_0 = \frac{a_c^2}{d_0^2}\begin{bmatrix} e\sin f\,f'^2\left(1 - \frac{\hat{w}_E}{f'}\cos i + \frac{1}{2}\left(\frac{\hat{w}_E}{f'}\right)^2\cos^2\theta \sin^2 i\right) \\ (1 - 2e\cos f)\,f'^2\left(1 - 2\frac{\hat{w}_E}{f'}\cos i + \left(\frac{\hat{w}_E}{f'}\right)^2\cos^2\theta\left(1 + \frac{\sin^2 i}{2}\right)\right) \\ (1 - 2e\cos f)\,f'^2\left(\frac{\hat{w}_E}{f'}\cos\theta \sin i - \left(\frac{\hat{w}_E}{f'}\right)^2\cos\theta \sin i \cos i\right) \end{bmatrix} \tag{3.97}$$

Then, the relative J_2 perturbation can be orbit-averaged as follows to keep only the secular components.

$$\frac{1}{2\pi}\int_0^{2\pi}\|\boldsymbol{V}_0\|\,\boldsymbol{V}_0 d\theta = \frac{a_c^2}{4d_0^2}\begin{bmatrix} 0 \\ 4f'^2 - 8\hat{w}_E f'\cos i + \hat{w}_E^2\sin^2 i + 4\hat{w}_E^2\cos^2 i \\ -4e\sin i\cos\omega(\hat{w}_E f' - \hat{w}_E^2\cos i) \end{bmatrix} \tag{3.98}$$

The term $||\boldsymbol{V}_0||\boldsymbol{V}_1$ in (3.95) can be calculated as

$$\|\boldsymbol{V}_0\|\,\boldsymbol{V}_1 = \frac{a_c}{d_0}(1 - e\cos f)\,f'\left(1 + \frac{1}{2}\left(\frac{\hat{w}_E}{f'}\right)^2\sin^2 i\cos^2\theta - \frac{\hat{w}_E}{f'}\cos i\right)\begin{bmatrix}\hat{x}' \\ \hat{y}' \\ \hat{z}'\end{bmatrix} + \frac{a_c}{d_0}\begin{bmatrix} 0 & -\tilde{\varphi}_1 & -\tilde{\varphi}_2 \\ \tilde{\varphi}_1 & 0 & \tilde{\varphi}_3 \\ \tilde{\varphi}_2 & -\tilde{\varphi}_3 & 0 \end{bmatrix}\begin{bmatrix}\hat{x} \\ \hat{y} \\ \hat{z}\end{bmatrix} \tag{3.99}$$

where

$$\tilde{\varphi}_1 = (1 - e\cos f)f'^2\left(1 - 2\frac{\hat{w}_E}{f'}\cos i + \left(\frac{\hat{w}_E}{f'}\right)^2\cos^2\theta\left(1 + \frac{\sin^2 i}{2}\right)\right) \tag{3.100}$$

$$\tilde{\varphi}_2 = (1 - e\cos f)(\hat{w}_E f'\cos\theta\sin i - \hat{w}_E^2\cos\theta\sin i\cos i) \tag{3.101}$$

$$\tilde{\varphi}_3 = (1 - e\cos f)(\hat{w}_E f'\sin\theta\sin i - \hat{w}_E^2\sin\theta\sin i\cos i) \tag{3.102}$$

Orbit-averaging these terms results in

$$\frac{1}{2\pi}\int_0^{2\pi}\|\boldsymbol{V}_0\|\,\boldsymbol{V}_1 d\theta = \frac{a_c}{d_0}f'\left(1 + \frac{1}{4}\left(\frac{\hat{w}_E}{f'}\right)^2\sin^2 i - \frac{\hat{w}_E}{f'}\cos i\right)\begin{bmatrix}\hat{x}' \\ \hat{y}' \\ \hat{z}'\end{bmatrix} + \frac{a_c}{d_0}\begin{bmatrix} 0 & -\varphi_1 & -\varphi_2 \\ \varphi_1 & 0 & \varphi_3 \\ \varphi_2 & -\varphi_3 & 0 \end{bmatrix}\begin{bmatrix}\hat{x} \\ \hat{y} \\ \hat{z}\end{bmatrix} \tag{3.103}$$

with $\varphi_1 = f'^2 - 2\hat{w}_E f'\cos i + \hat{w}_E^2\cos^2 i + \frac{1}{4}\hat{w}_E^2\sin^2 i$, $\varphi_2 = -\frac{1}{2}e\sin i \times \cos\omega(\hat{w}_E f' - \hat{\omega}_e^2\cos i)$, and $\varphi_3 = -\frac{1}{2}e\sin i\sin\omega\left(\hat{w}_E - \frac{\hat{w}_E^2}{f'}\cos i\right)$.

For the term $\frac{V_0^T V_1}{\|V_0\|} V_0$, the followings can be calculated

$$\begin{aligned} v_{0,x} v_{1,x} = \frac{a_c}{d_0} \Big(& \hat{x}' e f' \sin f - \hat{y} e \sin f (f'^2 - \hat{w}_E f' \cos i) \\ & - \hat{z} e \sin f \sin i \cos\theta \hat{w}_E f' \Big) \end{aligned} \tag{3.104}$$

$$\begin{aligned} v_{0,y} v_{1,y} = & \frac{a_c}{d_0} \hat{y} (1 - e \cos f)(f' - \hat{w}_E \cos i) \\ & + \frac{a_c}{d_0} \hat{x} (1 - e \cos f) f'^2 \left(1 - \frac{\hat{w}_E}{f'} \cos i\right)^2 \\ & + \frac{a_c}{d_0} \hat{z} (1 - e \cos f) \sin i \sin\theta \hat{w}_E (f' - \hat{w}_E \cos i) \end{aligned} \tag{3.105}$$

$$\begin{aligned} v_{0,z} v_{1,z} = \frac{a_c}{d_0} \Big(& \hat{z}' (1 - e \cos f) \hat{w}_E \cos\theta \sin i + \hat{x} \hat{w}_E^2 \sin^2 i \cos^2\theta \\ & - \hat{y} (1 - e \cos f) \hat{w}_E^2 \sin\theta \cos\theta \sin^2 i \Big) \end{aligned} \tag{3.106}$$

$$\|V_0\|^{-1} = \frac{d_0}{a_c} (1 + e \cos f) \frac{1}{f'} \left(1 + \frac{\hat{w}_E}{f'} \cos i - \frac{1}{2} \left(\frac{\hat{w}_E}{f'}\right)^2 \sin^2 i \cos^2\theta\right) \tag{3.107}$$

Based on (3.104)–(3.107), it follows that

$$\begin{aligned} \frac{V_0 V_1}{\|V_0\|} = & e \sin f \left(1 + \frac{\hat{w}_E}{f'} \cos i - \frac{1}{2} \left(\frac{\hat{w}_E}{f'}\right)^2 \sin^2 i \cos^2 \theta\right) \hat{x}' \\ & + \left(1 - \frac{1}{2} \left(\frac{\hat{w}_E}{f'}\right)^2 \sin^2 i \cos^2\theta - \left(\frac{\hat{w}_E}{f'}\right)^2 \cos^2 i\right) \hat{y}' \\ & + \left(\frac{\hat{w}_E}{f'} \cos\theta \sin i + \left(\frac{\hat{w}_E}{f'}\right)^2 \cos\theta \sin i \cos i\right) \hat{z}' \\ & + f' \left(1 - \frac{\hat{w}_E}{f'} \cos i + \frac{1}{2} \left(\frac{\hat{w}_E}{f'}\right)^2 \sin^2 i \cos^2\theta - \left(\frac{\hat{w}_E}{f'}\right)^2 \cos^2 i\right) \hat{x} \\ & - e \sin f f' \left(1 - \frac{1}{2} \left(\frac{\hat{w}_E}{f'}\right)^2 \sin^2 i \cos^2\theta - \left(\frac{\hat{w}_E}{f'}\right)^2 \cos^2 i\right) \hat{y} \\ & + \frac{\hat{w}_E^2}{f'} \sin\theta \cos\theta \sin^2 i \, \hat{y} + \hat{w}_E \sin\theta \sin i \, \hat{z} \\ & - e \sin f \sin i \cos\theta \left(\hat{w}_E + \frac{\hat{w}_E^2}{f'} \cos i\right) \hat{z} \end{aligned} \tag{3.108}$$

Multiplying (3.108) by the zeroth order and orbit-averaging it results in

$$\frac{1}{2\pi}\int_0^{2\pi}\left(\frac{\boldsymbol{V}_0\boldsymbol{V}_1}{\|\boldsymbol{V}_0\|}\right)\boldsymbol{V}_0 d\theta = \frac{a_c}{d_0}\begin{bmatrix} 0 & 0 & \psi_{13} \\ 0 & \psi_{22} & \psi_{23} \\ \psi_{13} & \psi_{23} & \psi_{33} \end{bmatrix}\begin{bmatrix} \hat{x}' \\ \hat{y}' \\ \hat{z}' \end{bmatrix} + \frac{a_c}{d_0}\begin{bmatrix} 0 & 0 & \psi'_{13} \\ \psi'_{21} & 0 & \psi'_{23} \\ \psi'_{31} & \psi'_{32} & 0 \end{bmatrix}\begin{bmatrix} \hat{x} \\ \hat{y} \\ \hat{z} \end{bmatrix} \tag{3.109}$$

where

$$\psi_{13} = -\frac{1}{2} e \sin i \sin\omega f' \hat{w}_E - \frac{1}{2f'} e \sin i \sin\omega f' \hat{w}_E^2 \cos i \tag{3.110}$$

$$\psi_{22} = f' - \frac{1}{4}\frac{\hat{w}_E^2}{f'}\sin^2 i - \frac{\hat{w}_E^2}{f'}\cos^2 i - \hat{w}_E \cos i \tag{3.111}$$

$$\psi_{23} = -\frac{1}{2} e\hat{w}_E \sin i \cos\omega \tag{3.112}$$

$$\psi_{33} = \frac{1}{2}\frac{\hat{w}_E^2}{f'}\sin^2 i \tag{3.113}$$

$$\psi'_{13} = \frac{1}{2} e \sin i \cos\omega \hat{w}_E f' \tag{3.114}$$

$$\psi'_{21} = f'^2 - 2\hat{w}_E f' \cos i + \frac{1}{4}\hat{w}_E^2 \sin^2 i \tag{3.115}$$

$$\psi'_{23} = -\frac{1}{2} e \sin i \sin\omega(\hat{w}_E f' - \hat{w}_E^2 \cos i) \tag{3.116}$$

$$\psi'_{31} = -\frac{1}{2} e \sin i \cos\omega(\hat{w}_E f' - \hat{w}_E^2 \cos i) \tag{3.117}$$

$$\psi'_{32} = \frac{1}{2} e \sin i \sin\omega f' \hat{w}_E \tag{3.118}$$

Combining (3.98), (3.103), and (3.109), the orbit-averaged dimensionless acceleration due to the atmospheric drag perturbation can be finally written as

$$\hat{f}_{drag,c} = -\frac{\hat{\beta}\hat{a}_c^2}{2}\begin{bmatrix} 0 \\ d_2^0 \\ d_3^0 \end{bmatrix} - \frac{\hat{\beta}\hat{a}_c}{2}\begin{bmatrix} 0 & d_{12}^1 & d_{13}^1 \\ d_{21}^1 & 0 & d_{23}^1 \\ d_{31}^1 & d_{32}^1 & 0 \end{bmatrix}\hat{\boldsymbol{r}}_e - \frac{\hat{\beta}\hat{a}_c}{2}\begin{bmatrix} d_{11}^2 & 0 & d_{13}^2 \\ 0 & d_{22}^2 & d_{23}^2 \\ d_{13}^2 & d_{23}^2 & d_{33}^2 \end{bmatrix}\hat{\boldsymbol{r}}_e' \tag{3.119}$$

where

$$d_2^0 = f'^2 - 2\hat{w}_E f' \cos i + \frac{1}{4}\hat{w}_E^2 \sin^2 i + \hat{w}_E^2 \cos^2 i \tag{3.120}$$

$$d_3^0 = -e \sin i \cos\omega(\hat{w}_E f' - \hat{w}_E^2 \cos i) \tag{3.121}$$

$$d_{12}^1 = -f'^2 + 2\hat{w}_E f' \cos i - \frac{1}{4}\hat{w}_E^2 \sin^2 i - \hat{w}_E^2 \cos^2 i \tag{3.122}$$

$$d_{13}^1 = \frac{1}{2} e \sin i \cos\boldsymbol{\omega}(\hat{w}_E f' - \hat{w}_E^2 \cos i) \tag{3.123}$$

$$d_{21}^1 = 2f'^2 - 4\hat{w}_E f' \cos i + \frac{1}{4}\hat{w}_E^2 \sin^2 i + \frac{1}{2}\hat{w}_E^2 \cos^2 i \tag{3.124}$$

$$d_{23}^1 = -e \sin i \sin\omega(\hat{w}_E f' - \hat{w}_E^2 \cos i) \tag{3.125}$$

$$d_{32}^1 = e \sin i \sin\omega\hat{w}_E f' - \frac{1}{2} e \sin i \sin\omega\hat{w}_E^2 \cos i \tag{3.126}$$

$$d_{11}^2 = f' - \hat{w}_E f' \cos i + \frac{\hat{w}_E^2}{4f'} \sin^2 i \tag{3.127}$$

$$d_{13}^2 = -\frac{1}{2} e \sin i \sin\omega\hat{w}_E - \frac{1}{2f'} e \sin i \sin\omega\hat{w}_E^2 \cos i \tag{3.128}$$

$$d_{22}^2 = 2f' - 2\hat{w}_E \cos i - \frac{\hat{w}_E}{f'} \cos^2 i \tag{3.129}$$

$$d_{23}^2 = -\frac{1}{2} e \sin i \cos\omega f' \hat{w}_E \tag{3.130}$$

$$d_{33}^2 = f' - \hat{w}_E f' \cos i + \frac{3\hat{w}_E^2}{4f'} \sin^2 i \tag{3.131}$$

Subtracting the acceleration of the chief microsatellite to that of the deputy microsatellite, we can obtain the acceleration of the deputy microsatellite relative to the chief microsatellite as

$$\begin{aligned}\Delta\hat{\boldsymbol{f}}_{drag} &= \hat{\boldsymbol{f}}_{drag,d} - \hat{\boldsymbol{f}}_{drag,c} \\ &= -\frac{(\beta_d - \beta_c)\hat{a}_c^2 d_0}{2}\begin{bmatrix} 0 \\ d_2^0 \\ d_3^0 \end{bmatrix} - \frac{\beta_d \hat{a}_c d_0}{2}\begin{bmatrix} 0 & d_{12}^1 & d_{13}^1 \\ d_{21}^1 & 0 & d_{23}^1 \\ d_{31}^1 & d_{32}^1 & 0 \end{bmatrix}\hat{\boldsymbol{r}}_e \\ &\quad - \frac{\beta_d \hat{a}_c d_0}{2}\begin{bmatrix} d_{11}^2 & 0 & d_{13}^2 \\ 0 & d_{22}^2 & d_{23}^2 \\ d_{31}^2 & d_{32}^2 & d_{33}^2 \end{bmatrix}\hat{\boldsymbol{r}}_e'\end{aligned} \tag{3.132}$$

where $\beta_c = \beta = \frac{c_D A_c}{m_c}\rho_c$, $\beta_d = \frac{c_D^d A_d}{m_d}\rho_c$, $c_D^d \in \mathbb{R}$ is the deputy microsatellite's drag coefficient, $A_d \in \mathbb{R}$ is the cross sectional area of the deputy microsatellite, and $m_d \in \mathbb{R}$ is the deputy microsatellite's mass.

3.4.6 System equation of relative motion

Assuming that the chief microsatellite has a small eccentricity elliptical orbit, the system equation of the relative motion between the deputy and the chief microsatellite is presented in face of the J_2 and the atmospheric drag perturbations.

3.4.6.1 Calculation of orbital angular velocity

From the view point of angular momentum (Sabatini and Palmerini, 2008), under the effects of the J_2 and the atmospheric drag perturbations, we can compute the orbital angular velocity as

$$\hat{\boldsymbol{\omega}}_0 = \frac{\boldsymbol{\omega}_0}{v_{o0}} = \begin{bmatrix} \hat{\omega}_{0x} \\ 0 \\ \hat{\omega}_{0z} \end{bmatrix} = \begin{bmatrix} \frac{r_c}{\hat{h}}(\hat{\boldsymbol{f}}_{J_2,z} + \hat{\boldsymbol{f}}_{drag,z}) \\ 0 \\ \frac{\hat{h}}{\hat{r}_c^2} \end{bmatrix} \tag{3.133}$$

Invoking (3.23), (3.32), (3.90), and (3.119) yields

$$\begin{aligned}\hat{\omega}_{0x} &= -\frac{3J_2\hat{v}_o^2\hat{R}_e^2}{2\hat{h}}(1 + 3e\cos f)\sin\theta\sin 2i \\ &\quad - \frac{\hat{\beta}\hat{a}_c^2}{2\hat{h}}(1 - 3e\cos f)(f'\hat{w}_E\cos\theta\sin i - \hat{w}_E^2\cos\theta\sin i\cos i)\end{aligned} \tag{3.134}$$

$$\hat{\omega}_{0z} = \frac{\hat{h}}{\hat{a}_c^2}(1 + 2e\cos f) \tag{3.135}$$

Therefore, the dimensionless variables $\hat{h}'$, f', θ', and r_c' are deduced first to calculate the derivative of the angular velocity. In fact, the derivative of $\hat{h}$ can be expressed as follows due to the J_2 and the atmospheric drag effects based on the assumption of small eccentricity (Sabatini and Palmerini, 2008).

$$\begin{aligned} \hat{h}' &= \frac{\hat{r}_c(\boldsymbol{f}_{J_2,y} + \boldsymbol{f}_{drag,y})}{v_{o0}^2 d_0} \\ &= -\frac{3J_2\hat{v}_o^2\hat{R}_e^2}{2}(1 + 3e\cos f)\sin^2 i \sin 2\theta \\ &\quad - \frac{1}{4}\hat{\beta}\hat{a}_c^3(1 - 3e\cos f)\left(2f'^2 - 4f'\hat{w}_E\cos i \right. \\ &\quad \left. + \hat{w}_E^2\cos^2\theta\sin^2 i + 2\hat{w}_E^2\cos^2 i\right) \end{aligned} \tag{3.136}$$

Using the chain rule, the derivative of $\hat{r}_c$ is

$$\hat{r}_c' = \frac{d\hat{r}_c}{df}f' = \hat{a}_c e\sin f(1 - 2e\cos f)f' \tag{3.137}$$

Because the orbital rate $\hat{\omega}_{0z}$ can be rewritten into a function of the true anomaly f, the right ascension of ascending node Ω, and the argument of perigee ω, we have

$$\hat{\omega}_{0z} = \frac{\hat{h}}{\hat{r}_c^2} = f' + \omega' + \Omega'\cos i \tag{3.138}$$

where ω' and Ω' can be computed by using (3.80) and (3.81), respectively. Then, the derivative of the true anomaly is given by

$$\begin{aligned} f' &= \frac{\hat{h}}{\hat{r}_c^2} - (\omega' + \Omega'\cos i) \\ &= \frac{\hat{h}}{\hat{a}_c^2}(1 + 2e\cos f) - \hat{v}_o s + \frac{5\hat{a}_c^3\hat{\beta}}{8\hat{h}}\hat{w}_E^2\sin^2 i\sin\omega\cos\omega \end{aligned} \tag{3.139}$$

$$\begin{aligned} f'' = (f')' &= \left(\frac{\hat{h}}{\hat{a}_c^2} - 2\frac{\hat{h}}{\hat{a}_c^3}\hat{a}_c'\right)(1 + 2e\cos f) - \frac{2\hat{h}e\sin f f'}{\hat{a}_c^2} - (\hat{v}_o s)' \\ &\quad + \left(\frac{5\hat{a}_c^3\hat{\beta}}{8\hat{h}}\hat{w}_E^2\sin^2 i\sin\omega\cos\omega\right)' \end{aligned}$$

$$
\begin{aligned}
&= \left(\frac{\hat{h}}{\hat{a}_c^2} - 2\frac{\hat{h}}{\hat{a}_c^3}\hat{a}'_c\right)(1+2e\cos f) - \frac{2\hat{h}e\sin f f'}{\hat{a}_c^2} \\
&\quad - \frac{3J_2\hat{R}_e^2\sqrt{\mu_0}}{8}\left(-\frac{7}{2}\hat{a}_c^{-\frac{9}{2}}(1+3\cos 2i)\hat{a}'_c - 6\hat{a}_c^{-\frac{7}{2}}\sin 2i i'\right) \\
&\quad + 5\frac{3\hat{\beta}\hat{\omega}_e^2}{16\hat{h}^2}\left(3\hat{h}\hat{a}_c^2\sin^2 i\sin 2\omega a' + 2\hat{h}\hat{a}_c^3\sin i\cos i\sin 2\omega i'\right. \\
&\quad \left. + 2\hat{h}\hat{a}_c^3\sin^2 i\cos 2\omega\omega' - \hat{a}_c^3\sin^2 i\sin 2\omega\hat{h}'\right)
\end{aligned}
\tag{3.140}
$$

Inserting (3.139) into (3.137) leads to

$$
\hat{r}'_c = \hat{a}_c e\sin f\left(\frac{\hat{h}}{\hat{a}_c^2} - \hat{v}_O s + \frac{5\hat{a}_c^3\hat{\beta}}{8\hat{h}}\hat{w}_E^2\sin^2 i\sin\omega\cos\omega\right) \tag{3.141}
$$

The derivative of the true latitude is therefore given by

$$
\theta' = f' + \omega' = \frac{\hat{h}}{\hat{r}_c^2} - \Omega'\cos i = \frac{\hat{h}}{\hat{a}_c^2}(1+2e\cos f) + \frac{3}{2}J_2\hat{v}_o\frac{\hat{R}_e^2}{\hat{a}_c^2}\cos^2 i \tag{3.142}
$$

Based on the above results, the derivative of angular velocity can be obtained from (3.133) as

$$
\begin{aligned}
\hat{\boldsymbol{\omega}}'_x &= \frac{\hat{r}'_c(\hat{\boldsymbol{f}}_{J_2,z} + \hat{\boldsymbol{f}}_{drag,z}) + \hat{r}_c(\hat{\boldsymbol{f}}'_{J_2,z} + \hat{\boldsymbol{f}}'_{drag,z})}{\hat{h}} \\
&\quad - \frac{\hat{r}_c(\hat{\boldsymbol{f}}_{J_2,z} + \hat{\boldsymbol{f}}_{drag,z})\hat{h}'}{\hat{h}^2}
\end{aligned}
\tag{3.143}
$$

$$
\hat{\boldsymbol{\omega}}'_z = \left(\frac{\hat{h}}{\hat{r}_c^2}\right)' = \frac{\hat{h}'}{\hat{r}_c^2} - \frac{2\hat{h}\hat{r}'_c}{\hat{r}_c^3} = \frac{\hat{h}'(1+2e\cos f)}{\hat{a}_c^2} - \frac{2\hat{h}'(1+3e\cos f)\hat{r}'_c}{\hat{a}_c^3} \tag{3.144}
$$

where

$$
\begin{aligned}
\hat{\boldsymbol{f}}'_{J_2,z} &= -\frac{3J_2\hat{v}_o^2\hat{R}_e^2(1+4e\cos f)}{2}\left(\frac{2\cos 2i\sin\theta i'}{\hat{a}_c}\right. \\
&\quad \left. + \frac{\sin 2i\cos\theta\theta'}{\hat{a}_c} - \frac{4\sin 2i\sin\theta\hat{r}'_c}{\hat{a}_c^2}\right)
\end{aligned}
\tag{3.145}
$$

$$\begin{aligned}\hat{f}'_{drag,z} &= \hat{\beta}\hat{a}_c(1-e\cos f)r_c'(\hat{w}_E^2\cos\theta\sin i\cos i - f'\hat{w}_E\cos\theta\sin i)\\ &- \frac{\hat{\beta}\hat{a}_c^2(1-2e\cos f)}{2}\Big(\hat{w}_E\left(f''\cos\theta - f'\sin\theta\theta'\right)\sin i\\ &+\hat{w}_E f'\cos\theta\cos i i' + \frac{1}{2}\hat{w}_E^2\sin\theta\sin 2i\theta'\\ &-\hat{w}_E^2\cos\theta\cos 2i i'\Big)\end{aligned} \tag{3.146}$$

3.4.6.2 Summary of the relative motion equation

Based on the result in Section 3.4.6.1, the dimensionless equations of relative motion without control force can be finally established as

$$\hat{\boldsymbol{r}}_e'' + \boldsymbol{N}\hat{\boldsymbol{r}}_e' + \boldsymbol{K}\hat{\boldsymbol{r}}_e = \boldsymbol{F} \tag{3.147}$$

with $\boldsymbol{N} = \begin{bmatrix} n_{11} & n_{12} & n_{13} \\ n_{21} & n_{22} & n_{23} \\ n_{31} & n_{32} & n_{33} \end{bmatrix} \in \mathbb{R}^{3\times3}$, $\boldsymbol{K} = \begin{bmatrix} k_{11} & k_{12} & k_{13} \\ k_{21} & k_{22} & k_{23} \\ k_{31} & k_{32} & k_{33} \end{bmatrix} \in \mathbb{R}^{3\times3}$, and $\boldsymbol{F} = [F_1 \quad F_2 \quad F_3]^{\mathrm{T}} \in \mathbb{R}^3$, where

$$n_{11} = 0.5\hat{\beta}_d\hat{a}_c d_{11}^2 \tag{3.148}$$

$$n_{12} = -2\hat{\boldsymbol{\omega}}_{0z} \tag{3.149}$$

$$n_{13} = 0.5\hat{\beta}_d\hat{a}_c d_{13}^2 \tag{3.150}$$

$$n_{21} = 2\hat{\boldsymbol{\omega}}_{0z} \tag{3.151}$$

$$n_{22} = 0.5\hat{\beta}_d\hat{a}_c d_{22}^2 \tag{3.152}$$

$$n_{23} = -2\hat{\omega}_{0x} + 0.5\hat{\beta}_d\hat{a}_c d_{23}^2 \tag{3.153}$$

$$n_{31} = 0.5\hat{\beta}_d\hat{a}_c d_{31}^2 \tag{3.154}$$

$$n_{32} = 2\hat{\omega}_{0x} + 0.5\hat{\beta}_d\hat{a}_c d_{32}^2 \tag{3.155}$$

$$n_{33} = 0.5\hat{\beta}_d\hat{a}_c d_{33}^2 \tag{3.156}$$

$$k_{11} = -\hat{\omega}_{0z}^2 - 2\hat{v}_o^2(1+3e\cos f) - 4\hat{v}_o^2 s \tag{3.157}$$

$$k_{12} = -\hat{\omega}_{0z}' + 0.5\hat{\beta}_d\hat{a}_c d_{12}^1 \tag{3.158}$$

$$k_{13} = \hat{\omega}_{0x}\hat{\omega}_{0z} - 4\hat{v}_o^2 e\tilde{s}\sin\omega + 0.5\hat{\beta}_d\hat{a}_c d_{13}^1 \tag{3.159}$$

$$k_{21} = \hat{\omega}_{0z}' + 0.5\hat{\beta}_d\hat{a}_c d_{21}^1 \tag{3.160}$$

$$k_{22} = -\hat{\omega}_{0x}^2 - \hat{\omega}_{0z}^2 + \hat{v}_o^2(1 + 3e\cos f) + \hat{v}_o^2 s \tag{3.161}$$

$$k_{23} = -\hat{\omega}_{0x}' + \hat{v}_o^2 e\tilde{s}\cos\omega + 0.5\hat{\beta}_d\hat{a}_c d_{23}^1 \tag{3.162}$$

$$k_{31} = \hat{\omega}_{0x}\hat{\omega}_{0z} - 4\hat{v}_o^2 e\tilde{s}\sin\omega + 0.5\hat{\beta}_d\hat{a}_c d_{31}^1 \tag{3.163}$$

$$k_{32} = \hat{\omega}_{0x}' + \hat{v}_o^2 e\tilde{s}\cos\omega + 0.5\hat{\beta}_d\hat{a}_c d_{32}^1 \tag{3.164}$$

$$k_{33} = -\hat{\omega}_{0x}^2 + \hat{v}_o^2\,(1 + 3e\cos f + 3s) \tag{3.165}$$

$$F_1 = 0 \tag{3.166}$$

$$F_2 = -0.5(\hat{\beta}_d - \hat{\beta}_c)\hat{a}_c d_2^0 \tag{3.167}$$

$$F_3 = -0.5(\hat{\beta}_d - \hat{\beta}_c)\hat{a}_c d_3^0 \tag{3.168}$$

According to (3.147), the orbit-averaged technique can be used to obtain the constant linearized equation of relative motion without any numerical integration. This model is more convenient for practical engineering. Moreover, the constant linearized equation for circular orbit can also be established based on the small eccentricity assumption. To achieve this, all the time-varying parameters in (3.147) should be orbit-averaged to obtain constant parameters. Hence, the dimensionless orbital angular velocity $\hat{\omega}_0$ is given by

$$\hat{\omega}_{0x} = \frac{3\hat{\beta}\hat{a}_c^2 e}{4\hat{h}}\hat{w}_E\sin i\left(f' - \hat{w}_E\cos i\right) - \frac{9J_2\hat{v}_o^2\hat{R}_e^2 e}{4\hat{h}}\sin\omega\sin 2i \tag{3.169}$$

$$\hat{\omega}_{0z} = \frac{\hat{h}}{\hat{a}_c^2} \tag{3.170}$$

According to (3.136), (3.137), (3.139), and (3.142), the followings can be obtained after conducting orbit-averaged method.

$$\hat{h}' = \frac{1}{2}\hat{\beta}\hat{a}_c^3\left(f'^2 - 2f'\hat{w}_E\cos i + \frac{1}{2}\hat{w}_E^2\sin^2 i + \hat{w}_E^2\cos^2 i\right) \tag{3.171}$$

$$\hat{r}_c' = 0 \tag{3.172}$$

$$f' = \frac{\hat{h}}{\hat{r}_c^2} - \hat{v}_o s + \frac{5\hat{a}_c^3\hat{\beta}}{8\hat{h}}\hat{w}_E^2\sin^2 i\sin\omega\cos\omega \tag{3.173}$$

$$\theta' = \frac{\hat{h}}{\hat{r}_c^2} + \frac{3}{2}J_2\hat{v}_o\frac{\hat{R}_e^2}{\hat{a}_c^2}\cos^2 i \tag{3.174}$$

Based on (3.139), the second derivative of the true anomaly satisfies

$$
\begin{aligned}
f'' = \frac{\hat{h}}{\hat{a}_c^2} - \frac{2\hat{h}}{\hat{a}_c^3}\hat{a}_c' + \frac{3J_2\hat{R}_e^2\sqrt{\mu_0}}{8}\Bigg(\frac{7}{2}\hat{a}_c^{-\frac{9}{2}}(1+3\cos 2i)\hat{a}_c' \\
+ 6\hat{a}_c^{-\frac{7}{2}}\sin 2ii'\Bigg) + \frac{5\hat{\beta}\hat{w}_E^2}{16\hat{h}^2}\Big(2\hat{h}\hat{a}_c^3 \sin i\cos i\sin 2\omega i' \\
+ 3\hat{h}\hat{a}_c^2\sin^2 i\sin 2\omega a' + 2\hat{h}\hat{a}_c^3\sin^2 i\cos 2\omega\omega' \\
- \hat{a}_c^3\sin^2 i\sin 2\omega\hat{h}'\Big)
\end{aligned}
\tag{3.175}
$$

Then, we can calculate that

$$
\begin{aligned}
\hat{\omega}_x' = \frac{\hat{r}_c'(\hat{f}_{J_2,z} + \hat{f}_{drag,z}) + \hat{r}_c(\hat{f}'_{J_2,z} + \hat{f}'_{drag,z})}{\hat{h}} \\
- \frac{\hat{r}_c(\hat{f}_{J_2,z} + \hat{f}_{drag,z})\hat{h}'}{\hat{h}^2}
\end{aligned}
\tag{3.176}
$$

$$
\hat{\omega}_z' = \frac{\hat{h}'}{\hat{a}_c^2} \tag{3.177}
$$

where

$$
\hat{f}'_{J_2,z} = -\frac{3J_2\hat{v}_o^2\hat{R}_e^2 e}{\hat{a}_c}(4\cos 2i\sin\omega i' + \sin 2i\cos\omega\theta') \tag{3.178}
$$

$$
\begin{aligned}
\hat{f}'_{drag,z} = -\frac{\hat{\beta}\hat{a}_c^2 e}{2}\Big(\hat{w}_E\left(f''\cos\omega - f'\sin\omega\theta'\right)\sin i \\
+ \hat{w}_E f'\cos\omega\cos ii' + \frac{1}{2}\hat{w}_E^2\sin\omega\sin 2i\theta' \\
- \hat{w}_E^2\cos\omega\cos 2ii'\Big)
\end{aligned}
\tag{3.179}
$$

Inserting (3.169)–(3.179) into (3.147), we finally establish the constant linearized equation of the relative motion. If the eccentricity is zero, the model becomes the constant linearized equation of the relative motion for circular orbit.

3.4.6.3 Verification of the established relative motion equation

In this section, numerical simulation is carried out to demonstrate the effectiveness of the new linearized equation (3.147). The satellite's physical parameters are specified as same as the physical parameters of the TECSAS mission led by the German Space Organization in collaboration with the Canadian Space Agency and the Federal Space Agency of Russia. The TECSAS was a technology demonstration mission for on-orbit servicing such as the rendezvous maneuver. The mass, the area, and the drag coefficient of the TECSAS are 175 kg, 2.22 m^2, and 2.3 kg, respectively. Its initial orbit parameters are $\omega = 0$ degree, $i = 78$ degrees, $\Omega = 320$ degrees, and $M = 0$ degree. Moreover, the effects of the J_2 geopotential and the atmospheric drag perturbations on a formation with small eccentricity are numerically analyzed. The impact of the eccentricity on a formation flying is further evaluated with simulation conducted by using different altitude and eccentricity.

A projected circular formation flying is simulated. The motion of the deputy microsatellite relative to the chief microsatellite forms an ellipse centered on the chief microsatellite. The ellipse's projection on the y–z plane is a circle subject to the constraint $y^2 + z^2 = d^2$, where $d \in \mathbb{R}_+$ is the radius of the projected circle. In simulation, $d = 100$ m is set. The desired motion of the deputy microsatellite relative to the chief microsatellite is planned as $\boldsymbol{X}_d = [x_d \quad y_d \quad z_d]^{\mathrm{T}}$ m:

$$\boldsymbol{X}_d = [0.5d\sin(cv_o t + \alpha) \quad d\cos(cv_o t + \alpha) \quad d\sin(cv_o t + \alpha)]^{\mathrm{T}} \tag{3.180}$$

$$\dot{\boldsymbol{X}}_d = dcv_o\,[0.5\cos(cv_o t + \alpha) \quad -\sin(cv_o t + \alpha) \quad \cos(cv_o t + \alpha)]^{\mathrm{T}} \tag{3.181}$$

where $c = \sqrt{1+s}$ and α is the initial phase angle in the y–z plane.

Setting $c = 0$ and $\alpha = 0$ and non-dimensionalizing (3.180)–(3.181) lead to

$$[\hat{x}_d \quad \hat{y}_d \quad \hat{z}_d]^{\mathrm{T}} = \left[0.5\sin(c\hat{v}_o\tau) \quad \cos(c\hat{v}_o\tau) \quad \sin(c\hat{v}_o\tau)\right]^{\mathrm{T}} \tag{3.182}$$

$$\dot{\boldsymbol{X}}_d = \left[0.5c\hat{v}_o\cos(c\hat{v}_o\tau) \quad -c\hat{v}_o(c\hat{v}_o\tau) \quad c\hat{v}_o\cos(c\hat{v}_o\tau)\right]^{\mathrm{T}} \tag{3.183}$$

Hence, the initial non-dimensionalized conditions for the considered formation flying are given as follows by setting $\tau = 0$ for (3.182)–(3.183).

$$[\hat{x}_0 \quad \hat{y}_0 \quad \hat{z}_0]^{\mathrm{T}} = [0 \quad 1 \quad 0]^{\mathrm{T}} \tag{3.184}$$

$$[\dot{x}' \quad \dot{y}' \quad \dot{z}']^{\mathrm{T}} = [0.5c \quad 0 \quad c]^{\mathrm{T}} \tag{3.185}$$

It is worth mentioning that different altitude results in different atmospheric drag. Because low orbit leads to severe atmospheric drag, the nominal altitude is set as 350 km in simulation to clearly show the effect of atmospheric drag. Moreover, since the proposed equation (3.147) is valid when the eccentricity is less than or equal to 0.001, a small eccentricity, *i.e.*, $e = 0.0001$, is selected for the linearized equation (3.147) in comparison with the true nonlinear model.

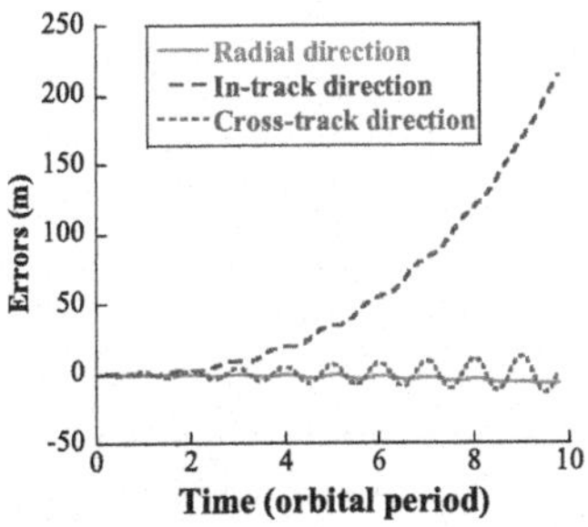

FIGURE 3.2 Relative position error between the linear model (3.147) and the nonlinear model for $e = 0.0001$.

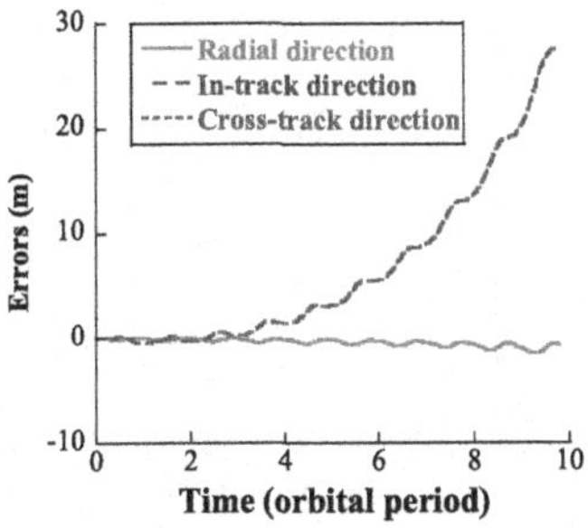

FIGURE 3.3 Relative position error between (3.147) and the nonlinear model for $e = 0.0001$ after orbit re-initialization.

For better showing the physical parameters, the non-dimensional simulation results are converted back to the dimensional values. In the case of the eccentricity $e = 0.0001$, Fig. 3.2 illustrates the errors of the relative motion between the true nonlinear model and the proposed linear model (3.147). It is seen that the initial errors between that two models are quite small. However, after two orbital periods' formation flying, the errors become large and large. That is because the secular perturbation of the atmospheric drag accumulates to affect the relative motion and the orbital elements of the chief satellite along with time. The accuracy of the proposed linear model (3.147) will deteriorate gradually. Fortunately, this drawback can be avoided by ground intervention. For example, the linear model (3.147) can be re-initialized after each orbital period. This can be achieved by uploading correction command to satellite through ground communication. By doing this, orbit correction and high-precision of the linearized model (3.147) could be guaranteed with the accumulated errors reduced. This can be validated in Fig. 3.3. It is observed that the error is less than 30 m even after ten orbital periods' flying. The linear model (3.147) can approximate the true nonlinear model with high precision. Hence, the linear model (3.147) is appropriate for control system and state estimation design.

To further show the practical application potential of the established linear model (3.147), simulation is further conducted by using the constant version of the linearized model (3.147) with the eccentricity $e = 0.0001$. Fig. 3.4 shows the error of the relative motion between the true nonlinear model and the constant version of the linear model (3.147) after orbit re-initialization. The error is governed to be less than 50 m. This accuracy is acceptable, because the constant version is an analytical solution without any numerical integration. Fig. 3.5 shows that the relative position error achieved by the constant version of the linearized model (3.147) for the case of the eccentricity $e = 0$ is better than 30 m.

To this end, it can be concluded from the above simulation results that the presented linear equation (3.147) accurately describes the relative motion of

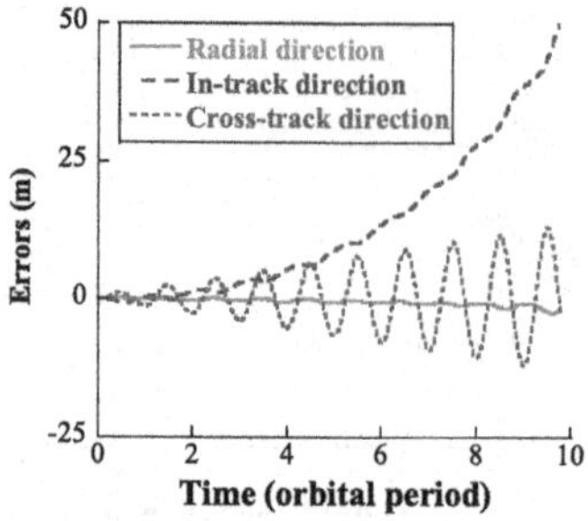

FIGURE 3.4 Relative position error between the constant version of the linearized model (3.147) and the nonlinear model for $e = 0.0001$ after orbit re-initialization.

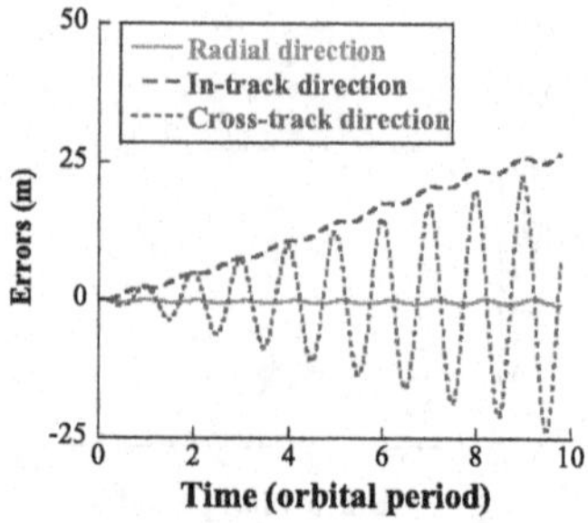

FIGURE 3.5 Relative position error between the constant linearized model (3.147) and the nonlinear model for $e = 0$ after orbit re-initialization.

formation flying microsatellite with small eccentricity. This model is applicable in practice.

3.4.7 Modeling of microsatellite formation flying control system

The control system of two microsatellites (one is the chief and the other is the deputy) formation flying is mathematically modeled in this section, which consists of the relative position control system and the relative attitude control system.

3.4.7.1 *Modeling of relative position control system*

In the system equation (3.147) of relative motion for the microsatellite formation flying, it is seen that there is no control force exerting on the chief and the deputy microsatellites. When investigating formation flying control problem, it should have control force. Let the control force exert on the deputy microsatellite, based on (3.147), the mathematical model of the relative position control system for microsatellite formation flying can be established as

$$\hat{\boldsymbol{r}}_e'' + \boldsymbol{N}\hat{\boldsymbol{r}}_e' + \boldsymbol{K}\hat{\boldsymbol{r}}_e - \boldsymbol{F} = \frac{1}{m_d}\boldsymbol{u}_T \tag{3.186}$$

where $\boldsymbol{u}_T \in \mathbb{R}^3$ is the control force acting on the deputy microsatellite.

3.4.7.2 *Modeling of relative attitude control system*

For the formation flying considered, the body-fixed coordinate frame of the deputy microsatellite can be denoted as $\mathcal{F}_{Bd}$. Let the unit quaternion $q_e = [q_{e1} \;\; \boldsymbol{q}_{e24}^{\mathrm{T}}]^{\mathrm{T}} \in \mathbb{R}^4$, $q_{e1} \in \mathbb{R}$, $\boldsymbol{q}_{e24} = [q_{e2} \;\; q_{e3} \;\; q_{e4}]^{\mathrm{T}} \in \mathbb{R}^3$, represent the relative attitude orientation of the deputy microsatellite in $\mathcal{F}_{Bd}$ with respect to the LVLH frame $\mathcal{F}_r$. $\boldsymbol{\omega}_e \in \mathbb{R}^3$ denotes the relative angular velocity of $\mathcal{F}_{Bd}$ with respect to the LVLH frame. It follows that $\boldsymbol{\omega}_e = \boldsymbol{\omega}_d - \mathbb{C}(\boldsymbol{q}_e)\boldsymbol{\omega}_0$, where

$\omega_d \in \mathbb{R}^3$ is the deputy microsatellite's angular velocity with respect to $\mathcal{F}_I$, and $\mathbb{C}(\boldsymbol{q}_e) \in \mathbb{R}^{3\times 3}$ is the rotation matrix from the LVLH frame $\mathcal{F}_r$ to the body-fixed frame $\mathcal{F}_{Bd}$. Then, the relative attitude control system can be modeled as

$$\dot{\boldsymbol{q}}_e = \boldsymbol{\Psi}(\omega_e)\boldsymbol{q}_e = \boldsymbol{E}(\boldsymbol{q}_e)\omega_e \tag{3.187}$$

$$\begin{aligned} \boldsymbol{J}_d\dot{\omega}_e &= \boldsymbol{u}_f + \boldsymbol{u}_{df} - (\omega_e + \mathbb{C}(\boldsymbol{q}_e)\omega_0)^{\times} \boldsymbol{J}_d \left(\omega_e + \mathbb{C}(\boldsymbol{q}_e)\omega_0\right) \\ &+ \boldsymbol{J}_d((\omega_e)^{\times}\mathbb{C}(\boldsymbol{q}_e)\omega_0 - \mathbb{C}(\boldsymbol{q}_e)\dot{\omega}_0) \end{aligned} \tag{3.188}$$

where $\boldsymbol{J}_d \in \mathbb{R}^{3\times 3}$ is the total inertia matrix of the deputy microsatellite. $\boldsymbol{u}_f \in \mathbb{R}^3$ is total control torque acting on the deputy microsatellite. $\boldsymbol{u}_{df} \in \mathbb{R}^3$ denotes the uncertain torque induced by external disturbance and modeling errors. The matrices $\boldsymbol{\Psi}(\omega_e) \in \mathbb{R}^{4\times 4}$ and $\boldsymbol{E}(\boldsymbol{q}_e) \in \mathbb{R}^{4\times 3}$ are defined as

$$\boldsymbol{\Psi}(\omega_e) = \frac{1}{2}\begin{bmatrix} 0 & -\omega_e^{\mathrm{T}} \\ \omega_e & -\omega_e^{\times} \end{bmatrix}, \quad \boldsymbol{E}(\boldsymbol{q}_e) = \frac{1}{2}\begin{bmatrix} -\boldsymbol{q}_{e24}^{\mathrm{T}} \\ q_{e1}\boldsymbol{I}_3 + \boldsymbol{q}_{e24}^{\times} \end{bmatrix} \tag{3.189}$$

3.5 Modeling of distributed microsatellite attitude control system

For formation flying with the number of satellites $N_0 \geq 3$, it is constantly called as the distributed formation flying. The desired attitude of the satellite formation flying is labeled with the number "0", the microsatellite communicating with the ground station is represented by the number "1", and the other microsatellites are labeled with the sequential numbers "$2, 3, \cdots$", respectively. Let $\omega_{i\to j} \in \mathbb{R}^3$ denote the relative angular velocity between the ith microsatellite and the jth microsatellite, $i, j = 0, 1, 2, \cdots, N_0$. The unit quaternion $\boldsymbol{q}_{i\to j} = [q_{i\to j1} \;\; \boldsymbol{q}_{i\to j24}^{\mathrm{T}}]^{\mathrm{T}} \in \mathbb{R}^4$, $q_{i\to j1} \in \mathbb{R}$, $\boldsymbol{q}_{i\to j24} = [q_{i\to j2} \;\; q_{i\to j3} \;\; q_{i\to j4}]^{\mathrm{T}} \in \mathbb{R}^3$, represents the relative attitude of the ith microsatellite with respect to the jth microsatellite, $i, j = 0, 1, 2, \cdots, N_0$.

Based on the coordinate frames defined in Fig. 3.1 and (3.187)–(3.188), the mathematical model of the relative rotational dynamics between any two microsatellites in the distributed formation flying is determined by

$$\dot{\boldsymbol{q}}_{i\to j} = \boldsymbol{\Psi}(\omega_{i\to j})\boldsymbol{q}_{i\to j} = \boldsymbol{E}(\boldsymbol{q}_{i\to j})\omega_{i\to j} \tag{3.190}$$

$$\begin{aligned} \boldsymbol{J}_i\dot{\omega}_{i\to j} &= \boldsymbol{u}_{fi} + \boldsymbol{u}_{dfi} - (\omega_{i\to j} + \mathbb{C}(\boldsymbol{q}_{i\to j})\omega_0)^{\times} \boldsymbol{J}_i(\omega_{i\to j} + \mathbb{C}(\boldsymbol{q}_{i\to j})\omega_0) \\ &+ \boldsymbol{J}_i(\omega_{i\to j}^{\times}\mathbb{C}(\boldsymbol{q}_{i\to j})\omega_0 - \mathbb{C}(\boldsymbol{q}_{i\to j})\dot{\omega}_0) \end{aligned} \tag{3.191}$$

where $\boldsymbol{J}_i \in \mathbb{R}^{3\times 3}$ is the total inertia matrix of the ith microsatellite. $\boldsymbol{u}_{fi} \in \mathbb{R}^3$ is total control torque acting on the ith microsatellite. $\boldsymbol{u}_{dfi}$ denotes the uncertain torque acting on the ith microsatellite induced by external disturbance. The matrices $\boldsymbol{\Psi}(\boldsymbol{\omega}_{i\to j}) \in \mathbb{R}^{4\times 4}$ and $\boldsymbol{E}(\boldsymbol{q}_{i\to j}) \in \mathbb{R}^{4\times 3}$ are defined as

$$\boldsymbol{\Psi}(\boldsymbol{\omega}_{i\to j}) = \frac{1}{2}\begin{bmatrix} 0 & -\boldsymbol{\omega}_{i\to j}^{\mathrm{T}} \\ \boldsymbol{\omega}_{i\to j} & -\boldsymbol{\omega}_{i\to j}^{\times} \end{bmatrix} \tag{3.192}$$

$$\boldsymbol{E}(\boldsymbol{q}_{i\to j}) = \frac{1}{2}\begin{bmatrix} -\boldsymbol{q}_{i\to j24}^{\mathrm{T}} \\ q_{i\to j1}\boldsymbol{I}_3 + \boldsymbol{q}_{i\to j24}^{\times} \end{bmatrix} \tag{3.193}$$

3.6 Summary

The work of modeling the microsatellite control system was finished in this chapter. Following the definitions of coordinate frames in the first, the mathematical model of a single microsatellite control system was presented. Moreover, the relative control system of two formation flying microsatellites in an elliptical orbit was mathematically modeled. Under the assumption of small eccentricity, a linear dynamic model was established to describe their relative motion perturbed by the J_2 and the atmospheric drag forces. The chief microsatellite's orbit was propagated by using the mean orbital elements with the main effects of J_2 and atmospheric drag perturbations explicitly captured. In comparison with the existing linear models, one of the main contributions of this modeling approach was that the negative influence of J_2 and atmospheric drag perturbations on the true anomaly and the orbit angular velocity was considered. Hence, the established model was more complete and general. In addition, the relative position variable is the state of the presented model of the relative motion control systems. This lets that model be convenient and appropriate for filter-based state estimation and control system design. This chapter provides the subsequent filtering approaches deign for the microsatellite control system with a mathematical model.

Part II

Sigma-point predictive filtering for microsatellite control system

Chapter 4

Unscented predictive filter

4.1 Introduction

The early results of the real-time predictive filtering (PF) were seen in Crassidis and Markley (1997c); Crassidis et al. (1993); Crassidis and Markley (1997a). As stated in Section 1.4.2, the predictive filter provides an efficient and real-time solution to determine the optimal state estimation online even in the presence of severe system modeling errors. For the method, the modeling errors are not restricted to Gaussian noise only. It is thus manifest that the PF is more general than other nonlinear estimation approaches including the extended Kalman filter. Hence, this solution has been widely adopted in many application areas, such as spacecraft attitude determination (Crassidis and Markley, 1997b; Crassidis, 1999; Ji and Yang, 2010), chaotic synchronization system (Kurian and Puthusserypady, 2006), and inertial alignment system, etc. However, the PF still has certain disadvantages that need to be addressed. For example, the estimation of the modeling error is only accurate to the first order of the posterior mean. This will introduce serious errors in the state estimation with the estimated modeling error. Therefore, it will inevitably result in the deterioration of the estimation precision and slow rate of convergence, etc. Moreover, if the estimated modeling error of the PF deviates largely from its actual value, the error will be propagated by the state equation to amplify the error's effect, resulting in filtering divergence.

Motivated by tackling with the above shortcomings, an alternative solution, *i.e.*, the sigma-point unscented predictive filtering (UPF), is developed in this chapter. The Unscented Transformation (UT) (Julier, 2002, 2003) and the classical PF are integrated to develop this filter. In the UPF, the minimal set of the deterministically chosen sigma points is applied. These sigma points completely have the same mean and covariance to the system state distribution. When these points are propagated through the nonlinear model error functions, they can guarantee a higher estimation accuracy for the modeling error. It is theoretically proved that the estimated modeling error of the UPF can capture the posterior mean accurately up to the third order for nonlinear Gaussian systems, with estimation errors only introduced in the fourth and higher orders. At the same time, the estimated modeling error can capture the posterior mean accurately to the second-order for any nonlinearity. This is better than the accuracy of the classi-

Predictive Filtering for Microsatellite Control System. https://doi.org/10.1016/B978-0-12-821865-5.00018-8

cal PF. Moreover, in comparison with the PF, the UPF has more robustness. It avoids the drawbacks of the PF.

4.2 Unscented predictive filter

In the classical predictive filtering Algorithm 2.1, the first-order Taylor approximation is used, but provides an insufficiently accurate approximation in many cases. Significant bias or even slow rate of convergence of the estimation may result. When the modeling error $\boldsymbol{d}_k$ is quite large, the estimation performance of the PF will be deteriorated. Hence, new filtering approaches should be developed to improve its state estimation performance. Motivated by this, invoking the UT technique (Julier et al., 2000), a sigma-point unscented predictive filtering (UPF) approach is developed for the nonlinear discrete system (2.32)–(2.33) with its estimation procedures listed in Algorithm 4.1, where $N \in \mathbb{N}$ is the number of sigma points, W_i^m and W_i^c to be determined by the designer are the weighted proportion for generating these sigma pints, $i = 0, 1, \cdots, N$, $\hat{\boldsymbol{x}}_k$ is the estimation of the state $\boldsymbol{x}_k$, and $\hat{\boldsymbol{d}}_k$ is the estimation of the modeling error $\boldsymbol{d}_k$. Moreover, the numerical integral method is adopted to update the state estimation in Algorithm 4.1.

Remark 4.1. In Algorithm 4.1, different UT schemes lead to different N and different weights W_i^m and W_i^c. For example, if the symmetric UT scheme in Julier and Uhlmann (1997) is chosen, then it follows that $N = 2n + 1$ and

$$W_0^m = W_0^c = \frac{\kappa}{n+\kappa} \tag{4.1}$$

$$W_i^c = \frac{1}{2(n+\kappa)}, \quad i = 1, 2, \cdots, 2n \tag{4.2}$$

$$W_i^c = \frac{1}{2(n+\kappa)}, \quad i = 1, 2, \cdots, 2n \tag{4.3}$$

where $\kappa \in \mathbb{R}$ is a constant. Suppose the UT scheme is selected as the scaled UT (Julier et al., 2000), it leads to $N = 2n + 1$ and

$$W_0^m = \frac{\kappa}{n+\kappa} \tag{4.4}$$

$$W_0^c = \frac{\kappa}{n+\kappa} + (1 - \kappa_\alpha^2 + \kappa_\beta) \tag{4.5}$$

$$W_i^m = \frac{1}{2(n+\kappa)}, \quad i = 1, 2, \cdots, 2n \tag{4.6}$$

$$W_i^c = \frac{1}{2(n+\kappa)}, \quad i = 1, 2, \cdots, 2n \tag{4.7}$$

with constants $\kappa = \kappa_\alpha^2(n + \kappa_0) - n$, κ_0, κ_α, and κ_β.

Algorithm 4.1: Unscented predictive filter.

Input Data: Initial estimation $\hat{\boldsymbol{x}}_0$, $\hat{\boldsymbol{y}}_0$, and measurement $\boldsymbol{y}_k$
Result: State estimation $\hat{\boldsymbol{x}}_k$
begin
Initializing the state and the covariance matrix as
$\hat{\boldsymbol{x}}_0 = \mathbb{E}(\boldsymbol{x}_0)$
$\boldsymbol{P}_{\tilde{x},0} = \mathrm{Var}(\boldsymbol{x}_0) = \mathbb{E}\left((\boldsymbol{x}_0 - \hat{\boldsymbol{x}}_0)(\boldsymbol{x}_0 - \hat{\boldsymbol{x}}_0)^{\mathrm{T}}\right)$
Choosing UT scheme
for $k = 1, 2, \cdots$ **do**
Using $\hat{\boldsymbol{x}}_{k-1}$, $\boldsymbol{P}_{\tilde{x},k-1}$, and the chosen UT to generate sigma points
$\boldsymbol{\xi}_{i,k-1}, i = 0, 1, \cdots, N$
Updating the modeling error's estimation $\hat{\boldsymbol{d}}_{k-1}$ as
$\boldsymbol{\beta}_{i,k-1} = \boldsymbol{M}(\boldsymbol{\xi}_{i,k-1}), i = 0, 1, \cdots, N$
$\boldsymbol{\gamma}_{i,k-1} = \boldsymbol{\beta}_{i,k-1}(\boldsymbol{h}_{k-1}(\boldsymbol{\xi}_{i,k-1}) - \boldsymbol{y}_k), i = 0, 1, \cdots, N$
$\boldsymbol{\alpha}_{i,k-1} = \boldsymbol{\beta}_{i,k-1}\underline{\boldsymbol{Z}}(\boldsymbol{\xi}_{i,k-1}, \Delta t), i = 0, 1, \cdots, N$
$\overline{\boldsymbol{h}}_{k-1} = \sum_{i=0}^{N} W_i^m \boldsymbol{\gamma}_{i,k-1} = \sum_{i=0}^{N} W_i^m \underline{\boldsymbol{h}}(\boldsymbol{\xi}_{i,k-1})$
$\overline{\boldsymbol{Z}}_{k-1} = \sum_{i=0}^{N} W_i^m \boldsymbol{\alpha}_{i,k} = \sum_{i=0}^{N} W_i^m \underline{\boldsymbol{Z}}(\boldsymbol{\xi}_{i,k-1}, \Delta t)$
$\hat{\boldsymbol{d}}_{i,k-1} = -(\boldsymbol{\alpha}_{i,k-1} + \boldsymbol{\gamma}_{i,k-1}), i = 0, 1, \cdots, N$
$\hat{\boldsymbol{d}}_{k-1} = -(\overline{\boldsymbol{Z}}_{k-1} + \overline{\boldsymbol{h}}_{k-1})$
Updating the state estimation
$\boldsymbol{\delta}_{i,k|k-1} = \boldsymbol{f}(\boldsymbol{\xi}_{i,k-1}), i = 0, 1, \cdots, N$
$\hat{\boldsymbol{x}}_{k|k-1} = \sum_{i=0}^{N} W_i^c \boldsymbol{\delta}_{i,k|k-1}$
$\hat{\boldsymbol{x}}_k = \hat{\boldsymbol{x}}_{k|k-1} + \boldsymbol{g}(\hat{\boldsymbol{x}}_{k|k-1})\hat{\boldsymbol{d}}_{k-1}$
Updating the state estimation error covariance
$\boldsymbol{\chi}_{i,k} = \boldsymbol{\delta}_{i,k|k-1} + \boldsymbol{g}(\boldsymbol{\xi}_{i,k-1})\hat{\boldsymbol{d}}_{i,k-1}, i = 0, 1, \cdots, N$
$\boldsymbol{P}_{\tilde{x},k} = \sum_{i=0}^{N} W_i^c (\boldsymbol{\chi}_{i,k} - \hat{\boldsymbol{x}}_k)(\boldsymbol{\chi}_{i,k} - \hat{\boldsymbol{x}}_k)^{\mathrm{T}}$
end
end

4.3 Estimation performance of UPF

As an example, the symmetric UT is chosen to implement Algorithm 4.1 in this section, *i.e.*, $N = 2n + 1$. The state estimation performance of the corresponding UPF is analyzed. If Algorithm 4.1 is implemented by using other UT schemes, the corresponding state estimation performance can be analyzed by using the same procedures given in this section.

4.3.1 Estimation accuracy of modeling error

As a stepping stone, we apply the Cholesky decomposition method to calculate the matrix square root of $\boldsymbol{P}_{\tilde{x},k}$. Based on this, we have

$$\boldsymbol{P}_{\tilde{x},k} = \boldsymbol{S}_k \boldsymbol{S}_k^{\mathrm{T}} \tag{4.8}$$

where $\boldsymbol{S}_k = [\boldsymbol{\sigma}_{k_1} \quad \boldsymbol{\sigma}_{k_2} \quad \cdots \quad \boldsymbol{\sigma}_{k_n}]$, $\boldsymbol{\sigma}_{k_i} \in \mathbb{R}^n$, $i = 1, 2, \cdots, n$. Then, it follows that $\boldsymbol{P}_{\tilde{x},k} = \boldsymbol{S}_k \boldsymbol{S}_k^{\mathrm{T}} = \sum_{i=1}^{n} \boldsymbol{\sigma}_{k_i} \boldsymbol{\sigma}_{k_i}^{\mathrm{T}}$. Correspondingly, the set of sigma-points are deterministically chosen as

$$\begin{cases} \boldsymbol{\xi}_{0,k} = \hat{\boldsymbol{x}}_k \\ \boldsymbol{\xi}_{i,k} = \hat{\boldsymbol{x}}_k + \sqrt{n+\kappa}\,\boldsymbol{\sigma}_{k_i} = \hat{\boldsymbol{x}}_k + \hat{\boldsymbol{\sigma}}_{k_i} \\ \boldsymbol{\xi}_{i+n,k} = \hat{\boldsymbol{x}}_k - \sqrt{n+\kappa}\,\boldsymbol{\sigma}_{k_i} = \hat{\boldsymbol{x}}_k - \hat{\boldsymbol{\sigma}}_{k_i} \end{cases} \tag{4.9}$$

with $i = 1, 2, \cdots, n$ and $\hat{\boldsymbol{\sigma}}_{k_i} = \sqrt{n+\kappa}\,\boldsymbol{\sigma}_{k_i}$.

The sigma points given in (4.9) are now propagated through the nonlinear function to generate

$$\begin{cases} \boldsymbol{\gamma}_{0,k} = \underline{\boldsymbol{h}}(\boldsymbol{\xi}_{0,k}) = \underline{\boldsymbol{h}}(\hat{\boldsymbol{x}}_k) \\ \boldsymbol{\gamma}_{i,k} = \underline{\boldsymbol{h}}(\boldsymbol{\xi}_{i,k}) = \underline{\boldsymbol{h}}(\hat{\boldsymbol{x}}_k) + \sum_{j=1}^{\infty} \dfrac{\boldsymbol{D}_{\hat{\sigma}_{k_i}}^{j} \underline{\boldsymbol{h}}}{j!} \\ \boldsymbol{\gamma}_{i+n,k} = \underline{\boldsymbol{h}}(\boldsymbol{\xi}_{i+n,k}) = \underline{\boldsymbol{h}}(\hat{\boldsymbol{x}}_k) + \sum_{j=1}^{\infty} \dfrac{\boldsymbol{D}_{-\hat{\sigma}_{k_i}}^{j} \underline{\boldsymbol{h}}}{j!} \end{cases} \tag{4.10}$$

where $i = 1, 2, \cdots, n$.

Similarly, for $i = 1, 2, \cdots, n$ one has

$$\begin{cases} \boldsymbol{\alpha}_{0,k} = \underline{\boldsymbol{Z}}(\boldsymbol{\xi}_{0,k}) = \underline{\boldsymbol{Z}}(\hat{\boldsymbol{x}}_k) \\ \boldsymbol{\alpha}_{i,k} = \underline{\boldsymbol{Z}}(\boldsymbol{\xi}_{i,k}) = \underline{\boldsymbol{Z}}(\hat{\boldsymbol{x}}_k) + \sum_{j=1}^{\infty} \dfrac{\boldsymbol{D}_{\hat{\sigma}_{k_i}}^{j} \underline{\boldsymbol{Z}}}{j!} \\ \boldsymbol{\alpha}_{i+n,k} = \underline{\boldsymbol{Z}}(\boldsymbol{\xi}_{i+n,k}) = \underline{\boldsymbol{Z}}(\hat{\boldsymbol{x}}_k) + \sum_{j=1}^{\infty} \dfrac{\boldsymbol{D}_{-\hat{\sigma}_{k_i}}^{j} \underline{\boldsymbol{Z}}}{j!} \end{cases} \tag{4.11}$$

Due to the vector differential operator (2.54)–(2.55), the following equations hold

$$\boldsymbol{D}_{-\hat{\sigma}_{k_i}}^{l_1} \underline{\boldsymbol{h}} = (-\hat{\boldsymbol{\sigma}}_{k_i}^{\mathrm{T}} \nabla)^{l_1} \boldsymbol{h}(\boldsymbol{x}) \Big|_{x=\hat{x}_k} = -\boldsymbol{D}_{\hat{\sigma}_{k_i}}^{l_1} \underline{\boldsymbol{h}} \tag{4.12}$$

$$\boldsymbol{D}_{-\hat{\sigma}_{k_i}}^{l_2} \underline{\boldsymbol{h}} = (-\hat{\boldsymbol{\sigma}}_{k_i}^{\mathrm{T}} \nabla)^{l_2} \boldsymbol{h}(\boldsymbol{x}) \Big|_{x=\hat{x}_k} = \boldsymbol{D}_{\hat{\sigma}_{k_i}}^{l_2} \underline{\boldsymbol{h}} \tag{4.13}$$

$$D^{l_1}_{-\hat{\sigma}_{k_i}} \underline{Z} = (-\hat{\sigma}^{\mathrm{T}}_{k_i} \nabla)^{l_1} Z(x) \Big|_{x=\hat{x}_k} = -D^{l_1}_{\hat{\sigma}_{k_i}} \underline{Z} \tag{4.14}$$

$$D^{l_2}_{-\hat{\sigma}_{k_i}} \underline{Z} = (-\hat{\sigma}^{\mathrm{T}}_{k_i} \nabla)^{l_2} Z(x) \Big|_{x=\hat{x}_k} = D^{l_2}_{\hat{\sigma}_{k_i}} \underline{Z} \tag{4.15}$$

where $l_1 \in \mathbb{N}$ is odd number and $l_2 \in \mathbb{N}$ is even number satisfying $0 < l_1 \leq 2n$, $0 < l_2 \leq 2n$, respectively.

Note that

$$\frac{1}{n+\kappa} \sum_{i=1}^{n} \frac{D^2_{\hat{\sigma}_{k_i}} \underline{h}}{2!} = \frac{1}{n+\kappa} \mathbb{E}\left(\frac{D_{\hat{\sigma}_{k_i}} (D_{\hat{\sigma}_{k_i}} \underline{h})}{2!} \right) = \frac{(\nabla^{\mathrm{T}} P_{\tilde{x},k} \nabla) \underline{h}(x)|_{x=\hat{x}_k}}{2!} \tag{4.16}$$

$$\frac{1}{n+\kappa} \sum_{i=1}^{n} \frac{D^2_{\hat{\sigma}_{k_i}} \underline{Z}}{2!} = \mathbb{E}\left(\frac{D^2_{\Delta x} \underline{Z}}{2!} \right) \tag{4.17}$$

$$\frac{1}{n+\kappa} \sum_{i=1}^{n} (D_{\hat{\sigma}_{k_i}} \underline{h})(D_{\hat{\sigma}_{k_i}} \underline{h})^{\mathrm{T}} = G_{\underline{h}} \left(\sum_{i=1}^{n} \hat{\sigma}_{k_i} \hat{\sigma}^{\mathrm{T}}_{k_i} \right) G^{\mathrm{T}}_{\underline{h}} = G_{\underline{h}} P_{\tilde{x},k} G^{\mathrm{T}}_{\underline{h}} \tag{4.18}$$

$$\frac{1}{n+\kappa} \sum_{i=1}^{n} (D_{\hat{\sigma}_{k_i}} \underline{Z})(D_{\hat{\sigma}_{k_i}} \underline{Z})^{\mathrm{T}} = G_{\underline{Z}} \left(\sum_{i=1}^{n} \hat{\sigma}_{k_i} \hat{\sigma}^{\mathrm{T}}_{k_i} \right) G^{\mathrm{T}}_{\underline{Z}} = G_{\underline{Z}} P_{\tilde{x},k} G^{\mathrm{T}}_{\underline{Z}} \tag{4.19}$$

$$\frac{1}{n+\kappa} \sum_{i=1}^{n} (D_{\hat{\sigma}_{k_i}} \underline{h})(D_{\hat{\sigma}_{k_i}} \underline{Z})^{\mathrm{T}} = G_{\underline{h}} P_{\tilde{x},k} G^{\mathrm{T}}_{\underline{Z}} \tag{4.20}$$

$$\frac{1}{n+\kappa} \sum_{i=1}^{n} (D_{\hat{\sigma}_{k_i}} \underline{Z})(D_{\hat{\sigma}_{k_i}} \underline{h})^{\mathrm{T}} = G_{\underline{Z}} P_{\tilde{x},k} G^{\mathrm{T}}_{\underline{h}} \tag{4.21}$$

Then, after the unscented transformation, the posterior distribution of nonlinear functions $\overline{h}_k$ and $\overline{Z}_k$ are represented by

$$\begin{aligned}
\bar{h}_{\mathrm{UPF}} &= \sum_{i=0}^{2n} W^m_i \gamma_{i,k} = W^m_0 \gamma_{0,k} + \sum_{i=1}^{n} (W^m_i \gamma_{i,k} + W^m_{i+n} \gamma_{i+n,k}) \\
&= \frac{\kappa}{n+\kappa} \underline{h}(\hat{x}_k) + \frac{1}{n+\kappa} \sum_{i=1}^{n} \left(\underline{h}(\hat{x}_k) + \frac{D^2_{\hat{\sigma}_{k_i}} \underline{h}}{2!} + \frac{D^4_{\hat{\sigma}_{k_i}} \underline{h}}{4!} + \cdots \right) \\
&= \underline{h}(\hat{x}_k) + \mathbb{E}\left(\frac{D^2_{\Delta x} \underline{h}}{2} \right) + \frac{1}{n+\kappa} \sum_{i=1}^{n} \left(\frac{D^4_{\hat{\sigma}_{k_i}} \underline{h}}{4!} + \frac{D^6_{\hat{\sigma}_{k_i}} \underline{h}}{6!} + \cdots \right)
\end{aligned} \tag{4.22}$$

$$
\begin{aligned}
\bar{\boldsymbol{Z}}_{\mathrm{UPF}} &= \sum_{i=0}^{2n} \boldsymbol{W}_i^m \boldsymbol{\alpha}_{i,k} = \boldsymbol{W}_0^m \boldsymbol{\alpha}_{0,k} + \sum_{i=1}^{n} (\boldsymbol{W}_i^m \boldsymbol{\alpha}_{i,k} + \boldsymbol{W}_{i+n}^m \boldsymbol{\alpha}_{i+n,k}) \\
&= \frac{\kappa}{n+\kappa} \underline{\boldsymbol{Z}}(\hat{\boldsymbol{x}}_k) + \frac{1}{n+\kappa} \sum_{i=1}^{n} \left(\underline{\boldsymbol{Z}}(\hat{\boldsymbol{x}}_k) + \frac{\boldsymbol{D}_{\hat{\sigma}_{k_i}}^2 \underline{\boldsymbol{Z}}}{2!} + \frac{\boldsymbol{D}_{\hat{\sigma}_{k_i}}^4 \underline{\boldsymbol{Z}}}{4!} + \cdots \right) \\
&= \underline{\boldsymbol{Z}}(\hat{\boldsymbol{x}}_k) + \mathbb{E}\left(\frac{\boldsymbol{D}_{\Delta x}^2 \underline{\boldsymbol{Z}}}{2} \right) + \frac{1}{n+\kappa} \sum_{i=1}^{n} \left(\frac{\boldsymbol{D}_{\hat{\sigma}_{k_i}}^4 \underline{\boldsymbol{Z}}}{4!} + \frac{\boldsymbol{D}_{\hat{\sigma}_{k_i}}^6 \underline{\boldsymbol{Z}}}{6!} + \cdots \right)
\end{aligned} \tag{4.23}
$$

Applying (4.16)–(4.23), it is obtained from Algorithm 4.1 that the posterior mean of the estimated modeling error provided by the UPF can be calculated as

$$
\begin{aligned}
\hat{\boldsymbol{d}}_{\mathrm{UPF}} &= -(\underline{\boldsymbol{h}}(\hat{\boldsymbol{x}}_k) + \underline{\boldsymbol{Z}}(\hat{\boldsymbol{x}}_k)) - \left(\mathbb{E}\left(\frac{\boldsymbol{D}_{\Delta x}^2 \underline{\boldsymbol{h}}}{2} \right) + \mathbb{E}\left(\frac{\boldsymbol{D}_{\Delta x}^2 \underline{\boldsymbol{Z}}}{2} \right) \right) \\
&\quad - \frac{1}{n+\kappa} \sum_{i=1}^{n} \sum_{j=1}^{\infty} \frac{\boldsymbol{D}_{\hat{\sigma}_{k_i}}^{(2j+2)} \underline{\boldsymbol{h}}}{(2j+2)!} - \frac{1}{n+\kappa} \sum_{i=1}^{n} \sum_{j=1}^{\infty} \frac{\boldsymbol{D}_{\hat{\sigma}_{k_i}}^{(2j+2)} \underline{\boldsymbol{Z}}}{(2j+2)!}
\end{aligned} \tag{4.24}
$$

The covariance of the estimated modeling error obtained from the UPF is calculated as

$$
\begin{aligned}
(\boldsymbol{P}_{d,k})_{\mathrm{UPF}} &= \sum_{i=0}^{2n} \mathbb{E}(\boldsymbol{A}\boldsymbol{A}^{\mathrm{T}}) \\
&= \frac{1}{n+\kappa} \sum_{i=1}^{n} \mathbb{E}(\boldsymbol{B}\boldsymbol{B}^{\mathrm{T}} + \boldsymbol{C}\boldsymbol{C}^{\mathrm{T}}) \\
&= \boldsymbol{G}_{\underline{h}} \boldsymbol{P}_{\tilde{x},k} \boldsymbol{G}_{\underline{h}}^{\mathrm{T}} + \boldsymbol{G}_{\underline{Z}} \boldsymbol{P}_{\tilde{x},k} \boldsymbol{G}_{\underline{Z}}^{\mathrm{T}} + \boldsymbol{G}_{\underline{h}} \boldsymbol{P}_{\tilde{x},k} \boldsymbol{G}_{\underline{Z}}^{\mathrm{T}} + \boldsymbol{G}_{\underline{Z}} \boldsymbol{P}_{\tilde{x},k} \boldsymbol{G}_{\underline{h}}^{\mathrm{T}} \\
&\quad - \mathbb{E}\left(\frac{\boldsymbol{D}_{\hat{\sigma}_{k_i}}^2 (\underline{\boldsymbol{Z}} + \underline{\boldsymbol{h}})}{2!} \right) \mathbb{E}\left(\left(\frac{\boldsymbol{D}_{\hat{\sigma}_{k_i}}^2 (\underline{\boldsymbol{Z}} + \underline{\boldsymbol{h}})}{2!} \right)^{\mathrm{T}} \right) + \boldsymbol{\Sigma}_1 + \boldsymbol{\Sigma}_2
\end{aligned} \tag{4.25}
$$

with

$$
\boldsymbol{A} = \boldsymbol{\gamma}_{i,k} + \boldsymbol{\alpha}_{i,k} - \bar{\boldsymbol{h}}_{\mathrm{UPF}} - \bar{\boldsymbol{Z}}_{\mathrm{UPF}} \tag{4.26}
$$

$$
\boldsymbol{B} = \boldsymbol{D}_{\hat{\sigma}_{k_i}} \underline{\boldsymbol{h}} + \frac{\boldsymbol{D}_{\hat{\sigma}_{k_i}}^3 \underline{\boldsymbol{h}}}{3!} + \boldsymbol{D}_{\hat{\sigma}_{k_i}} \underline{\boldsymbol{Z}} + \frac{\boldsymbol{D}_{\hat{\sigma}_{k_i}}^3 \underline{\boldsymbol{Z}}}{3!} + \cdots \tag{4.27}
$$

$$
\boldsymbol{C} = \frac{\boldsymbol{D}_{\hat{\sigma}_{k_i}}^2 \underline{\boldsymbol{h}}}{2!} + \frac{\boldsymbol{D}_{\hat{\sigma}_{k_i}}^4 \underline{\boldsymbol{h}}}{4!} + \frac{\boldsymbol{D}_{\hat{\sigma}_{k_i}}^2 \underline{\boldsymbol{Z}}}{2!} + \frac{\boldsymbol{D}_{\hat{\sigma}_{k_i}}^2 \underline{\boldsymbol{Z}}}{4!} + \cdots \tag{4.28}
$$

$$\boldsymbol{\Sigma}_1 = \frac{1}{n+\kappa}\underbrace{\sum_{m=1}^{\infty}\sum_{i=1}^{\infty}\sum_{j=1}^{\infty}\frac{1}{i!j!}\mathbb{E}((\boldsymbol{D}^i_{\hat{\sigma}_{km}}(\underline{\boldsymbol{Z}}+\underline{\boldsymbol{h}}))(\boldsymbol{D}^j_{\hat{\sigma}_{km}}(\underline{\boldsymbol{Z}}+\underline{\boldsymbol{h}}))^{\mathrm{T}})}_{\text{condition\_1}} \tag{4.29}$$

$$\boldsymbol{\Sigma}_2 = \bar{\kappa}\underbrace{\sum_{i=1}^{\infty}\sum_{j=1}^{\infty}\sum_{l=1}^{n}\sum_{m=1}^{n}\frac{1}{i!j!}\mathbb{E}(\boldsymbol{D}^{2i}_{\hat{\sigma}_{k_l}}(\underline{\boldsymbol{Z}}+\underline{\boldsymbol{h}}))\mathbb{E}(\boldsymbol{D}^{2j}_{\hat{\sigma}_{km}}(\underline{\boldsymbol{Z}}+\underline{\boldsymbol{h}}))^{\mathrm{T}}}_{\text{condition\_2}} \tag{4.30}$$

where $\bar{\kappa} = \frac{1}{(2i)!(2j)!(n+\kappa)^2}$.

Theorem 4.1. *When the UPF in Algorithm 4.1 is applied to the nonlinear discrete system (2.32)–(2.33) with the modeling error $\boldsymbol{d}_k$ and the measurement noise $\boldsymbol{v}_k$, the estimation $\hat{\boldsymbol{d}}_k$ can capture the posterior mean of the modeling error accurately to the third order for nonlinear Gaussian discrete system through the use of the symmetrically-distributed set of points, with the approximation errors only introduced in the fourth or higher-order moments. Moreover, the UPF completely capture the posterior mean of the modeling error accurately to the second order for any nonlinear function. The estimation performance for the modeling error achieved by the UPF is better than the performance guaranteed by the classical PF presented in Algorithm 2.1.*

Proof. Comparing (2.63)–(2.64) with (4.24)–(4.25), the conclusion in Theorem 4.1 can be proved. □

Remark 4.2. The classical PF presented in Algorithm 2.1 only calculates the posterior mean and covariance accurately to the first order with all the high-order moments truncated. It can be concluded the UPF achieves higher estimation accuracy for the modeling error than the classical PF. Moreover, the higher accuracy of the covariance $(\boldsymbol{P}_{d,k})_{\text{UPF}}$ of the modeling error will increase the estimation rate of the modeling error.

4.3.2 Estimation accuracy of system states

Let $(\hat{\boldsymbol{x}}_k)_{\text{UPF}}$ denote the state estimation provided by the UPF, then the set of sigma points $\boldsymbol{\xi}_{k,i}$, $i = 0, 1, \cdots, 2n$, can be represented as follows by symmetrically-distributed.

$$\begin{cases} \boldsymbol{\xi}_{k,0} = (\hat{\boldsymbol{x}}_k)_{\text{UPF}} \\ \boldsymbol{\xi}_{k,i} = (\hat{\boldsymbol{x}}_k)_{\text{UPF}} + \sqrt{n+\kappa}\boldsymbol{\sigma}_{k_i} = \hat{\boldsymbol{x}}_k + \hat{\boldsymbol{\sigma}}_{\boldsymbol{x}_{k_i}} \\ \boldsymbol{\xi}_{k,i+n} = (\hat{\boldsymbol{x}}_k)_{\text{UPF}} - \sqrt{n+\kappa}\boldsymbol{\sigma}_{k_i} = \hat{\boldsymbol{x}}_k - \hat{\boldsymbol{\sigma}}_{\boldsymbol{x}_{k_i}} \end{cases} \tag{4.31}$$

where $i = 1, 2, \cdots, n$ and $\hat{\boldsymbol{\sigma}}_{\boldsymbol{x}_{k_i}} = \sqrt{n+\kappa}\boldsymbol{\sigma}_{\boldsymbol{x}_{k_i}}$. Here, $\boldsymbol{\sigma}_{\boldsymbol{x}_{k_i}}$ satisfies $(\boldsymbol{P}_{\tilde{\boldsymbol{x}},k})_{\mathrm{UPF}} = \sum_{i=1}^{n} \boldsymbol{\sigma}_{\boldsymbol{x}_{k_i}} \boldsymbol{\sigma}_{\boldsymbol{x}_{k_i}}^{\mathrm{T}}$, where $(\boldsymbol{S}_k)_{\mathrm{UPF}} = [\boldsymbol{\sigma}_{\boldsymbol{x}_{k_1}} \quad \boldsymbol{\sigma}_{\boldsymbol{x}_{k_2}} \quad \cdots \quad \boldsymbol{\sigma}_{\boldsymbol{x}_{k_n}}]$, $\boldsymbol{\sigma}_{\boldsymbol{x}_{k_i}} \in \mathbb{R}^n$, $i = 1, 2, \cdots, n$, is the Cholesky decomposition of $(\boldsymbol{P}_{\tilde{\boldsymbol{x}},k})_{\mathrm{UPF}}$.

Applying these points (4.31) to propagate through the nonlinear transformation and yields

$$\begin{cases} \boldsymbol{\delta}_{k,0} = \boldsymbol{f}(\boldsymbol{\xi}_{k,0}) = \boldsymbol{f}((\hat{\boldsymbol{x}}_k)_{\mathrm{UPF}}) \\ \boldsymbol{\delta}_{k,i} = \boldsymbol{f}(\boldsymbol{\xi}_{k,i}) = \boldsymbol{f}((\hat{\boldsymbol{x}}_k)_{\mathrm{UPF}}) + \boldsymbol{D}_{\hat{\boldsymbol{\sigma}}_{\boldsymbol{x}_{k_i}}} \boldsymbol{f} + \dfrac{\boldsymbol{D}^2_{\hat{\boldsymbol{\sigma}}_{\boldsymbol{x}_{k_i}}} \boldsymbol{f}}{2!} + \cdots \\ \boldsymbol{\delta}_{k,i+n} = \boldsymbol{f}(\boldsymbol{\xi}_{k,i+n}) = \boldsymbol{f}((\hat{\boldsymbol{x}}_k)_{\mathrm{UPF}}) + \boldsymbol{D}_{-\hat{\boldsymbol{\sigma}}_{\boldsymbol{x}_{k_i}}} \boldsymbol{f} + \dfrac{\boldsymbol{D}^2_{-\hat{\boldsymbol{\sigma}}_{\boldsymbol{x}_{k_i}}} \boldsymbol{f}}{2!} \cdots \end{cases} \tag{4.32}$$

$$\begin{cases} \boldsymbol{\rho}_{k,0} = \boldsymbol{\rho}(\boldsymbol{\xi}_{k,0}) = \boldsymbol{\eta}((\hat{\boldsymbol{x}}_k)_{\mathrm{UPF}}) \\ \boldsymbol{\rho}_{k,i} = \boldsymbol{\rho}(\boldsymbol{\xi}_{k,i}) = \boldsymbol{\eta}((\hat{\boldsymbol{x}}_k)_{\mathrm{UPF}}) + \boldsymbol{D}_{\hat{\boldsymbol{\sigma}}_{\boldsymbol{x}_{k_i}}} \boldsymbol{\eta} + \dfrac{\boldsymbol{D}^2_{\hat{\boldsymbol{\sigma}}_{\boldsymbol{x}_{k_i}}} \boldsymbol{\eta}}{2!} + \cdots \\ \boldsymbol{\rho}_{k,i+n} = \boldsymbol{\rho}(\boldsymbol{\xi}_{k,i+n}) = \boldsymbol{\eta}((\hat{\boldsymbol{x}}_k)_{\mathrm{UPF}}) + \boldsymbol{D}_{-\hat{\boldsymbol{\sigma}}_{\boldsymbol{x}_{k_i}}} \boldsymbol{\eta} + \dfrac{\boldsymbol{D}^2_{-\hat{\boldsymbol{\sigma}}_{\boldsymbol{x}_{k_i}}} \boldsymbol{\eta}}{2!} \cdots \end{cases} \tag{4.33}$$

Then, it follows from (2.68) that

$$\boldsymbol{\xi}_{k+1,i} = \boldsymbol{\xi}_{k,i} + \Delta t \boldsymbol{\delta}_{k,i} + \Delta t \boldsymbol{\rho}_{k,i} + \boldsymbol{\mu}_{k,i} \tag{4.34}$$

where $\boldsymbol{\mu}_{k,i} = \boldsymbol{\mu}(\boldsymbol{\xi}_{k,i}, \hat{\boldsymbol{d}}_{i,k})$ is the higher-order discretization error and $i = 0, 1, \cdots, 2n+1$.

To this end, the application the UPF leads to

$$\begin{aligned} (\hat{\boldsymbol{x}}_{k+1})_{\mathrm{UPF}} &= \sum_{i=0}^{2n} W_i^c \boldsymbol{\xi}_{k+1,i} = \sum_{i=0}^{2n} W_i^c (\boldsymbol{\xi}_{k,i} + \Delta t \boldsymbol{\delta}_{k,i} + \Delta t \boldsymbol{\rho}_{k,i} + \boldsymbol{\mu}_{k,i}) \\ &= (\hat{\boldsymbol{x}}_k)_{\mathrm{UPF}} + \Delta t \boldsymbol{f}(\hat{\boldsymbol{x}}_k)_{\mathrm{UPF}} + \Delta t \boldsymbol{\eta}(\hat{\boldsymbol{x}}_k)_{\mathrm{UPF}} + \sum_{i=0}^{2n} W_i^c \boldsymbol{\mu}_{k,i} \\ &\quad + \frac{\Delta t}{n+\kappa} \sum_{i=1}^{n} \sum_{j=1}^{\infty} \frac{\boldsymbol{D}^{2j}_{\hat{\boldsymbol{\sigma}}_{k_i}} \boldsymbol{f}}{(2j)!} + \frac{\Delta t}{n+\kappa} \sum_{i=1}^{n} \sum_{j=1}^{\infty} \frac{\boldsymbol{D}^{2j}_{\hat{\boldsymbol{\sigma}}_{k_i}} \boldsymbol{\eta}}{(2j)!} \end{aligned} \tag{4.35}$$

Taking (2.69) and (4.35) into consideration, the estimation error of the states achieved by the UPF is denoted by $(\tilde{\boldsymbol{x}}_{k+1})_{\mathrm{UPF}}$, and it is specified by

$$\begin{aligned}(\tilde{\boldsymbol{x}}_{k+1})_{\mathrm{UPF}} &= \boldsymbol{x}_{k+1} - (\hat{\boldsymbol{x}}_{k+1})_{\mathrm{UPF}} \\ &= (\tilde{\boldsymbol{x}}_k)_{\mathrm{UPF}} + \Delta t(\boldsymbol{G}_f(\tilde{\boldsymbol{x}}_k)_{\mathrm{UPF}} + \boldsymbol{G}_\eta(\tilde{\boldsymbol{x}}_k)_{\mathrm{UPF}}) \\ &\quad + (\bar{\boldsymbol{\mu}}_k)_{\mathrm{UPF}} + (\boldsymbol{s}_k)_{\mathrm{UPF}} \\ &= \boldsymbol{A}_k(\tilde{\boldsymbol{x}}_k)_{\mathrm{UPF}} + (\bar{\boldsymbol{\mu}}_k)_{\mathrm{UPF}} + (\boldsymbol{s}_k)_{\mathrm{UPF}}\end{aligned} \tag{4.36}$$

where

$$\boldsymbol{A}_k = \boldsymbol{I}_n + \Delta t \boldsymbol{G}_f + \Delta t \boldsymbol{G}_\eta \tag{4.37}$$

$$(\boldsymbol{s}_k)_{\mathrm{UPF}} = -\Delta t \bar{\boldsymbol{B}}((\hat{\boldsymbol{x}}_k)_{\mathrm{UPF}})\boldsymbol{v}_k = -(\boldsymbol{B}_k)_{\mathrm{UPF}}\boldsymbol{v}_k \tag{4.38}$$

$$\begin{aligned}(\bar{\boldsymbol{\mu}}_k)_{\mathrm{UPF}} &= \boldsymbol{\mu}(\boldsymbol{x}_k, \boldsymbol{d}_k)\boldsymbol{v}_k - \sum_{i=1}^{2n} W_i^c \boldsymbol{\mu}_{k,i} - \boldsymbol{\varepsilon}(\bar{\boldsymbol{B}}((\hat{\boldsymbol{x}}_k)_{\mathrm{UPF}}))\boldsymbol{v}_k \\ &\quad + \Delta t(\Delta \boldsymbol{f}(\boldsymbol{x}_k, (\hat{\boldsymbol{x}}_k)_{\mathrm{UPF}}) + \Delta \boldsymbol{\eta}(\boldsymbol{x}_k, \boldsymbol{d}_k, (\hat{\boldsymbol{x}}_k)_{\mathrm{UPF}}, (\hat{\boldsymbol{d}}_k)_{\mathrm{UPF}}))\end{aligned} \tag{4.39}$$

with

$$\Delta \boldsymbol{f}(\boldsymbol{x}_k, (\hat{\boldsymbol{x}}_k)_{\mathrm{UPF}}) = \sum_{i=2}^{\infty} \frac{1}{i!} \boldsymbol{D}_{\hat{x}_k}^i \boldsymbol{f} - \frac{1}{n+\kappa} \sum_{j=1}^{n} \sum_{l=1}^{\infty} \frac{\boldsymbol{D}_{\hat{\sigma}_{k_j}}^{2l} \boldsymbol{f}}{(2l)!} \tag{4.40}$$

$$\Delta \boldsymbol{\eta}(\boldsymbol{x}_k, \boldsymbol{d}_k, (\hat{\boldsymbol{x}}_k)_{\mathrm{UPF}}, (\hat{\boldsymbol{d}}_k)_{\mathrm{UPF}}) = \sum_{i=2}^{\infty} \frac{1}{i!} \boldsymbol{D}_{\hat{x}_k}^i \boldsymbol{\eta} - \frac{1}{n+\kappa} \sum_{j=1}^{n} \sum_{l=1}^{\infty} \frac{\boldsymbol{D}_{\hat{\sigma}_{k_j}}^{2l} \boldsymbol{\eta}}{(2l)!} \tag{4.41}$$

Consequently, the state estimation error transition (4.36) is in a recursive form. According to the definition of the covariance of the state estimation error, we can calculate the covariance achieved by the UPF as

$$\begin{aligned}(\boldsymbol{P}_{\tilde{x},k+1})_{\mathrm{UPF}} &= \sum_{i=0}^{2n} W_i^c (\boldsymbol{\xi}_{k+1,i} - (\hat{\boldsymbol{x}}_{k+1})_{\mathrm{UPF}})(\boldsymbol{\xi}_{k+1,i} - (\hat{\boldsymbol{x}}_{k+1})_{\mathrm{UPF}})^{\mathrm{T}} \\ &= (\boldsymbol{P}_{\tilde{x},k})_{\mathrm{UPF}} + \bar{\boldsymbol{\psi}}(\boldsymbol{v}_k) + \bar{o}(\boldsymbol{\delta}_{k,i}, \boldsymbol{\mu}_{k,i}) \\ &\quad + \Delta t \sum_{i=0}^{2n} W_i^c (\boldsymbol{\xi}_{k,i} - (\hat{\boldsymbol{x}}_k)_{\mathrm{UPF}})(\boldsymbol{\delta}_{k,i} - \boldsymbol{f}((\hat{\boldsymbol{x}}_k)_{\mathrm{UPF}}))^{\mathrm{T}}\end{aligned}$$

$$
\begin{aligned}
&+\Delta t\sum_{i=0}^{2n} W_i^c(\boldsymbol{\delta}_{k,i}-\boldsymbol{f}((\hat{\boldsymbol{x}}_k)_{\mathrm{UPF}}))(\boldsymbol{\xi}_{k,i}-(\hat{\boldsymbol{x}}_k)_{\mathrm{UPF}})^{\mathrm{T}}\\
&+\Delta t\sum_{i=0}^{2n} W_i^c(\boldsymbol{\xi}_{k,i}-(\hat{\boldsymbol{x}}_k)_{\mathrm{UPF}})(\boldsymbol{\rho}_{k,i}-\boldsymbol{\eta}((\hat{\boldsymbol{x}}_k)_{\mathrm{UPF}}))^{\mathrm{T}}\\
&+\Delta t\sum_{i=0}^{2n} W_i^c(\boldsymbol{\rho}_{k,i}-\boldsymbol{\eta}((\hat{\boldsymbol{x}}_k)_{\mathrm{UPF}}))(\boldsymbol{\xi}_{k,i}-(\hat{\boldsymbol{x}}_k)_{\mathrm{UPF}})^{\mathrm{T}}\\
&+\Delta t^2\sum_{i=0}^{2n} W_i^c(\boldsymbol{\delta}_{k,i}-\boldsymbol{f}((\hat{\boldsymbol{x}}_k)_{\mathrm{UPF}}))(\boldsymbol{\delta}_{k,i}-\boldsymbol{f}((\hat{\boldsymbol{x}}_k)_{\mathrm{UPF}}))^{\mathrm{T}}\\
&+\Delta t^2\sum_{i=0}^{2n} W_i^c(\boldsymbol{\rho}_{k,i}-\boldsymbol{\eta}((\hat{\boldsymbol{x}}_k)_{\mathrm{UPF}}))(\boldsymbol{\rho}_{k,i}-\boldsymbol{\eta}((\hat{\boldsymbol{x}}_k)_{\mathrm{UPF}}))^{\mathrm{T}}\\
&+\Delta t^2\sum_{i=0}^{2n} W_i^c(\boldsymbol{\delta}_{k,i}-\boldsymbol{f}((\hat{\boldsymbol{x}}_k)_{\mathrm{UPF}}))(\boldsymbol{\rho}_{k,i}-\boldsymbol{\eta}((\hat{\boldsymbol{x}}_k)_{\mathrm{UPF}}))^{\mathrm{T}}\\
&+\Delta t^2\sum_{i=0}^{2n} W_i^c(\boldsymbol{\rho}_{k,i}-\boldsymbol{\eta}((\hat{\boldsymbol{x}}_k)_{\mathrm{UPF}}))(\boldsymbol{\delta}_{k,i}-\boldsymbol{f}((\hat{\boldsymbol{x}}_k)_{\mathrm{UPF}}))^{\mathrm{T}}
\end{aligned}
\tag{4.42}
$$

where $\bar{\boldsymbol{o}}(\boldsymbol{\delta}_{k,i},\boldsymbol{\mu}_{k,i})$ is the sum of the polynomial of $\boldsymbol{\mu}_{k,i}-\sum_{i=0}^{2n} W_i^c\boldsymbol{\mu}_{k,i}$ and other polynomials. $\bar{\boldsymbol{\psi}}(\boldsymbol{v}_k)$ is the sum of the polynomial of $\boldsymbol{v}_k$.

The set of sigma points are then used to expand the terms of (4.42), it yields

$$
\begin{aligned}
&\sum_{i=0}^{n} W_i^c(\boldsymbol{\xi}_{k,i}-(\hat{\boldsymbol{x}}_k)_{\mathrm{UPF}})(\boldsymbol{\delta}_{k,i}-\boldsymbol{f}((\hat{\boldsymbol{x}}_k)_{\mathrm{UPF}}))^{\mathrm{T}}\\
&=\sum_{i=0}^{n} W_i^c((\boldsymbol{\xi}_{k,i}-(\hat{\boldsymbol{x}}_k)_{\mathrm{UPF}})\boldsymbol{\delta}_i^{\mathrm{T}}+(\boldsymbol{\xi}_{k,i+n}-(\hat{\boldsymbol{x}}_k)_{\mathrm{UPF}})\boldsymbol{\delta}_{i+n}^{\mathrm{T}})\\
&=\frac{1}{n+\kappa}\sum_{i=1}^{n}\hat{\boldsymbol{\sigma}}_{x_{k_i}}\left(\sum_{j=1}^{\infty}\frac{\boldsymbol{D}_{\hat{\boldsymbol{\sigma}}_{x_{k_i}}}^{(2j-1)}\boldsymbol{f}}{(2j-1)!}\right)^{\mathrm{T}}\\
&=\boldsymbol{P}_{\tilde{x},k}\boldsymbol{G}_f^{\mathrm{T}}+\sum_{i=1}^{n}\sum_{j=1}^{\infty}\frac{1}{(n+\kappa)(2j+1)!}\hat{\boldsymbol{\sigma}}_{x_{k_i}}\boldsymbol{D}_{\hat{\boldsymbol{\sigma}}_{x_{k_i}}}^{2j+1}\boldsymbol{f}
\end{aligned}
\tag{4.43}
$$

$$\sum_{i=0}^{2n} W_i^c(\boldsymbol{\delta}_{k,i} - \boldsymbol{f}((\hat{\boldsymbol{x}}_k)_{\text{UPF}}))(\boldsymbol{\delta}_{k,i} - \boldsymbol{f}(\hat{\boldsymbol{x}}_k)_{\text{UPF}})^{\text{T}}$$
$$= \sum_{i=0}^{2n} W_i^c \boldsymbol{\delta}_{k,i}\boldsymbol{\delta}_{k,i}^{\text{T}} - \boldsymbol{f}((\hat{\boldsymbol{x}}_k)_{\text{UPF}})(\boldsymbol{f}((\hat{\boldsymbol{x}}_k)_{\text{UPF}}))^{\text{T}}$$
$$= \boldsymbol{G}_f \boldsymbol{P}_{\tilde{x},k}\boldsymbol{G}_f^{\text{T}} - \mathbb{E}\left(\frac{\boldsymbol{D}_{\Delta x}^2 \boldsymbol{f}}{2!}\right)\mathbb{E}\left(\left(\frac{\boldsymbol{D}_{\Delta x}^2 \boldsymbol{f}}{2!}\right)^{\text{T}}\right)$$
$$+ \frac{1}{n+\kappa}\sum_{k=1}^{n}\underbrace{\sum_{i=1}^{\infty}\sum_{j=1}^{\infty}\frac{1}{i!j!}(\boldsymbol{D}_{\hat{\sigma}_{x_{k_i}}}^{i}\boldsymbol{f})(\boldsymbol{D}_{\hat{\sigma}_{x_{k_i}}}^{j}\boldsymbol{f})^{\text{T}}}_{\text{condition\_1}}$$
$$\underbrace{-\sum_{i=1}^{\infty}\sum_{j=1}^{\infty}\frac{1}{(2i)!(2j)!(n+\kappa)^2}\sum_{p=1}^{n}\sum_{m=1}^{n}(\boldsymbol{D}_{\hat{\sigma}_{x_{kp}}}^{2i}\boldsymbol{f})(\boldsymbol{D}_{\hat{\sigma}_{x_{km}}}^{2j}\boldsymbol{f})^{\text{T}}}_{\text{condition\_2}} \tag{4.44}$$

$$\sum_{i=0}^{2n} W_i^c(\boldsymbol{\xi}_{k,i} - (\hat{\boldsymbol{x}}_k)_{\text{UPF}})(\boldsymbol{\rho}_{k,i} - \boldsymbol{\eta}(\hat{\boldsymbol{x}}_k)_{\text{UPF}})^{\text{T}}$$
$$= \boldsymbol{P}_{\tilde{x},k}\boldsymbol{G}_\eta^{\text{T}} + \sum_{i=1}^{n}\sum_{j=1}^{\infty}\frac{1}{(n+\kappa)(2j+1)!}\hat{\boldsymbol{\sigma}}_{x_{k_i}}\boldsymbol{D}_{\hat{\sigma}_{x_{k_i}}}^{2j+1}\boldsymbol{\eta} \tag{4.45}$$

$$\sum_{i=0}^{2n} W_i^c(\boldsymbol{\delta}_{k,i} - \boldsymbol{f}((\hat{\boldsymbol{x}}_k)_{\text{UPF}}))(\boldsymbol{\xi}_{k,i} - (\hat{\boldsymbol{x}}_k)_{\text{UPF}})^{\text{T}}$$
$$= \boldsymbol{G}_f \boldsymbol{P}_{\tilde{x},k} + \sum_{i=1}^{n}\sum_{j=1}^{\infty}\frac{1}{(n+\kappa)(2j+1)!}\boldsymbol{D}_{\hat{\sigma}_{x_{k_i}}}^{2j+1}\boldsymbol{f}\hat{\boldsymbol{\sigma}}_{x_{k_i}} \tag{4.46}$$

$$\sum_{i=0}^{2n} W_i^c(\boldsymbol{\rho}_{k,i} - \boldsymbol{\eta}(\hat{\boldsymbol{x}}_k)_{\text{UPF}})(\boldsymbol{\xi}_{k,i} - (\hat{\boldsymbol{x}}_k)_{\text{UPF}})^{\text{T}}$$
$$= \boldsymbol{G}_\eta \boldsymbol{P}_{\tilde{x},k} + \sum_{i=1}^{n}\sum_{j=1}^{\infty}\frac{1}{(n+\kappa)(2j+1)!}\boldsymbol{D}_{\hat{\sigma}_{x_{k_i}}}^{2j+1}\boldsymbol{\eta}\hat{\boldsymbol{\sigma}}_{x_{k_i}} \tag{4.47}$$

$$\sum_{i=0}^{2n} W_i^c(\boldsymbol{\rho}_{k,i} - \boldsymbol{\eta}(\hat{\boldsymbol{x}}_k)_{\text{UPF}})(\boldsymbol{\rho}_{k,i} - \boldsymbol{\eta}(\hat{\boldsymbol{x}}_k)_{\text{UPF}})^{\text{T}}$$
$$= \boldsymbol{G}_\eta \boldsymbol{P}_{\tilde{x},k}\boldsymbol{G}_\eta^{\text{T}} - \mathbb{E}\left(\frac{\boldsymbol{D}_{\Delta x}^2 \boldsymbol{\eta}}{2!}\right)\mathbb{E}\left(\left(\frac{\boldsymbol{D}_{\Delta x}^2 \boldsymbol{\eta}}{2!}\right)^{\text{T}}\right)$$

$$
+\frac{1}{n+\kappa}\underbrace{\sum_{l=1}^{n}\sum_{i=1}^{\infty}\sum_{j=1}^{\infty}\frac{1}{i!j!}(\boldsymbol{D}^{i}_{\hat{\sigma}x_{k_l}}\boldsymbol{\eta})(\boldsymbol{D}^{j}_{\hat{\sigma}x_{k_l}}\boldsymbol{\eta})^{\mathrm{T}}}_{\text{condition\_1}}
$$
$$
-\underbrace{\sum_{i=1}^{\infty}\sum_{j=1}^{\infty}\frac{1}{(2i)!(2j)!}\sum_{p=1}^{n}\sum_{m=1}^{n}(\boldsymbol{D}^{2i}_{\hat{\sigma}x_{kp}}\boldsymbol{\eta})(\boldsymbol{D}^{2j}_{\hat{\sigma}x_{km}}\boldsymbol{\eta})^{\mathrm{T}}}_{\text{condition\_2}} \tag{4.48}
$$

$$
\sum_{i=0}^{2n}W_i^c(\boldsymbol{\delta}_{k,i}-\boldsymbol{f}((\hat{\boldsymbol{x}}_k)_{\mathrm{UPF}}))(\boldsymbol{\rho}_{k,i}-\boldsymbol{\eta}(\hat{\boldsymbol{x}}_k)_{\mathrm{UPF}})^{\mathrm{T}}
$$
$$
=\boldsymbol{G}_f\boldsymbol{P}_{\tilde{x},k}\boldsymbol{G}_\eta^{\mathrm{T}}-\mathbb{E}\left(\frac{\boldsymbol{D}^2_{\Delta x}\boldsymbol{f}}{2!}\right)\mathbb{E}\left(\left(\frac{\boldsymbol{D}^2_{\Delta x}\boldsymbol{\eta}}{2!}\right)^{\mathrm{T}}\right)
$$
$$
+\frac{1}{n+\kappa}\underbrace{\sum_{k=1}^{n}\sum_{i=1}^{\infty}\sum_{j=1}^{\infty}\frac{1}{i!j!}(\boldsymbol{D}^{i}_{\hat{\sigma}x_{k_i}}\boldsymbol{f})(\boldsymbol{D}^{j}_{\hat{\sigma}x_{k_i}}\boldsymbol{\eta})^{\mathrm{T}}}_{\text{condition\_1}} \tag{4.49}
$$
$$
-\underbrace{\sum_{i=1}^{\infty}\sum_{j=1}^{\infty}\frac{1}{(2i)!(2j)!(n+\kappa)^2}\sum_{p=1}^{n}\sum_{m=1}^{n}(\boldsymbol{D}^{2i}_{\hat{\sigma}x_{kp}}\boldsymbol{f})(\boldsymbol{D}^{2j}_{\hat{\sigma}x_{km}}\boldsymbol{\eta})^{\mathrm{T}}}_{\text{condition\_2}}
$$

$$
\sum_{i=0}^{2n}W_i^c(\boldsymbol{\rho}_{k,i}-\boldsymbol{\eta}(\hat{\boldsymbol{x}}_k)_{\mathrm{UPF}})(\boldsymbol{\delta}_{k,i}-\boldsymbol{f}((\hat{\boldsymbol{x}}_k)_{\mathrm{UPF}}))^{\mathrm{T}}
$$
$$
=\boldsymbol{G}_\eta\boldsymbol{P}_{\tilde{x},k}\boldsymbol{G}_f^{\mathrm{T}}-\mathbb{E}\left(\frac{\boldsymbol{D}^2_{\Delta x}\boldsymbol{\eta}}{2!}\right)\mathbb{E}\left(\left(\frac{\boldsymbol{D}^2_{\Delta x}\boldsymbol{f}}{2!}\right)^{\mathrm{T}}\right)
$$
$$
+\frac{1}{n+\kappa}\underbrace{\sum_{k=1}^{n}\sum_{i=1}^{\infty}\sum_{j=1}^{\infty}\frac{1}{i!j!}(\boldsymbol{D}^{i}_{\hat{\sigma}x_{k_i}}\boldsymbol{\eta})(\boldsymbol{D}^{j}_{\hat{\sigma}x_{k_i}}\boldsymbol{f})^{\mathrm{T}}}_{\text{condition\_1}} \tag{4.50}
$$
$$
-\underbrace{\sum_{i=1}^{\infty}\sum_{j=1}^{\infty}\frac{1}{(2i)!(2j)!(n+\kappa)^2}\sum_{p=1}^{n}\sum_{m=1}^{n}(\boldsymbol{D}^{2i}_{\hat{\sigma}x_{kp}}\boldsymbol{\eta})(\boldsymbol{D}^{2j}_{\hat{\sigma}x_{km}}\boldsymbol{f})^{\mathrm{T}}}_{\text{condition\_2}}
$$

Then, invoking (4.43)–(4.50) results in

$$
\begin{aligned}
&(\boldsymbol{P}_{\tilde{x},k+1})_{\mathrm{UPF}} = \\
&-(\Delta t)^2\left(\mathbb{E}\left(\frac{\boldsymbol{D}_{\Delta x}^2 \boldsymbol{f}}{2!}\right)+\mathbb{E}\left(\frac{\boldsymbol{D}_{\Delta x}^2 \boldsymbol{\eta}}{2!}\right)\right)\left(\mathbb{E}\left(\frac{\boldsymbol{D}_{\Delta x}^2 \boldsymbol{f}}{2!}\right)+\mathbb{E}\left(\frac{\boldsymbol{D}_{\Delta x}^2 \boldsymbol{\eta}}{2!}\right)\right)^{\mathrm{T}} \\
&+\boldsymbol{A}_k(\boldsymbol{P}_{\tilde{x},k})_{\mathrm{UPF}}\boldsymbol{A}_k^{\mathrm{T}}+\tilde{\boldsymbol{\Sigma}}_1+\tilde{\boldsymbol{\Sigma}}_2+\bar{o}_k+\bar{\boldsymbol{\psi}}_k
\end{aligned}
\tag{4.51}
$$

where

$$
\begin{aligned}
\tilde{\boldsymbol{\Sigma}}_1 = &\sum_{i=1}^{n}\sum_{j=1}^{\infty}\frac{\Delta t}{(n+\kappa)(2j+1)!}\hat{\boldsymbol{\sigma}}_{x_{k_i}}\boldsymbol{D}_{\hat{\sigma}_{x_{ki}}}^{2j+1}(\boldsymbol{f}+\boldsymbol{\eta}) \\
&+\sum_{i=1}^{n}\sum_{j=1}^{\infty}\frac{\Delta t}{(n+\kappa)(2j+1)!}\boldsymbol{D}_{\hat{\sigma}_{x_{k_i}}}^{2j+1}(\boldsymbol{f}+\boldsymbol{\eta})\hat{\boldsymbol{\sigma}}_{x_{k_i}}
\end{aligned}
\tag{4.52}
$$

$$
\begin{aligned}
\tilde{\boldsymbol{\Sigma}}_2 = &\frac{(\Delta t)^2}{n+\kappa}\sum_{l=1}^{n}\underbrace{\sum_{i=1}^{\infty}\sum_{j=1}^{\infty}\frac{1}{i!j!}\boldsymbol{D}_{\hat{\sigma}_{x_{k_l}}}^{i}(\boldsymbol{f}+\boldsymbol{\eta})(\boldsymbol{D}_{\hat{\sigma}_{x_{k_l}}}^{i}(\boldsymbol{f}+\boldsymbol{\eta}))^{\mathrm{T}}-}_{\text{condition\_1}} \\
&\underbrace{\sum_{i=1}^{\infty}\sum_{j=1}^{\infty}\frac{(\Delta t)^2}{(2i)!(2j)!(n+\kappa)^2}\sum_{p=1}^{n}\sum_{m=1}^{n}\boldsymbol{D}_{\hat{\sigma}_{x_{k_p}}}^{2i}(\boldsymbol{f}+\boldsymbol{\eta})(\boldsymbol{D}_{\hat{\sigma}_{x_{km}}}^{2j}(\boldsymbol{f}+\boldsymbol{\eta}))^{\mathrm{T}}}_{\text{condition\_2}}
\end{aligned}
\tag{4.53}
$$

Moreover, one can rewrite (4.51) as

$$
(\boldsymbol{P}_{\tilde{x},k+1})_{\mathrm{UPF}} = \boldsymbol{A}_k(\boldsymbol{P}_{\tilde{x},k})_{\mathrm{UPF}}\boldsymbol{A}_k^{\mathrm{T}} - \boldsymbol{K}_k\boldsymbol{K}_k^{\mathrm{T}} + \boldsymbol{Q}_k \tag{4.54}
$$

with

$$
\boldsymbol{A}_k = \boldsymbol{I}_n + \Delta t\boldsymbol{G}_f + \Delta t\boldsymbol{G}_\eta \tag{4.55}
$$

$$
\begin{aligned}
\boldsymbol{K}_k &= \Delta t\frac{(\nabla^{\mathrm{T}}\boldsymbol{P}_{\tilde{x},k}\nabla)\boldsymbol{f}(\boldsymbol{x})|_{\boldsymbol{x}=\hat{\boldsymbol{x}}_k}}{2!}+\Delta t\frac{(\nabla^{\mathrm{T}}\boldsymbol{P}_{\tilde{x},k}\nabla)\boldsymbol{\eta}(\boldsymbol{x})|_{\boldsymbol{x}=\hat{\boldsymbol{x}}_k}}{2!} \\
&= \Delta t\mathbb{E}\left(\frac{\boldsymbol{D}_{\Delta x}^2 \boldsymbol{f}}{2!}\right)+\Delta t\mathbb{E}\left(\frac{\boldsymbol{D}_{\Delta x}^2 \boldsymbol{\eta}}{2!}\right)
\end{aligned}
\tag{4.56}
$$

$$
\boldsymbol{Q}_k = \tilde{\boldsymbol{\Sigma}}_1+\tilde{\boldsymbol{\Sigma}}_2+\bar{o}_k+\bar{\boldsymbol{\psi}}_k \tag{4.57}
$$

It is seen in (4.54) that $\boldsymbol{Q}_k$ is a matrix particularly introduced. If the value of $\boldsymbol{Q}_k$ approximates the sum of the last four terms of (4.54) and the initial value satisfies $(\boldsymbol{P}_{\tilde{x},0})_{\mathrm{UPF}} = \mathbb{E}(\tilde{\boldsymbol{x}}_0\tilde{\boldsymbol{x}}_0^{\mathrm{T}})$, then $(\boldsymbol{P}_{\tilde{x},k})_{\mathrm{UPF}}$ calculated from (4.54) will approximate its actual value, that is $(\boldsymbol{P}_{\tilde{x},k})_{\mathrm{UPF}} \approx \mathbb{E}(\tilde{\boldsymbol{x}}_k\tilde{\boldsymbol{x}}_k^{\mathrm{T}})$. Therefore, (4.54) can be used to estimate the covariance of the state estimation error.

Theorem 4.2. *When applying the UPF to the state estimation problem of the nonlinear discrete system* (2.32)–(2.33)*, if the initial value satisfies* $(\boldsymbol{P}_{\tilde{x},0})_{UPF} = (\boldsymbol{P}_{\tilde{x},0})_{PF} = \boldsymbol{P}_0$*, then* $(\boldsymbol{P}_{\tilde{x},k})_{UPF} \leq (\boldsymbol{P}_{\tilde{x},k})_{PF}$ *is valid for every* $k \geq 0$*. The state estimation error's covariance of the UPF has faster convergence rate than the classical PF. Better filtering performance is achieved by the UPF than the classical PF.*

Proof. Comparing (2.74) with (4.54), we can easily prove the conclusion in Theorem 4.2. □

Remark 4.3. The exact value of $\boldsymbol{Q}_k$ can not be obtained in practice, it is determined based on experience information. Hence, the estimation obtained by using (4.54) would not be precise. However, this does not affect the estimation performance. It will not be introduced into updating the algorithm of UPF. It has an influence on the stability region during the filtering process. This will be analyzed in the subsequent stability analysis of the UPF.

4.4 Covariance constraint analysis

For the classical PF, the optimal state estimation is determined based on the assumption that the consistent estimations of the states must match the available measurements with a residual error covariance which is approximately equal to the known measurement error covariance. This necessary condition is hereafter referred to as the "covariance constraint". The covariance constraint is imposed by requiring the following approximation to be satisfied:

$$\mathbb{E}((\boldsymbol{y}_k - \boldsymbol{h}(\hat{\boldsymbol{x}}_k))(\boldsymbol{y}_k - \boldsymbol{h}(\hat{\boldsymbol{x}}_k))^{\mathrm{T}}) = \hat{\boldsymbol{R}}_k \approx \boldsymbol{R}_k,\ k = 1, 2, \cdots, \varsigma \tag{4.58}$$

where $\varsigma \in \mathbb{N}$ denotes the number of the set of measurements. This means that the estimated output $\boldsymbol{h}(\hat{\boldsymbol{x}}_k)$ is required to fit the actual measurements $\boldsymbol{y}_k$ with approximately the same error covariance as the actual measurements fit the truth. Otherwise, the estimation is statistically inconsistent.

Theoretically, the optimal weighting matrix $\boldsymbol{W}_E$ of any filtering algorithms should be chosen with covariance constraint (4.58) satisfied. Hence, there is a balance between the modeling error and the residual error of estimation. When the covariance constraint is not satisfied, then choosing the optimal weighting matrix $\boldsymbol{W}_E$ and compensating for the modeling error should be done simultaneously to ensure $\hat{\boldsymbol{R}}_k \to \boldsymbol{R}_k$. In this case, it is difficult to determine the optimal weighting matrix $\boldsymbol{W}_E$. Note that $\boldsymbol{W}_E$ is constantly chosen based on the experience of engineers.

Based on (2.64) and (4.25), it follows that

$$\mathbb{E}((\boldsymbol{y}_k - \boldsymbol{h}(\hat{\boldsymbol{x}}_k))(\boldsymbol{y}_k - \boldsymbol{h}(\hat{\boldsymbol{x}}_k))^{\mathrm{T}}) = \boldsymbol{R}_k$$
$$= \mathbb{E}\left(\underbrace{\sum_{i=1}^{\infty}\sum_{j=1}^{\infty}\frac{1}{i!j!}(\boldsymbol{D}^i_{\Delta x}\boldsymbol{h})(\boldsymbol{D}^j_{\Delta x}\boldsymbol{h})^{\mathrm{T}}}_{\text{condition\_1}}\right) - \mathbb{E}\left(\frac{\boldsymbol{D}^2_{\Delta x}\boldsymbol{h}}{2!}\right)\mathbb{E}\left(\left(\frac{\boldsymbol{D}^2_{\Delta x}\boldsymbol{h}}{2!}\right)^{\mathrm{T}}\right)$$
$$- \mathbb{E}\left(\underbrace{\sum_{i=1}^{\infty}\sum_{j=1}^{\infty}\frac{1}{(2i)!(2j)!}\boldsymbol{D}^{2i}_{\Delta x}\boldsymbol{h}(\boldsymbol{D}^{2j}_{\Delta x}\boldsymbol{h})^{\mathrm{T}}}_{\text{condition\_2}}\right) + \boldsymbol{G_h}\boldsymbol{P}_{\tilde{x},k}\boldsymbol{G}_{\boldsymbol{h}}^{\mathrm{T}} \tag{4.59}$$

$$\mathbb{E}((\boldsymbol{y}_k - \boldsymbol{h}((\hat{\boldsymbol{x}}_k)_{\mathrm{PF}}))(\boldsymbol{y}_k - \boldsymbol{h}((\hat{\boldsymbol{x}}_k)_{PF}))^{\mathrm{T}}) = (\hat{\boldsymbol{R}}_k)_{\mathrm{PF}} = \boldsymbol{G_h}\boldsymbol{P}_{\tilde{x},k}\boldsymbol{G}_{\boldsymbol{h}}^{\mathrm{T}} \tag{4.60}$$

$$\mathbb{E}((\boldsymbol{y}_k - \boldsymbol{h}((\hat{\boldsymbol{x}}_k)_{\mathrm{UPF}}))(\boldsymbol{y}_k - \boldsymbol{h}((\hat{\boldsymbol{x}}_k)_{\mathrm{UPF}}))^{\mathrm{T}})$$
$$= (\hat{\boldsymbol{R}}_k)_{\mathrm{UPF}} = \boldsymbol{G_h}\boldsymbol{P}_{\tilde{x},k}\boldsymbol{G}_{\boldsymbol{h}}^{\mathrm{T}} - \mathbb{E}\left(\frac{\boldsymbol{D}^2_{\Delta x}\boldsymbol{h}}{2!}\right)\mathbb{E}\left(\left(\frac{\boldsymbol{D}^2_{\Delta x}\boldsymbol{h}}{2!}\right)^{\mathrm{T}}\right)$$
$$+ \frac{1}{n+\kappa}\sum_{m=1}^{n}\underbrace{\sum_{i=1}^{\infty}\sum_{j=1}^{\infty}\frac{1}{i!j!}(\boldsymbol{D}^i_{\hat{\sigma}_{km}}\boldsymbol{h})(\boldsymbol{D}^j_{\hat{\sigma}_{km}}\boldsymbol{h})^{\mathrm{T}}}_{\text{condition\_1}}$$
$$- \sum_{m=1}^{n}\underbrace{\sum_{i=1}^{\infty}\sum_{j=1}^{\infty}\frac{1}{(2i)!(2j)!(n+\kappa)^2}\sum_{l=1}^{n}\sum_{p=1}^{n}\frac{1}{(2i)!(2j)!}\boldsymbol{D}^{2i}_{\hat{\sigma}_{kl}}\boldsymbol{h}(\boldsymbol{D}^{2j}_{\hat{\sigma}_{kp}}\boldsymbol{h})^{\mathrm{T}}}_{\text{condition\_1}} \tag{4.61}$$

It is found in (4.59)–(4.61) that the covariance of the estimation error achieved by the PF is only accurate to the second-order accuracy of the real value. The covariance of the estimation error achieved by the UPF is accurate to the fourth-order accuracy of the real value. Hence, $(\hat{\boldsymbol{R}}_k)_{\mathrm{UPF}}$ is more consistent with $\boldsymbol{R}_k$ than $(\hat{\boldsymbol{R}}_k)_{\mathrm{PF}}$. This implies that the selection of the weighting matrix $\boldsymbol{W}_E$ for the UPF depends less on the covariance constraint. The UPF has more freedom to choose $\boldsymbol{W}_E$. Hence, the practical application value of the UPF is great than the classical PF.

4.5 Stochastic stability of UPF

In this section, the stability of achieved by the UPF is rigorously proved.

4.5.1 Boundedness analysis of state estimation covariance

Theorem 4.3. *When the UPF is applied to the nonlinear stochastic system described by* (2.32)–(2.33), *for every* $k \geq 0$, *if there exist positive real numbers* $0 < z \leq 1$, $\bar{k} > 0$, $\bar{r} \geq \underline{r} \geq 0$, *and* $\bar{q} \geq \underline{q} > 0$ *satisfying*

$$\mathbf{0} \leq \boldsymbol{A}_k \boldsymbol{A}_k^{\mathrm{T}} \leq (1-z)\boldsymbol{I}_n \tag{4.62}$$

$$\mathbf{0} \leq \frac{1}{4}(\Delta t)^2 \boldsymbol{\Upsilon}\boldsymbol{\Upsilon}^{\mathrm{T}} \leq \bar{k}\boldsymbol{I}_n \tag{4.63}$$

$$\underline{r}\boldsymbol{I}_n \leq \boldsymbol{R}_k \leq \bar{r}\boldsymbol{I}_n \tag{4.64}$$

$$\underline{q}\boldsymbol{I}_n \leq \boldsymbol{Q}_k \leq \bar{q}\boldsymbol{I}_n \tag{4.65}$$

and the initial condition $(\boldsymbol{P}_{\tilde{x},0})_{UPF}$ *satisfies* $\underline{p}\boldsymbol{I}_n \leq (\boldsymbol{P}_{\tilde{x},0})_{UPF} \leq \bar{p}\boldsymbol{I}_n$, *where* $\boldsymbol{\Upsilon} = (\nabla^{\mathrm{T}}\nabla)\boldsymbol{f}(\boldsymbol{x})|_{\boldsymbol{x}=\hat{\boldsymbol{x}}_k} + (\nabla^{\mathrm{T}}\nabla)\boldsymbol{\eta}(\boldsymbol{x})|_{\boldsymbol{x}=\hat{\boldsymbol{x}}_k}$, *then the recursion* (4.54) *is bounded as*

$$\underline{p}\boldsymbol{I}_n \leq (\boldsymbol{P}_{\tilde{x},k})_{UPF} \leq \bar{p}\boldsymbol{I}_n, k \geq 0 \tag{4.66}$$

Proof. When $\underline{q} = \underline{p}$ and $\bar{q} = \bar{p}z + \bar{p}^2\bar{k}$, then it follows from (4.63)–(4.64) that

$$\underline{p}\boldsymbol{A}_k\boldsymbol{A}_k^{\mathrm{T}} - \frac{1}{4}(\Delta t)^2\,\underline{p}^2\boldsymbol{\Upsilon}\boldsymbol{\Upsilon}^{\mathrm{T}} + \boldsymbol{Q}_k \geq \underline{p}\boldsymbol{I}_n \tag{4.67}$$

$$\bar{p}\boldsymbol{A}_k\boldsymbol{A}_k^{\mathrm{T}} - \frac{1}{4}(\Delta t)^2\,\bar{p}^2\boldsymbol{\Upsilon}\boldsymbol{\Upsilon}^{\mathrm{T}} + \boldsymbol{Q}_k \leq \bar{p}\boldsymbol{I}_n \tag{4.68}$$

If the initial condition $\underline{p}\boldsymbol{I}_n \leq (\boldsymbol{P}_{\tilde{x},0})_{\mathrm{UPF}} \leq \bar{p}\boldsymbol{I}_n$ is satisfied, then one has

$$\underline{p}\boldsymbol{I}_n \leq (\boldsymbol{P}_{\tilde{x},1})_{\mathrm{UPF}} = \boldsymbol{A}_0(\boldsymbol{P}_{\tilde{x},0})_{\mathrm{UPF}}\boldsymbol{A}_0^{\mathrm{T}} - \boldsymbol{K}_0\boldsymbol{K}_0^{\mathrm{T}} + \boldsymbol{Q}_0 \leq \bar{p}\boldsymbol{I}_n \tag{4.69}$$

Hence, applying (4.67)–(4.69), $\underline{p}\boldsymbol{I}_n \leq (\boldsymbol{P}_{\tilde{x},1})_{\mathrm{UPF}} \leq \bar{p}\boldsymbol{I}_n$ is proved.

Repeating the above procedure, $\underline{p}\boldsymbol{I}_n \leq (\boldsymbol{P}_{\tilde{x},k})_{\mathrm{UPF}} \leq \bar{p}\boldsymbol{I}_n$ can be proved to be valid for $k \geq 0$. The proof is thereby completed here. □

4.5.2 Boundedness analysis of state estimation errors

Once the recursion equation for the state estimation error and the corresponding state estimation error's covariance is derived in Section 4.3, the stochastic boundedness analysis of the UPF is analyzed in this subsection.

Assumption 4.1. For every $k \geq 0$, there exist positive real constants $\bar{a} \in \mathbb{R}_+$, $\underline{a} \in \mathbb{R}_+$, $\bar{p} \in \mathbb{R}_+$, $\underline{p} \in \mathbb{R}_+$, $\bar{q} \in \mathbb{R}_+$, $\underline{q} \in \mathbb{R}_+$, $\underline{r} \in \mathbb{R}_+$, $\bar{r} \in \mathbb{R}_+$, $\underline{\lambda} \in \mathbb{R}_+$, $\bar{\lambda} \in \mathbb{R}_+$, $\underline{s} \in \mathbb{R}_+$, $\bar{s} \in \mathbb{R}_+$, $\bar{k} \in \mathbb{R}_+$, $\delta \in \mathbb{R}_+$ satisfying

$$\underline{a} \leq ||\boldsymbol{A}_k|| \leq \bar{a} \tag{4.70}$$

$$\frac{1}{2} \|\Delta t \boldsymbol{\Upsilon}\| \leq \sqrt{\bar{k}} \tag{4.71}$$

$$\underline{q}\boldsymbol{I}_n \leq \boldsymbol{Q}_k \leq \bar{q}\boldsymbol{I}_n \tag{4.72}$$

$$\underline{r}\boldsymbol{I}_n \leq \boldsymbol{R}_k \tag{4.73}$$

$$\boldsymbol{0} \leq \boldsymbol{W}_E \leq \bar{w}\boldsymbol{I}_n \tag{4.74}$$

$$\underline{\lambda} \leq ||\boldsymbol{\Lambda}(\Delta t)|| \leq \bar{\lambda} \tag{4.75}$$

$$\underline{s} \leq \|(\boldsymbol{U}(\hat{\boldsymbol{x}}_k))_{\text{UPF}}\| \leq \bar{s} \tag{4.76}$$

$$\underline{p}\boldsymbol{I}_n \leq (\boldsymbol{P}_{\tilde{x},k})_{\text{UPF}} \leq \bar{p}\boldsymbol{I}_n \tag{4.77}$$

Assumption 4.2. $\boldsymbol{A}_k$ is invertible for every $k \geq 0$.

Assumption 4.3. Two positive real numbers $\kappa_{\bar{\mu}} > 0$ and $\varepsilon' > 0$ exist such that the nonlinear function given by (4.36) is bounded, *i.e.*,

$$\left\|(\bar{\boldsymbol{\mu}}_k)_{\text{UPF}}\right\| \leq \kappa_{\bar{\mu}} \left\|\boldsymbol{x}_k - (\hat{\boldsymbol{x}}_k)_{\text{UPF}}\right\|^2 \tag{4.78}$$

$$||\boldsymbol{x}_k - (\hat{\boldsymbol{x}}_k)_{\text{UPF}}|| \leq \varepsilon' \tag{4.79}$$

Lemma 4.1. *If Assumptions 4.1–4.3 are satisfied, then a real and positive scalar* $0 < \alpha < 1$ *exists with the following inequality be satisfied.*

$$(\boldsymbol{A}_k - \boldsymbol{K}_k\boldsymbol{C}_k)^{\mathrm{T}}\boldsymbol{\Pi}_{k+1}(\boldsymbol{A}_k - \boldsymbol{K}_k\boldsymbol{C}_k) \leq (1-\alpha)\boldsymbol{\Pi}_k, k \geq 0 \tag{4.80}$$

where $\boldsymbol{\Pi}_k = (\boldsymbol{P}_{\tilde{x},k}^{-1})_{UPF}$, $\boldsymbol{C}_k = ((\boldsymbol{P}_{\tilde{x},k}^{-1})_{UPF}\boldsymbol{A}_k^{-1}\boldsymbol{K}_k)^{\mathrm{T}}$, *and*

$$\alpha = 1 - \frac{1}{1 + \frac{\underline{q}}{\bar{p}(\bar{a}+\bar{p}^2\bar{k}/\underline{a}\underline{p})^2}} \tag{4.81}$$

Proof. From (4.54) and the definition of $\boldsymbol{C}_k$, one has

$$\begin{aligned}(\boldsymbol{P}_{\tilde{x},k+1})_{\text{UPF}} &= \boldsymbol{A}_k(\boldsymbol{P}_{\tilde{x},k})_{\text{UPF}}\boldsymbol{A}_k^{\mathrm{T}} + \boldsymbol{Q}_k - \boldsymbol{A}_k(\boldsymbol{P}_{\tilde{x},k})_{\text{UPF}}\boldsymbol{C}_k^{\mathrm{T}}\boldsymbol{K}_k^{\mathrm{T}} \\ &= (\boldsymbol{A}_k - \boldsymbol{K}_k\boldsymbol{C}_k)(\boldsymbol{P}_{\tilde{x},k})_{\text{UPF}}(\boldsymbol{A}_k - \boldsymbol{K}_k\boldsymbol{C}_k)^{\mathrm{T}} + \boldsymbol{Q}_k \\ &\quad + \boldsymbol{K}_k\boldsymbol{C}_k(\boldsymbol{P}_{\tilde{x},k})_{\text{UPF}}(\boldsymbol{A}_k - \boldsymbol{K}_k\boldsymbol{C}_k)^{\mathrm{T}}\end{aligned} \tag{4.82}$$

and

$$A_k^{-1}(A_k - K_k C_k)(P_{\tilde{x},k})_{\mathrm{UPF}} = (P_{\tilde{x},k})_{\mathrm{UPF}} - (P_{\tilde{x},k})_{\mathrm{UPF}} C_k^{\mathrm{T}} C_k (P_{\tilde{x},k})_{\mathrm{UPF}} \tag{4.83}$$

which is a symmetric matrix.

From (4.36)–(4.41) and (4.54), the linear term and the first-order term of the Taylor Series expansion of $(x_{k+1})_{\mathrm{UPF}}$ is $A_k(\tilde{x}_k)_{\mathrm{UPF}}$. K_k is the expectation of the second-order term of the Taylor Series expansion of $(x_{k+1})_{\mathrm{UPF}}$. It is obvious that the zeroth and the first-order terms of any nonlinear function account for the major than that of the residual terms. Then, we have

$$|K_k| < |A_k(\tilde{x}_k)_{\mathrm{UPF}}| \tag{4.84}$$

$$\frac{|K_k|}{\left|A_k(\tilde{x}_k)_{\mathrm{UPF}}\right|}\left|(\tilde{x}_k)_{\mathrm{UPF}}\right| < \left|(\tilde{x}_k)_{\mathrm{UPF}}\right| \tag{4.85}$$

Based on (4.85), it follows that

$$\begin{aligned} &A_k^{-1} K_k ((A_k^{-1}) K_k)^{\mathrm{T}} \\ &< (P_{\tilde{x},k})_{\mathrm{UPF}} (P_{\tilde{x},k}^{-1})_{\mathrm{UPF}} ((A_k^{-1}) K_k)((A_k^{-1}) K_k)^{\mathrm{T}} (P_{\tilde{x},k}^{-1})_{\mathrm{UPF}} \\ &< (P_{\tilde{x},k}^{-1})_{\mathrm{UPF}} \end{aligned} \tag{4.86}$$

Owing to $(P_{\tilde{x},k}^{-1})_{\mathrm{UPF}} = ((P_{\tilde{x},k}^{-1})_{\mathrm{UPF}})^{\mathrm{T}}$, one has

$$C_k C_k^{\mathrm{T}} < (P_{\tilde{x},k}^{-1})_{\mathrm{UPF}} \tag{4.87}$$

Combining (4.87) with (4.83), it can be obtained from $(P_{\tilde{x},k}^{-1})_{\mathrm{UPF}} > \mathbf{0}$ and $(P_{\tilde{x},k})_{\mathrm{UPF}} > \mathbf{0}$ that

$$\begin{aligned} &A_k^{-1}(A_k - K_k C_k)(P_{\tilde{x},k})_{\mathrm{UPF}} \\ &= (P_{\tilde{x},k})_{\mathrm{UPF}} - (P_{\tilde{x},k})_{\mathrm{UPF}} C_k^{\mathrm{T}} C_k (P_{\tilde{x},k})_{\mathrm{UPF}} > \mathbf{0} \end{aligned} \tag{4.88}$$

Since the following inequality is valid,

$$A_k^{-1} K_k C_k = (A_k^{-1} K_k)(A_k^{-1} K_k)^{\mathrm{T}} ((P_{\tilde{x},k}^{-1})_{\mathrm{UPF}})^{\mathrm{T}} \geq \mathbf{0} \tag{4.89}$$

it has

$$\begin{aligned} &K_k C_k (P_{\tilde{x},k})_{\mathrm{UPF}} (A_k - K_k C_k)^{\mathrm{T}} \\ &= A_k (A_k^{-1} K_k C_k)(A_k^{-1}(A_k - K_k C_k)(P_{\tilde{x},k})_{\mathrm{UPF}})^{\mathrm{T}} A_k^{\mathrm{T}} \geq \mathbf{0} \end{aligned} \tag{4.90}$$

Inserting (4.90) into (4.82) leads to

$$\begin{aligned}(\boldsymbol{P}_{\tilde{\boldsymbol{x}},k+1})_{\text{UPF}} &\geq (\boldsymbol{A}_k - \boldsymbol{K}_k\boldsymbol{C}_k)(\boldsymbol{P}_{\tilde{\boldsymbol{x}},k})_{\text{UPF}}(\boldsymbol{A}_k - \boldsymbol{K}_k\boldsymbol{C}_k)^{\text{T}} + \boldsymbol{Q}_k \\ &= (\boldsymbol{A}_k - \boldsymbol{K}_k\boldsymbol{C}_k)\Big((\boldsymbol{P}_{\tilde{\boldsymbol{x}},k})_{\text{UPF}}(\boldsymbol{A}_k - \boldsymbol{K}_k\boldsymbol{C}_k)^{\text{T}} \\ &\quad + (\boldsymbol{A}_k - \boldsymbol{K}_k\boldsymbol{C}_k)^{-1}\boldsymbol{Q}_k\Big)\end{aligned} \tag{4.91}$$

If Assumption 4.1 is satisfied, then one has

$$||\boldsymbol{K}_k|| \leq \bar{p}\sqrt{\bar{k}} \tag{4.92}$$

$$\|\boldsymbol{C}_k\| \leq \frac{\bar{p}\sqrt{\bar{k}}}{\underline{a}\underline{p}} \tag{4.93}$$

To this end, (4.91) can be rewritten as

$$(\boldsymbol{P}_{\tilde{\boldsymbol{x}},k+1})_{\text{UPF}} \geq (\boldsymbol{A}_k - \boldsymbol{K}_k\boldsymbol{C}_k)\left((\boldsymbol{P}_{\tilde{\boldsymbol{x}},k})_{\text{UPF}} + \frac{\underline{q}}{\left(\bar{a} + \frac{\bar{p}^2\bar{k}}{\underline{a}\underline{p}}\right)^2}\right)(\boldsymbol{A}_k - \boldsymbol{K}_k\boldsymbol{C}_k)^{\text{T}} \tag{4.94}$$

Since $(\boldsymbol{P}_{\tilde{\boldsymbol{x}},k})_{\text{UPF}} > \underline{p}\boldsymbol{I}_n$ and $\boldsymbol{A}_k - \boldsymbol{K}_k\boldsymbol{C}_k$ are nonsingular, taking the inverse of both sides and multiplying the inverse with $(\boldsymbol{A}_k - \boldsymbol{K}_k\boldsymbol{C}_k)^{\text{T}}$ and $(\boldsymbol{A}_k - \boldsymbol{K}_k\boldsymbol{C}_k)$, it can be proved from Assumption 4.1 that

$$(\boldsymbol{A}_k - \boldsymbol{K}_k\boldsymbol{C}_k)^{\text{T}}\boldsymbol{\Pi}_{k+1}(\boldsymbol{A}_k - \boldsymbol{K}_k\boldsymbol{C}_k) \leq \left(1 + \frac{\underline{q}}{\bar{p}(\bar{a} + \frac{\bar{p}^2\bar{k}}{\underline{a}\underline{p}})^2}\right)^{-1}\boldsymbol{\Pi}_k \tag{4.95}$$

Hence, the conclusion in Lemma 4.1 is proved. □

Lemma 4.2. *If Assumptions 4.1–4.3 are satisfied and* $\boldsymbol{\Pi}_k = (\boldsymbol{P}^{-1}_{\tilde{\boldsymbol{x}},k})$ *exists, then it follows that*

$$\begin{aligned}&\left(\boldsymbol{K}_k\boldsymbol{C}_k\left(\boldsymbol{x}_k - (\hat{\boldsymbol{x}}_k)_{UPF}\right)\right)^{\text{T}}\boldsymbol{\Pi}_k\begin{pmatrix}2(\boldsymbol{A}_k - \boldsymbol{K}_k\boldsymbol{C}_k)(\boldsymbol{x}_k - (\hat{\boldsymbol{x}}_k)_{UPF}) \\ +\boldsymbol{K}_k\boldsymbol{C}_k(\boldsymbol{x}_k - (\hat{\boldsymbol{x}}_k)_{UPF}) + (\bar{\boldsymbol{\mu}}_k)_{UPF}\end{pmatrix} \\ &\leq \kappa_1\left\|\boldsymbol{x}_k - (\hat{\boldsymbol{x}}_k)_{UPF}\right\|^2\end{aligned} \tag{4.96}$$

where $\kappa_1 = \frac{k_1}{\underline{p}}((2\bar{a} + k_1) + \kappa_{\bar{\mu}}\varepsilon')$.

Proof. It can be obtained from (4.92)–(4.93) in the proof of Lemma 4.1 that

$$\begin{aligned}\left\|\boldsymbol{K}_k\boldsymbol{C}_k\left(\boldsymbol{x}_k-(\hat{\boldsymbol{x}}_k)_{\mathrm{UPF}}\right)\right\| &\leq \frac{\bar{p}^2\bar{k}}{\underline{a}\underline{p}}\|\boldsymbol{x}_k-(\hat{\boldsymbol{x}}_k)_{\mathrm{UPF}}\| \\ &= k_1\|\boldsymbol{x}_k-(\hat{\boldsymbol{x}}_k)_{\mathrm{UPF}}\|\end{aligned} \tag{4.97}$$

where $k_1=\frac{\bar{p}^2\bar{k}}{\underline{a}\underline{p}}$.

Based on (4.78)–(4.79) and (4.92)–(4.93), using $||\boldsymbol{x}_k-(\hat{\boldsymbol{x}}_k)_{\mathrm{UPF}}||\leq\varepsilon'$, we prove that

$$\begin{aligned}&\left(\boldsymbol{K}_k\boldsymbol{C}_k\left(\boldsymbol{x}_k-(\hat{\boldsymbol{x}}_k)_{\mathrm{UPF}}\right)\right)^{\mathrm{T}}\boldsymbol{\Pi}_k\begin{pmatrix}2(\boldsymbol{A}_k-\boldsymbol{K}_k\boldsymbol{C}_k)(\boldsymbol{x}_k-(\hat{\boldsymbol{x}}_k)_{\mathrm{UPF}})\\+\boldsymbol{K}_k\boldsymbol{C}_k(\boldsymbol{x}_k-(\hat{\boldsymbol{x}}_k)_{\mathrm{UPF}})+(\bar{\boldsymbol{\mu}}_k)_{\mathrm{UPF}}\end{pmatrix}\\&\leq\frac{k_1}{\underline{p}}\|\boldsymbol{x}_k-(\hat{\boldsymbol{x}}_k)_{\mathrm{UPF}}\|\left((2\bar{a}+k_1+\kappa_{\bar{\mu}}\varepsilon')\|\boldsymbol{x}_k-(\hat{\boldsymbol{x}}_k)_{\mathrm{UPF}}\|\right)\\&=\kappa_1\left\|\boldsymbol{x}_k-(\hat{\boldsymbol{x}}_k)_{\mathrm{UPF}}\right\|^2\end{aligned} \tag{4.98}$$

The proof is completed here. □

Lemma 4.3. *If Assumptions 4.1–4.3 are satisfied and* $\boldsymbol{\Pi}_k=(\boldsymbol{P}_{\tilde{x},k}^{-1})$ *exists, then it follows that*

$$\begin{aligned}&(\bar{\boldsymbol{\mu}}_k)_{UPF}^{\mathrm{T}}\boldsymbol{\Pi}_k\begin{pmatrix}2(\boldsymbol{A}_k-\boldsymbol{K}_k\boldsymbol{C}_k)(\boldsymbol{x}_k-(\hat{\boldsymbol{x}}_k)_{UPF})\\+\boldsymbol{K}_k\boldsymbol{C}_k(\boldsymbol{x}_k-(\hat{\boldsymbol{x}}_k)_{UPF})+(\bar{\boldsymbol{\mu}}_k)_{UPF}\end{pmatrix}\\&\leq\kappa_2\left\|\boldsymbol{x}_k-(\hat{\boldsymbol{x}}_k)_{UPF}\right\|^3\end{aligned} \tag{4.99}$$

where $\kappa_2=\frac{\kappa_{\bar{\mu}}}{\underline{p}}((2\bar{a}+k_1)+\kappa_{\bar{\mu}}\varepsilon')$.

Proof. Because Assumptions 4.1–4.3 are satisfied, it can be obtained from (4.78), (4.79), and (4.98) in the proof of Lemma 4.2 that

$$\begin{aligned}&(\bar{\boldsymbol{\mu}}_k)_{\mathrm{UPF}}^{\mathrm{T}}\boldsymbol{\Pi}_k\begin{pmatrix}2(\boldsymbol{A}_k-\boldsymbol{K}_k\boldsymbol{C}_k)(\boldsymbol{x}_k-(\hat{\boldsymbol{x}}_k)_{\mathrm{UPF}})\\+\boldsymbol{K}_k\boldsymbol{C}_k(\boldsymbol{x}_k-(\hat{\boldsymbol{x}}_k)_{\mathrm{UPF}})+(\bar{\boldsymbol{\mu}}_k)_{\mathrm{UPF}}\end{pmatrix}\\&\leq\frac{k_1}{\underline{p}}\|\boldsymbol{x}_k-(\hat{\boldsymbol{x}}_k)_{\mathrm{UPF}}\|\left((2\bar{a}+k_1+\kappa_{\bar{\mu}}\varepsilon')\|\boldsymbol{x}_k-(\hat{\boldsymbol{x}}_k)_{\mathrm{UPF}}\|\right)\\&=\kappa_2\left\|\boldsymbol{x}_k-(\hat{\boldsymbol{x}}_k)_{\mathrm{UPF}}\right\|^3\end{aligned} \tag{4.100}$$

The proof is completed here. □

Lemma 4.4. *Suppose that Assumptions 4.1–4.3 are satisfied and* $\boldsymbol{\Pi}_k=(\boldsymbol{P}_{\tilde{x},k}^{-1})$ *exists, it has* $\mathbb{E}((\boldsymbol{s}_k)_{UPF}^{\mathrm{T}}\boldsymbol{\Pi}_k(\boldsymbol{s}_k)_{UPF})\leq\kappa_3\delta$, *where* $\kappa_3=\frac{\bar{\lambda}^2\bar{s}^2}{\underline{p}\underline{r}^2\underline{\lambda}^4\underline{s}^4}n$ *and* $\delta=\bar{r}^3$.

Proof. It is obtained from the definition of $(\boldsymbol{s}_k)_{\text{UPF}}$ that

$$\begin{aligned}(\boldsymbol{s}_k)_{\text{UPF}}^{\text{T}}\boldsymbol{\Pi}_k(\boldsymbol{s}_k)_{\text{UPF}} &= ((\boldsymbol{B}_k)_{\text{UPF}}\boldsymbol{v}_k)^{\text{T}}\boldsymbol{\Pi}_k((\boldsymbol{B}_k)_{\text{UPF}}\boldsymbol{v}_k)\\ &\le \frac{1}{\underline{p}}\boldsymbol{v}_k^{\text{T}}(\boldsymbol{B}_k)_{\text{UPF}}^{\text{T}}(\boldsymbol{B}_k)_{\text{UPF}}\boldsymbol{v}_k\\ &= \frac{1}{\underline{p}}tr(\boldsymbol{v}_k^{\text{T}}(\boldsymbol{B}_k)_{\text{UPF}}^{\text{T}}(\boldsymbol{B}_k)_{\text{UPF}}\boldsymbol{v}_k)\\ &= \frac{1}{\underline{p}}tr((\boldsymbol{B}_k)_{\text{UPF}}\boldsymbol{v}_k\boldsymbol{v}_k^{\text{T}}(\boldsymbol{B}_k)_{\text{UPF}}^{\text{T}})\end{aligned} \tag{4.101}$$

Computing the mean value of $(\boldsymbol{s}_k)_{\text{UPF}}^{\text{T}}\boldsymbol{\Pi}_k(\boldsymbol{s}_k)_{\text{UPF}}$ yields

$$\mathbb{E}((\boldsymbol{s}_k)_{\text{UPF}}^{\text{T}}\boldsymbol{\Pi}_k(\boldsymbol{s}_k)_{\text{UPF}}) \le \frac{1}{\underline{p}}tr((\boldsymbol{B}_k)_{\text{UPF}}\mathbb{E}(\boldsymbol{v}_k\boldsymbol{v}_k^{\text{T}})(\boldsymbol{B}_k)_{\text{UPF}}^{\text{T}}) \tag{4.102}$$

Because $\boldsymbol{v}_k$ is standard vector-valued white noise processes with $\mathbb{E}(\boldsymbol{v}_k{\boldsymbol{v}_k}^{\text{T}}) = \boldsymbol{R}_k$, it can be calculated from Assumption 4.1 and (4.102) that

$$\begin{aligned}\mathbb{E}((\boldsymbol{s}_k)_{\text{UPF}}^{\text{T}}\boldsymbol{\Pi}_k(\boldsymbol{s}_k)_{\text{UPF}}) &\le \frac{\bar{r}}{\underline{p}}tr((\boldsymbol{B}_k)_{\text{UPF}}(\boldsymbol{B}_k)_{\text{UPF}}^{\text{T}})\\ &\le \frac{\bar{r}^2\bar{\lambda}^2\bar{s}^2}{\underline{r}^2\underline{\lambda}^4\underline{s}^4}n\\ &= \kappa_3\delta\end{aligned} \tag{4.103}$$

The proof is hence ended here. □

Theorem 4.4. *Consider a nonlinear stochastic system described by* (2.32) *and* (2.33), *with the application of the UPF in Algorithm 4.1, provided that the system is subject to Assumptions 4.1–4.3 and the initial estimation errors satisfies*

$$||(\tilde{\boldsymbol{x}}_0)_{UPF}|| \le \varepsilon \tag{4.104}$$

$$\boldsymbol{R}_k \le \bar{r}\boldsymbol{I}_n \tag{4.105}$$

where $\varepsilon > 0$ *is a constant, then the state estimation error* $(\tilde{\boldsymbol{x}}_{k+1})_{UPF}$ *given in* (4.36) *is exponentially bounded in mean square and bounded with probability one.*

Proof. Choose a Lyapunov candidate function as

$$V_k((\tilde{\boldsymbol{x}}_k)_{\text{UPF}}) = ((\tilde{\boldsymbol{x}}_k)_{\text{UPF}})^{\text{T}}\boldsymbol{\Pi}_k(\tilde{\boldsymbol{x}}_k)_{\text{UPF}} \tag{4.106}$$

Since $\boldsymbol{\Pi}_k = (\boldsymbol{P}_{\tilde{x},k}^{-1})_{\mathrm{UPF}}$ exist and $(\boldsymbol{P}_{\tilde{x},k})_{\mathrm{UPF}}$ is positive-definite, (4.106) is bounded by

$$\frac{1}{\bar{p}}\|(\tilde{\boldsymbol{x}}_k)_{\mathrm{UPF}}\|^2 \leq \boldsymbol{V}_k((\tilde{\boldsymbol{x}}_k)_{\mathrm{UPF}}) \leq \frac{1}{\underline{p}}\|(\tilde{\boldsymbol{x}}_k)_{\mathrm{UPF}}\|^2 \tag{4.107}$$

From (4.36), we compute (4.106) as

$$\boldsymbol{V}_{k+1}\left((\tilde{\boldsymbol{x}}_{k+1})_{\mathrm{UPF}}\right) = \boldsymbol{\Upsilon}_1^{\mathrm{T}}\boldsymbol{\Pi}_{k+1}\boldsymbol{\Upsilon}_1 = \boldsymbol{\Upsilon}_2^{\mathrm{T}}\boldsymbol{\Pi}_{k+1}\boldsymbol{\Upsilon}_2 \tag{4.108}$$

where $\boldsymbol{\Upsilon}_1 = (\tilde{\boldsymbol{x}}_{k+1})_{\mathrm{UPF}}\boldsymbol{A}_k + (\bar{\boldsymbol{\mu}}_k)_{\mathrm{UPF}} + (\boldsymbol{s}_k)_{\mathrm{UPF}}$ and $\boldsymbol{\Upsilon}_2 = (\tilde{\boldsymbol{x}}_{k+1})_{\mathrm{UPF}}(\boldsymbol{A}_k - \boldsymbol{K}_k\boldsymbol{C}_k) + (\bar{\boldsymbol{\mu}}_k)_{\mathrm{UPF}} + (\tilde{\boldsymbol{x}}_{k+1})_{\mathrm{UPF}}(\boldsymbol{K}_k\boldsymbol{C}_k)^{\mathrm{T}} + (\boldsymbol{s}_k)_{\mathrm{UPF}}$.

Applying Lemma 4.1, it is found that (4.108) is further bounded by

$$\begin{aligned}
\boldsymbol{V}_{k+1}\left((\tilde{\boldsymbol{x}}_{k+1})_{\mathrm{UPF}}\right) \leq{}& (1-\alpha)\,\boldsymbol{V}_k\left((\tilde{\boldsymbol{x}}_k)_{\mathrm{UPF}}\right)\\
&+ \left(\boldsymbol{K}_k\boldsymbol{C}_k(\tilde{\boldsymbol{x}}_k)_{\mathrm{UPF}}\right)^{\mathrm{T}}\boldsymbol{\Pi}_k\left(\boldsymbol{K}_k\boldsymbol{C}_k(\tilde{\boldsymbol{x}}_k)_{\mathrm{UPF}}\right.\\
&\left.+ (\bar{\boldsymbol{\mu}}_k)_{\mathrm{UPF}} + 2\left(\boldsymbol{A}_k - \boldsymbol{K}_k\boldsymbol{C}_k\right)(\tilde{\boldsymbol{x}}_k)_{\mathrm{UPF}}\right)\\
&+ (\bar{\boldsymbol{\mu}}_k)_{\mathrm{UPF}}^{\mathrm{T}}\boldsymbol{\Pi}_k\left(2\left(\boldsymbol{A}_k - \boldsymbol{K}_k\boldsymbol{C}_k\right)(\tilde{\boldsymbol{x}}_k)_{\mathrm{UPF}}\right.\\
&\left.+ \boldsymbol{K}_k\boldsymbol{C}_k(\tilde{\boldsymbol{x}}_k)_{\mathrm{UPF}} + (\bar{\boldsymbol{\mu}}_k)_{\mathrm{UPF}}\right) + (\boldsymbol{s}_k)_{\mathrm{UPF}}^{\mathrm{T}}\boldsymbol{\Pi}_k(\boldsymbol{s}_k)_{\mathrm{UPF}}\\
&+ (\boldsymbol{s}_k)_{\mathrm{UPF}}^{\mathrm{T}}\boldsymbol{\Pi}_k\left(2\left(\boldsymbol{A}_k - \boldsymbol{K}_k\boldsymbol{C}_k\right)(\tilde{\boldsymbol{x}}_k)_{\mathrm{UPF}}\right.\\
&\left.+ 2\boldsymbol{K}_k\boldsymbol{C}_k(\tilde{\boldsymbol{x}}_k)_{\mathrm{UPF}} + 2(\bar{\boldsymbol{\mu}}_k)_{\mathrm{UPF}}\right)
\end{aligned} \tag{4.109}$$

Taking the conditional expectation $\mathbb{E}\left(\boldsymbol{V}_{k+1}\left((\tilde{\boldsymbol{x}}_{k+1})_{\mathrm{UPF}}\right)\middle|\,(\tilde{\boldsymbol{x}}_{k+1})_{\mathrm{UPF}}\right)$ and considering the white noise property, it is seen that the last term in (4.109) vanishes since it depend on $\boldsymbol{v}_k$. Therefore, it can be obtained from Lemma 4.2 and Lemma 4.3 that the following inequality is valid for $||(\tilde{\boldsymbol{x}}_k)_{\mathrm{UPF}}|| \leq \varepsilon'$.

$$\begin{aligned}
&\mathbb{E}\left(\boldsymbol{V}_{k+1}\left((\tilde{\boldsymbol{x}}_{k+1})_{\mathrm{UPF}}\right)\middle|\,(\tilde{\boldsymbol{x}}_{k+1})_{\mathrm{UPF}}\right) - \boldsymbol{V}_k\left((\tilde{\boldsymbol{x}}_k)_{\mathrm{UPF}}\right)\\
&\leq -\alpha\boldsymbol{V}_k\left((\tilde{\boldsymbol{x}}_k)_{\mathrm{UPF}}\right) + \kappa_1\left\|(\tilde{\boldsymbol{x}}_k)_{\mathrm{UPF}}\right\|^2 + \kappa_2\left\|(\tilde{\boldsymbol{x}}_k)_{\mathrm{UPF}}\right\|^3 + \kappa_3\delta
\end{aligned} \tag{4.110}$$

Defining $\varepsilon = \min\left(\varepsilon', \frac{\alpha}{2\bar{p}\kappa_2}\right)$ and using (4.106)–(4.107), it leads to

$$\kappa_2\|(\tilde{\boldsymbol{x}}_k)_{\mathrm{UPF}}\|\,\|(\tilde{\boldsymbol{x}}_k)_{\mathrm{UPF}}\|^2 \leq \frac{\alpha}{2\bar{p}}\|(\tilde{\boldsymbol{x}}_k)_{\mathrm{UPF}}\|^2 \leq \frac{\alpha}{2}\boldsymbol{V}_k((\tilde{\boldsymbol{x}}_k)_{\mathrm{UPF}}) \tag{4.111}$$

holds for $||(\tilde{\boldsymbol{x}}_k)_{\mathrm{UPF}}|| \leq \varepsilon$.

Inserting (4.111) into (4.110) yields

$$\begin{aligned}
&\mathbb{E}\left(\boldsymbol{V}_{k+1}\left((\tilde{\boldsymbol{x}}_{k+1})_{\mathrm{UPF}}\right)\middle|\,(\tilde{\boldsymbol{x}}_{k+1})_{\mathrm{UPF}}\right) - \boldsymbol{V}_k\left((\tilde{\boldsymbol{x}}_k)_{\mathrm{UPF}}\right)\\
&\leq -\frac{\alpha}{2}\boldsymbol{V}_k\left((\tilde{\boldsymbol{x}}_k)_{\mathrm{UPF}}\right) + \tilde{\kappa}_3\delta
\end{aligned} \tag{4.112}$$

for $||(\tilde{\boldsymbol{x}}_k)_{\mathrm{UPF}}|| \leq \varepsilon$, where $\tilde{\kappa}_3 = \kappa_3 + \frac{\kappa_1}{\delta}\|(\tilde{\boldsymbol{x}}_k)_{\mathrm{UPF}}\|^2$.

To this end, we are able to apply Lemma 2.1 with $||(\tilde{\boldsymbol{x}}_0)_{\text{UPF}}|| \leq \varepsilon$, $\underline{\upsilon} = \frac{1}{\bar{p}}$, $\bar{\upsilon} = \frac{1}{\underline{p}}$, and $\tilde{\mu} = \tilde{\kappa}_3\delta$. However, note that $\tilde{\varepsilon} \leq \|(\tilde{\boldsymbol{x}}_k)_{\text{UPF}}\| \leq \varepsilon$ with $\tilde{\varepsilon} < \varepsilon$, applying Definition 2.1, the supermartingale inequality

$$\begin{aligned}
&\mathbb{E}\left(\boldsymbol{V}_{k+1}\left((\tilde{\boldsymbol{x}}_{k+1})_{\text{UPF}}\right)\middle|\, (\tilde{\boldsymbol{x}}_{k+1})_{\text{UPF}}\right) - \boldsymbol{V}_k\left((\tilde{\boldsymbol{x}}_k)_{\text{UPF}}\right) \\
&\leq -\frac{\alpha}{2}\boldsymbol{V}_k\left((\tilde{\boldsymbol{x}}_k)_{\text{UPF}}\right) + \tilde{\kappa}_3\delta \leq \mathbf{0}
\end{aligned} \tag{4.113}$$

is fulfilled to guarantee the boundedness of the estimation error.

Choosing $\delta = \frac{\alpha\tilde{\varepsilon}^2}{2\bar{p}\tilde{\kappa}_3}$ with some $\tilde{\varepsilon} < \varepsilon$, we prove that $||(\tilde{\boldsymbol{x}}_k)_{\text{UPF}}|| \geq \tilde{\varepsilon}$ and

$$\begin{aligned}
\frac{\alpha}{2}\boldsymbol{V}_k((\tilde{\boldsymbol{x}}_k)_{\text{UPF}}) &\geq \frac{\alpha}{2\bar{p}}\|(\tilde{\boldsymbol{x}}_k)_{\text{UPF}}\|^2 \\
&\geq \frac{\alpha\tilde{\varepsilon}^2}{2\bar{p}} \\
&= \tilde{\kappa}_3\delta
\end{aligned} \tag{4.114}$$

Therefore, we conclude that the estimation error is always bounded if the initial error and the noise terms are bounded. Additionally, invoking Definition 2.2, it is further proved from (4.113) that the state estimation error is bounded with probability one. The proof is hence completed here. □

Remark 4.4. In the process of proving convergence for the covariance $(\boldsymbol{P}_{\tilde{x},k})_{\text{UPF}}$, the high-order error in the $\boldsymbol{Q}_k$ is included. Because the covariance $(\boldsymbol{P}_{\tilde{x},k})_{\text{UPF}}$ is verified to be convergent and bounded, $\boldsymbol{Q}_k$ is also bounded. It is obvious that the high-order error must be bounded. $\hat{\boldsymbol{d}}(\boldsymbol{x}_k)$ is the estimation of the modeling error in the UPF. If the UPF is convergent, this estimation must remains bounded. We can obtain the conclusion that if the UPF is convergent, the expectation of the actual modeling error must be bounded, so that the actual modeling error is also required to be bounded.

4.6 Application to microsatellite attitude control system

In this section, the UPF developed in Algorithm 4.1 is applied to solve the attitude control system's state estimation problem of a single microsatellite. The attitude and the angular velocity are to be estimated by using the UPF.

4.6.1 Measurement model of star sensors

The single microsatellite considered have two star sensors mounted in the $\mathcal{F}_B$ to measure its attitude. The attitude measurement obtained from those two sensors can be mathematically modeled as

$$\boldsymbol{y} = \begin{bmatrix} \mathbb{C}(\boldsymbol{q})\,\boldsymbol{S}_1 \\ \mathbb{C}(\boldsymbol{q})\,\boldsymbol{S}_2 \end{bmatrix} + \begin{bmatrix} \Delta\boldsymbol{S}_1 \\ \Delta\boldsymbol{S}_2 \end{bmatrix} \tag{4.115}$$

where $y \in \mathbb{R}^6$ is the attitude measurement value obtained from the star sensors. $\mathbb{C}(q)$ is the rotation matrix from the inertial frame $\mathcal{F}_I$ to the body-fixed frame $\mathcal{F}_B$. $S_1 \in \mathbb{R}^3$ and $S_2 \in \mathbb{R}^3$ are the star vector in the inertial frame $\mathcal{F}_I$. $\Delta S_1 \in \mathbb{R}^3$ and $\Delta S_2 \in \mathbb{R}^3$ are the sensor measurement noises. The angular error of star sensors can be normalized and adopted the Gaussian random distribution to obtain the measurement noises. The output y is normalized, which remains a unit vector.

4.6.2 Simulation results

Incorporating the measurement (4.105), the attitude control system described by (3.1)–(3.2) can be rewritten into the following nonlinear equation:

$$\begin{cases} \dot{x} = f(x) + g(x)d \\ y = h(x) + v \end{cases} \tag{4.116}$$

where $x = [q^{\mathrm{T}} \quad \omega_{BI}^{\mathrm{T}}]^{\mathrm{T}}, d = [0 \quad u_d^{\mathrm{T}}]^{\mathrm{T}}, v = [(\Delta S_1)^{\mathrm{T}} \quad (\Delta S_2)^{\mathrm{T}}]^{\mathrm{T}}, g(x) = J^{-1}$, and

$$h(x) = \begin{bmatrix} \mathbb{C}(q)\, S_1 \\ \mathbb{C}(q)\, S_2 \end{bmatrix} \tag{4.117}$$

$$f(x) = \begin{bmatrix} E(q)\omega_{BI} \\ -J^{-1}(\omega_{BI})^{\times} J\omega_{BI} + J^{-1}u_c \end{bmatrix} \tag{4.118}$$

Then, the nonlinear model (4.116) can be written into the form of (2.32)–(2.33) after discretization. As a result, the proposed UPF algorithm can be applied to solve the state's estimation problem of this microsatellite attitude control system.

Numerical simulations are carried out to validate the estimation performance of the UPF. The microsatellite simulated is with its inertia parameter $J = \mathrm{diag}([49.96 \quad 55.40 \quad 63.0]^{\mathrm{T}})$ kg·m^2. The modeling error is assumed to consist of the constant value $\Delta N_c \in \mathbb{R}^3$ and the periodic value $\Delta N_e \in \mathbb{R}^3$, *i.e.*,

$$d = \Delta N_c + \Delta N_e \tag{4.119}$$

To demonstrate that the UPF achieves better estimation performance than the classical PF in Algorithm 2.1 and the standard UKF. The PF in Algorithm 2.1 and the standard UKF are also applied to this microsatellite attitude control system with simulation results presented.

In simulations, q_{24} is used into the filtering algorithms to propagate and estimate the attitude, then to normalize the quaternion by using the constraint

$q^{\mathrm{T}}q = 1$ to obtain the final quaternion for each filter cycle. The initial attitude is $q = [1 \quad 0 \quad 0 \quad 0]^{\mathrm{T}}$ and the initial angular velocity is $\omega^{\mathrm{T}} = [0.1 \quad 0.1 \quad 0.1]^{\mathrm{T}}$ deg/sec. The initial estimation value of the attitude quaternion is chosen as

$$q(0) = \begin{bmatrix} \cos\frac{\Theta_0}{2} & \frac{\sqrt{3}}{3}\sin\frac{\Theta_0}{2} & \frac{\sqrt{3}}{3}\sin\frac{\Theta_0}{2} & \frac{\sqrt{3}}{3}\sin\frac{\Theta_0}{2} \end{bmatrix}^{\mathrm{T}} \quad (4.120)$$

where $\Theta_0 \in \mathbb{R}$ is a constant to be specified. The weighting matrix of the UPF and the PF is chosen as $W_E = 10^5 I_3$. The parameters of the symmetrically-distributed set of points are chosen as $n = 6$ and $\kappa = 0.8$ for the UPF. Moreover, the following two cases are considered:

1) Case #1: This is a nominal case, in which $\Theta_0 = 5$ degrees,

$$\Delta N_c = [-0.003 \quad -0.004 \quad -0.003]^{\mathrm{T}} \text{ Nm} \quad (4.121)$$

$$\Delta N_e = [0.003\cos(\omega_T t) \quad 0.001\cos(\omega_T t) \quad -0.002\cos(\omega_T t)]^{\mathrm{T}} \text{ Nm} \quad (4.122)$$

and $\omega_T = 0.66$ deg/sec are assumed. The initial estimate value of the angular velocity is $[0 \quad 0 \quad 0]^{\mathrm{T}}$ deg/sec. The standard deviation of that two star sensors' measurement noise is assumed as 20 arcseconds.

2) Case #2: In this case, severe measurement noise and modeling error are assumed. $\Theta_0 = 5$ degrees, $\Delta N_c = [-0.015 \quad -0.02 \quad -0.015]^{\mathrm{T}}$ Nm, $\Delta N_e = [0.015\cos(\omega_T t) \quad 0.005\cos(\omega_T t) \quad -0.01\cos(\omega_T t)]^{\mathrm{T}}$ Nm, and $\omega_T =$ 0.66 deg/sec are assumed. The initial estimate value of the angular velocity is $[0 \quad 0 \quad 0]^{\mathrm{T}}$ deg/sec. The standard deviation of that two star sensors' measurement noise is assumed as 100 arcseconds.

Remark 4.5. As defined in (2.69), $\tilde{x} = x - \hat{x}$ denotes the state estimation error. When applying state estimation approaches to the attitude control system of a single microsatellite, the corresponding state estimation error is specified as $\tilde{x} = [\tilde{q}^{\mathrm{T}} \quad \tilde{\omega}_{BI}^{\mathrm{T}}]^{\mathrm{T}}$. In satellite engineering, the unit-quaternion q can be mathematically transformed into the Euler attitude angles to explicitly describe the satellite's attitude, *i.e.*,

$$\Theta = [\phi \quad \theta \quad \psi]^{\mathrm{T}} = f_t(q) \quad (4.123)$$

where $f_t(\cdot)$ is the transformation function (Sidi, 1997). ϕ, θ, and ψ are the roll, the pitch, and the yaw attitude Euler angles with respect to the unit-quaternion q, respectively. Hence, the state estimation error described by the unit quaternion $\tilde{q}$ is transformed into the attitude Euler angles estimation error throughout the book, which is denoted as $\Theta_e = [\phi_e \quad \theta_e \quad \psi_e]^{\mathrm{T}}$. The angular velocity state estimation error is re-denoted as $\omega_E = [\omega_{Ex} \quad \omega_{Ey} \quad \omega_{Ez}]^{\mathrm{T}} = \tilde{\omega}_{BI}$. The estimation error of the modeling error u_d is represented as $d_e = [d_{e1} \quad d_{e2} \quad d_{e3}]^{\mathrm{T}}$.

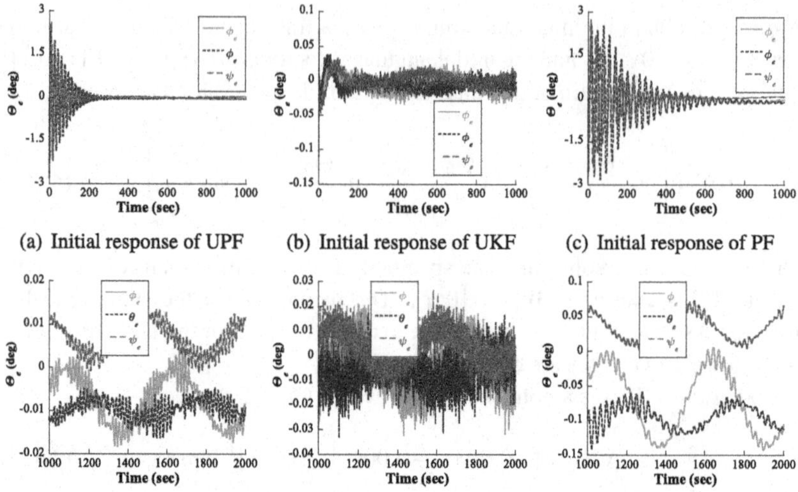

(a) Initial response of UPF (b) Initial response of UKF (c) Initial response of PF

(d) Steady behavior of UPF (e) Steady behavior of UKF (f) Steady behavior of PF

FIGURE 4.1 Attitude estimation error using UPF, UKF, and PF in Case #1.

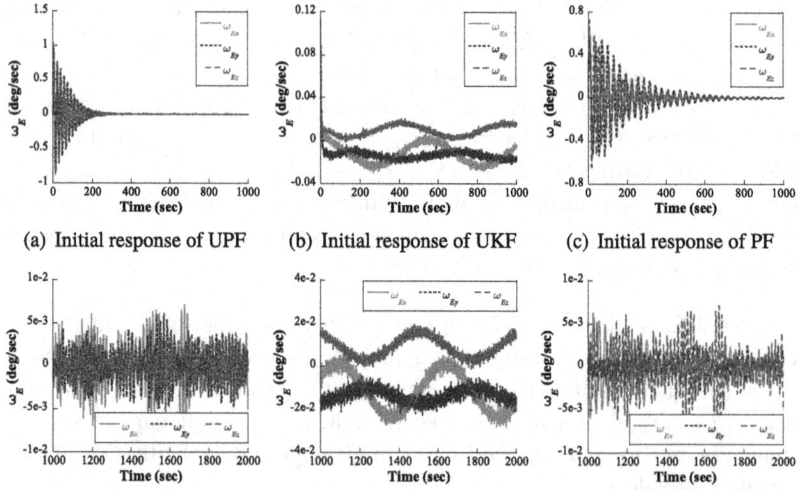

(a) Initial response of UPF (b) Initial response of UKF (c) Initial response of PF

(d) Steady behavior of UPF (e) Steady behavior of UKF (f) Steady behavior of PF

FIGURE 4.2 Angular velocity estimation error using UPF, UKF, and PF in Case #1.

When the UPF, the UKF, and the PF were applied in Case #1, the state estimation results were shown in Figs. 4.1–4.2. It was found in Fig. 4.1 that the UKF and the UPF achieved a faster rate of estimation for the attitude angle state. While the PF led to a slower state estimation and larger overshoot than the UKF and the UPF. It was seen in Fig. 4.1(c) that the states were estimated by the PF after 700 seconds. The attitude state estimation accuracy guaranteed by the UPF and the UKF was almost the same, which was higher than the attitude estimation accuracy achieved by the PF. However, the UPF and the PF resulted in better angular velocity state estimation accuracy than the UKF, as we can see in Fig. 4.2. The angular velocity accuracy ensured by the PF and the UPF was $|\omega_{Ex}| \leq 0.006$ deg/sec, $|\omega_{Ey}| \leq 0.006$ deg/sec, and $|\omega_{Ez}| \leq 0.006$ deg/sec. The corresponding angular velocity governed by the UKF was $|\omega_{Ex}| \leq 0.028$ deg/sec, $|\omega_{Ey}| \leq 0.022$ deg/sec, and $|\omega_{Ez}| \leq 0.02$ deg/sec. This was owing to the modeling error $\boldsymbol{d}$. The UKF is not capable of handling the modeling error $\boldsymbol{d}$ than the PF and the UPF. The PF and the UPF incorporated an estimation law for the modeling error and then compensated for it. It was seen in Figs. 4.3–4.4 that the UPF led to better estimation accuracy and faster estimation rate of modeling error than the PF. However, comparing with Fig. 4.3(a) and Fig. 4.4(a), the UPF achieved larger estimation of the modeling error than the PF in the initial response. This was because the initial estimation error was treated as the modeling error. The initial modeling error was thus larger than the actual modeling error. The steady behavior in Figs. 4.3–4.4 show that the estimation accuracy of the modeling error achieved by the UPF was higher than the PF. Hence, the UPF can compensate for the modeling error and improve the accuracy of the system model quickly. Due to this improved precision of the system model, better filtering performance was obtained from the UPF.

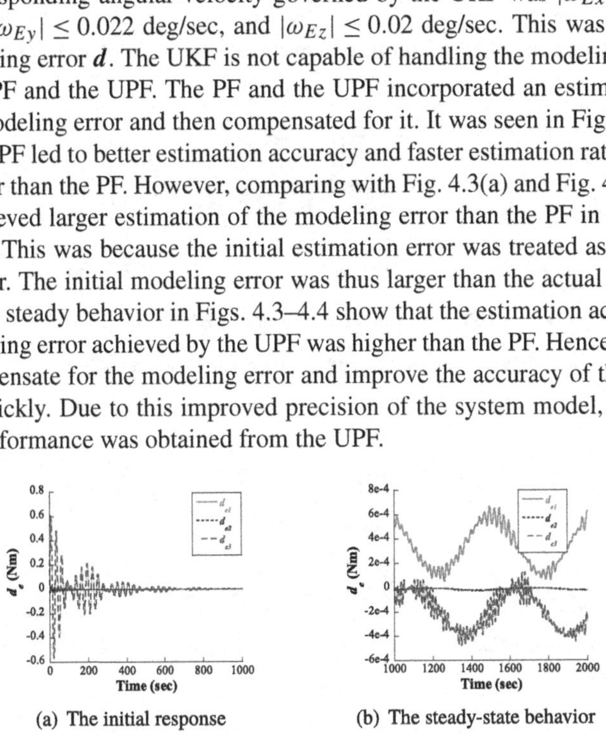

FIGURE 4.3 Estimation error of the modeling error ensured by UPF in Case #1.

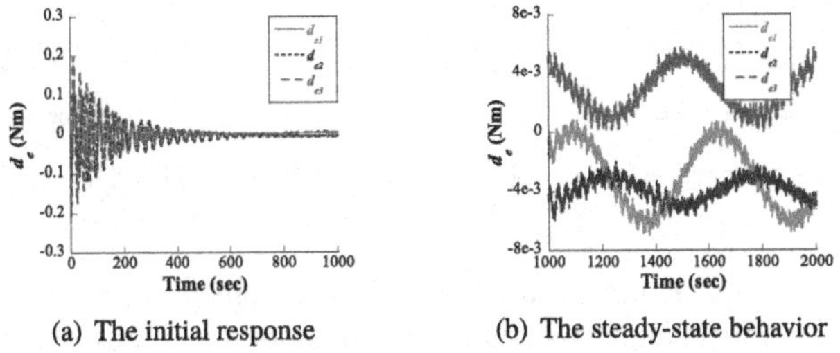

FIGURE 4.4 Estimation error of the modeling error ensured by PF in Case #1.

In Case #2, it was seen in Figs. 4.5–4.6 that the filtering performances achieved by UKF and the PF were deteriorated by the severe measurement noise

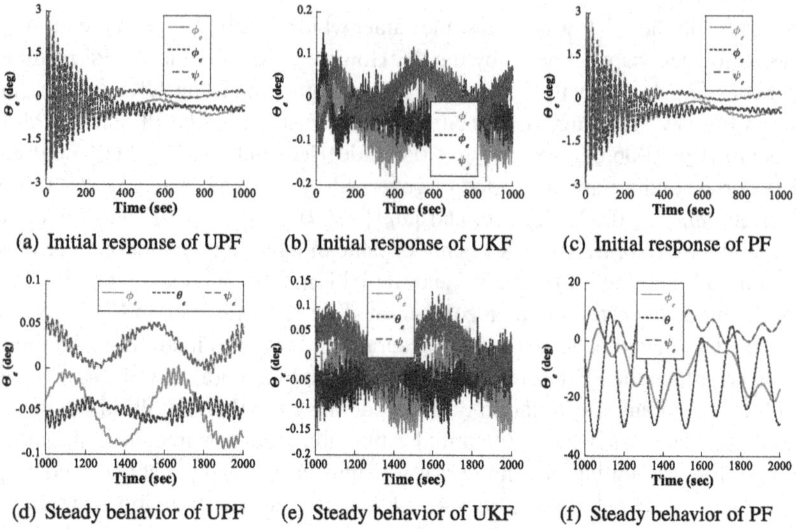

FIGURE 4.5 Attitude estimation error using UPF, UKF, and PF in Case #2.

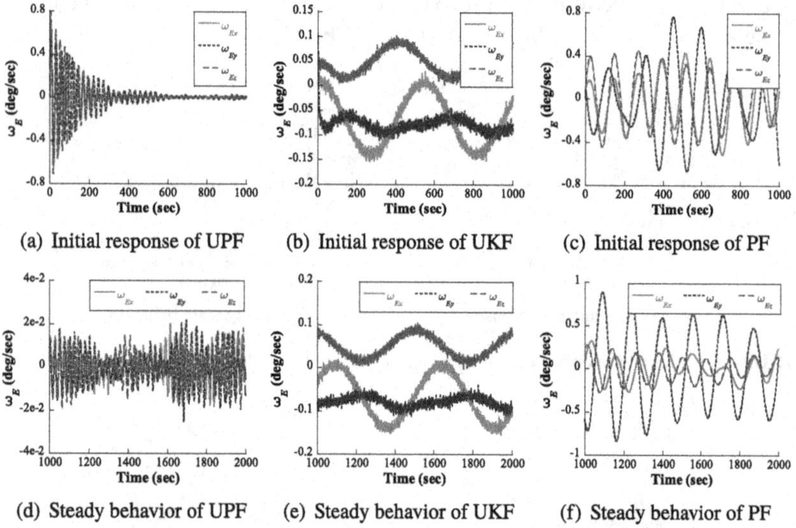

FIGURE 4.6 Angular velocity estimation error using UPF, UKF, and PF in Case #2.

and modeling error. However, the UPF still maintained good state estimation performance, which demonstrated the superiority of the proposed filter. The estimation of the modeling error obtained from the UPF and the PF were shown in Figs. 4.7–4.8, respectively.

Note that small initial estimation, weak measurement noise, and small modeling error were considered in Case #1 and Case #2, the estimation error

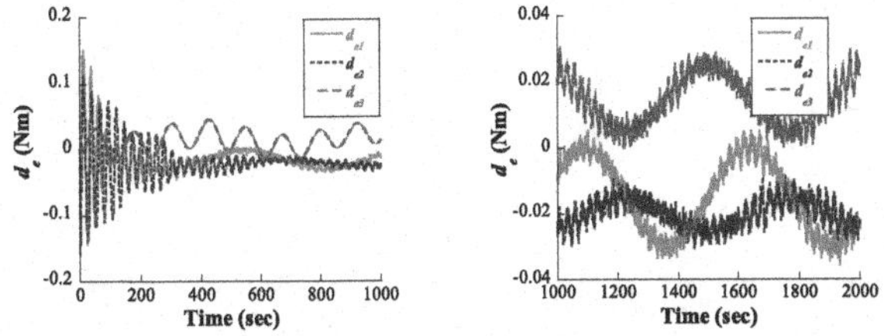

(a) The initial response (b) The steady-state behavior

FIGURE 4.7 Estimation error of the modeling error ensured by UPF in Case #2.

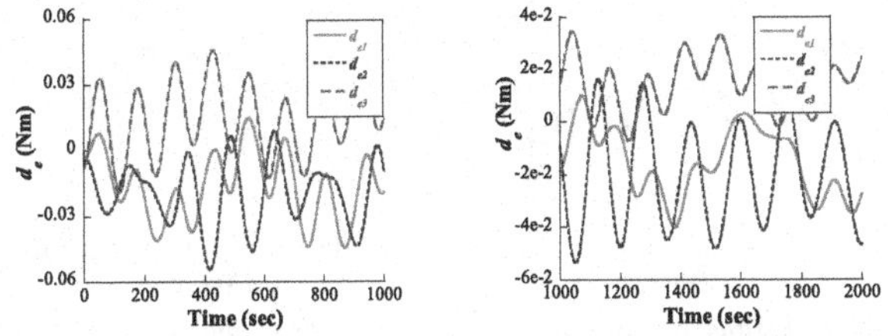

(a) The initial response (b) The steady-state behavior

FIGURE 4.8 Estimation error of the modeling error ensured by PF in Case #2.

achieved by the UPF was bounded and stable, as it was proved in Theorem 4.4 and validated in Figs. 4.1–4.2 and Figs. 4.5–4.6. To further verify that the initial estimation and measurement noise would affect state estimation performance, the following Case #3 and Case #4 were further simulated. Those two cases were with large initial estimation and severe measurement noise, respectively.

3) Case #3: This case is with a large initial estimation error. Almost all the parameters are the same as Case #1 except that the initial attitude estimation and the initial angular velocity estimation are assumed as $\Theta_0 = 80$ degrees and $[1.7 \quad 1.7 \quad 1.7]^{\mathrm{T}}$ deg/sec, respectively.

4) Case #4: Severe measurement noise is assumed with its standard deviation of 1760 arcseconds. The other parameters are set as same as Case #1.

The simulation results in Figs. 4.9–4.10 show that state estimation was not stable and diverged in the presence of large initial estimation and severe measurement noise. This was due to the high nonlinearities of the microsatellite attitude control system. The conclusion in Theorem 4.4 was proved. Note that the estimation error of states in Case #3 and Case #4 were more severe than Case #1, this was because the UPF has a good ability to estimate and deal with the modeling error. The large initial estimation and the severe noise were lumped as modeling error to be compensated. To this end, the conclusions in Theorems 4.1–4.4 were validated.

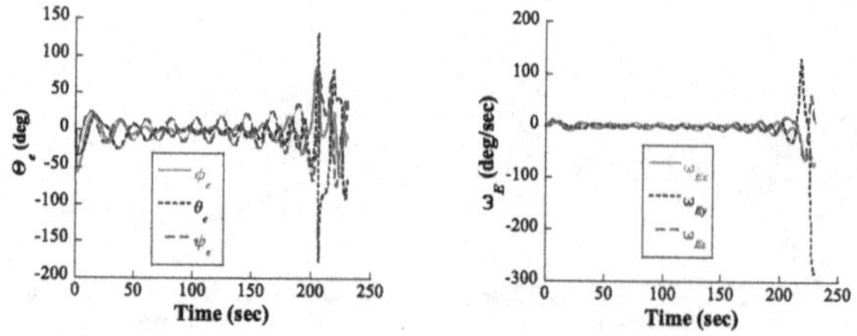

(a) Attitude estimation error (b) Estimation error $\boldsymbol{\Theta}_e$

FIGURE 4.9 State estimation error ensured by UPF in Case #3.

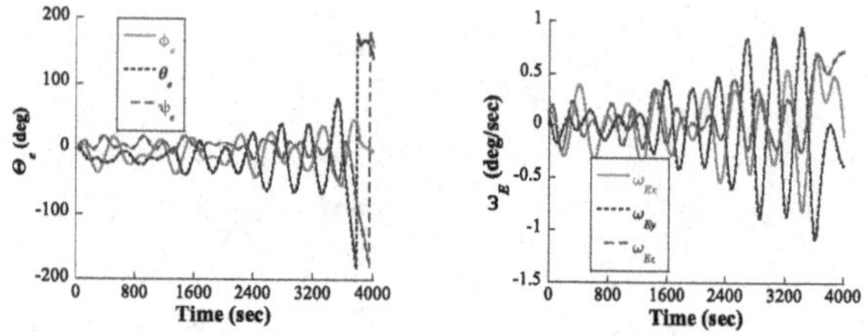

(a) Estimation error $\boldsymbol{\Theta}_e$ (b) Estimation error $\boldsymbol{\omega}_E$

FIGURE 4.10 State estimation error ensured by UPF in Case #4.

4.7 Application to microsatellite formation flying control system

In this section, the UPF developed in Algorithm 4.1 is applied to solve the control system's state estimation problem of two microsatellites flying in chief and deputy formation. More specifically, the relative position and the relative attitude between those two microsatellites will be estimated by using the UPF.

The mathematical model established in Section 3.4.7 can be used to describe the relative attitude and position motion of this formation flying. Although the dimensionless relative position control system mode is established in (3.186), the relative position control system without dimensionless is presented in this section to evaluate the state estimation performance, which is given as follows based on (3.186).

$$\ddot{\boldsymbol{r}}_{e0} + \boldsymbol{N}\dot{\boldsymbol{r}}_{e0} + \boldsymbol{K}\boldsymbol{r}_{e0} = \boldsymbol{F} + \frac{1}{m_d}\boldsymbol{u}_T \tag{4.124}$$

To measure the relative position, several Vision-based Digital Signal Processors (VDSP) are embedded as the navigation measurement sensor for this formation flying. It comprises an electro-optical sensor combined with specific beacons. This sensor is made up of a Position-Sensing-Diode (PSD) placed in the focal plane of a wide-angle lens (Gunnam et al., 2002). Let this PSD be fixed in the body-fixed frame of the deputy microsatellite with its location denoted as $\boldsymbol{p} = [p_x \quad p_y \quad p_z]^{\mathrm{T}}$. Hence, the following relationship holds

$$\boldsymbol{r}_e = \boldsymbol{r}_{e0} + \mathbb{C}^{\mathrm{T}}(\boldsymbol{q}_e)\boldsymbol{p} \tag{4.125}$$

where $\boldsymbol{r}_e = [x \quad y \quad z]^{\mathrm{T}}$ is the position vector of the PSD in the LVLH frame $\mathfrak{F}_r$. Then, it follows from (4.125) that

$$\dot{\boldsymbol{r}}_e = \boldsymbol{r}_{e0} + \boldsymbol{\omega}_d^{\times}(\mathbb{C}^{\mathrm{T}}(\boldsymbol{q}_e)\boldsymbol{p}) \tag{4.126}$$

$$\ddot{\boldsymbol{r}}_e = \ddot{\boldsymbol{r}}_{e0} + \dot{\boldsymbol{\omega}}_d^{\times}(\mathbb{C}^{\mathrm{T}}(\boldsymbol{q}_e)\boldsymbol{p}) + \boldsymbol{\omega}_d^{\times}(\boldsymbol{\omega}_d^{\times}(\mathbb{C}^{\mathrm{T}}(\boldsymbol{q}_e)\boldsymbol{p})) \tag{4.127}$$

Inserting (4.125)–(4.127) into (4.124) yields

$$\begin{aligned}\ddot{\boldsymbol{r}}_e = &(\mathbb{C}^{\mathrm{T}}(\boldsymbol{q}_e)\boldsymbol{\omega} + \boldsymbol{\omega}_o)^{\times}\left((\mathbb{C}^{\mathrm{T}}(\boldsymbol{q}_e)\boldsymbol{\omega} + \boldsymbol{\omega}_o)^{\times}(\mathbb{C}^{\mathrm{T}}(\boldsymbol{q}_e)\boldsymbol{p})\right) + \boldsymbol{F} + \boldsymbol{d}_2\\ &+ (\mathbb{C}^{\mathrm{T}}(\boldsymbol{q}_e)\dot{\boldsymbol{\omega}} + \boldsymbol{\omega}_o)^{\times}(\mathbb{C}^{\mathrm{T}}(\boldsymbol{q}_e)\boldsymbol{p}) - \boldsymbol{K}(\boldsymbol{r}_e - \mathbb{C}^{\mathrm{T}}(\boldsymbol{q}_e)\boldsymbol{p})\\ &- \boldsymbol{N}\left(\dot{\boldsymbol{r}}_e - (\mathbb{C}^{\mathrm{T}}(\boldsymbol{q}_e)\boldsymbol{\omega} + \boldsymbol{\omega}_o)^{\times}(\mathbb{C}^{\mathrm{T}}(\boldsymbol{q}_e)\boldsymbol{p})\right) + \frac{1}{m_d}\boldsymbol{u}_T\end{aligned} \tag{4.128}$$

where $\boldsymbol{d}_2$ denotes disturbance force due to the J_2 perturbation and the atmospheric drag perturbation modeled in Chapter 2.

4.7.1 Measurement model of PSD

In this subsection, the measurement model of PSD is introduced (Crassidis et al., 2000a; Xing et al., 2010). When the PSD is applied, the beacons are fixed on the chief satellite with their positions in the chief body-fixed frame known exactly. The sensor focal plane is known within the deputy frame. Suppose that the Z-axis of the sensor coordinate system is directed outward along the boresight. Because the coordinate axis of the sensor frame has the same direction as the body-fixed frame $\mathfrak{F}_{Bd}$, they have the same attitude with respect to the LVLH frame $\mathfrak{F}_r$. If the ith beacon at position $[X_i \quad Y_i \quad Z_i]$ is observed in the focal plane by the coordinate $[\chi_i \quad \gamma_i]$, the object to the image space projection transformation is determined by

$$\chi_i = \chi_0 - f\frac{A_{11}(X_i - x) + A_{12}(Y_i - y) + A_{13}(Z_i - z)}{A_{31}(X_i - x) + A_{32}(Y_i - y) + A_{33}(Z_i - z)} \tag{4.129}$$

$$\gamma_i = \gamma_0 - f\frac{A_{21}(X_i - x) + A_{22}(Y_i - y) + A_{23}(Z_i - z)}{A_{31}(X_i - x) + A_{32}(Y_i - y) + A_{33}(Z_i - z)} \tag{4.130}$$

where $i = 1, 2, \cdots, N$, $[\chi_0 \quad \gamma_0]$ is the optical axis offset, f is the focal length, N is the total number of beacons, and A_{kl} are the coefficients of the attitude matrix $\mathbb{C}(\boldsymbol{q}_e)$, $k = 1, 2$, $l = 1, 2, 3$.

The ideal observation can be expressed in a unit vector form as

$$\boldsymbol{b}_i = \mathbb{C}(\boldsymbol{q}_e)\boldsymbol{r}_i,\ i = 1, 2, \cdots, N \tag{4.131}$$

where $\boldsymbol{b}_i \in \mathbb{R}^3$ and $\boldsymbol{r}_i \in \mathbb{R}^3$ are the unit vectors expressed in the body-fixed frame of the sensors and the LVLH frame $\mathfrak{F}_r$, respectively, and they are given by

$$\boldsymbol{b}_i = \frac{1}{\sqrt{\underline{f}^2 + (\chi_i - \chi_0)^2 + (\gamma_i - \gamma_0)^2}} \begin{bmatrix} -(\chi_i - \chi_0) \\ -(\gamma_i - \gamma_0) \\ \underline{f} \end{bmatrix} \tag{4.132}$$

$$\boldsymbol{r}_i = \frac{1}{\sqrt{(X_i - x)^2 + (Y_i - y)^2 + (Z_i - z)^2}} \begin{bmatrix} X_i - x \\ Y_i - y \\ Z_i - z \end{bmatrix} \tag{4.133}$$

However, the measurement is coupled with noise. Hence, the real measurement with observation noise can be expressed as

$$\tilde{\boldsymbol{b}}_i = \boldsymbol{A}\boldsymbol{r}_i + \boldsymbol{\varepsilon}_i \tag{4.134}$$

where $\tilde{\boldsymbol{b}}_i \in \mathbb{R}^3$, $i = 1, 2, \cdots, N$, denotes the ith measurement, and $\boldsymbol{\varepsilon}_i \in \mathbb{R}^3$ is the measurement noise.

To this end, the observation or the measurement model can finally be established as

$$\boldsymbol{y} = \begin{bmatrix} \boldsymbol{A}\boldsymbol{r}_1 \\ \boldsymbol{A}\boldsymbol{r}_2 \\ \vdots \\ \boldsymbol{A}\boldsymbol{r}_N \end{bmatrix} + \begin{bmatrix} \boldsymbol{\varepsilon}_1 \\ \boldsymbol{\varepsilon}_2 \\ \vdots \\ \boldsymbol{\varepsilon}_N \end{bmatrix} \tag{4.135}$$

where $\boldsymbol{y} = [\tilde{\boldsymbol{b}}_1^{\mathrm{T}}\ \ \tilde{\boldsymbol{b}}_2^{\mathrm{T}}\ \ \cdots\ \ \tilde{\boldsymbol{b}}_N^{\mathrm{T}}]^{\mathrm{T}}$.

4.7.2 Transformed microsatellite formation flying control system

Combining the relative attitude control model (3.187)–(3.188), the rewritten relative position dynamics (4.128), and measurement model (4.135), the formation flying control system of two microsatellites can be transformed into the following nonlinear equation.

$$\begin{cases} \dot{\boldsymbol{x}} = \boldsymbol{f}(\boldsymbol{x}) + \boldsymbol{g}(\boldsymbol{x})\boldsymbol{d} \\ \boldsymbol{y} = \boldsymbol{h}(\boldsymbol{x}) + \boldsymbol{v} \end{cases} \tag{4.136}$$

where $x = [q_e^T \ \ r_e^T \ \ \omega^T \ \ \dot{r}_e]^T$, $d = [0 \ \ u_{df}^T \ \ 0 \ \ (F + d_2)^T]^T$, $h(x) = [(Ar_1)^T \ \ (Ar_2)^T \ \ \cdots \ \ (Ar_N)^T]^T$, $v = [\varepsilon_1^T \ \ \varepsilon_2^T \ \ \cdots \ \ \varepsilon_N^T]^T$,

$$g(x) = \begin{bmatrix} 0 & 0 & 0 & 0 \\ 0 & J_d^{-1} & 0 & 0 \\ 0 & 0 & 0 & 0 \\ 0 & 0 & 0 & I_3 \end{bmatrix} \tag{4.137}$$

and $f(x) = [F_1^T(x) \ \ F_2^T(x)]^T$. Here, $\dot{r}_e = [v_x \ \ v_y \ \ v_z]^T$ is the relative velocity. $F_1(x)$ and $F_2(x)$ are defined as

$$F_1(x) = \begin{bmatrix} f_1(x) \\ f_2(x) \end{bmatrix} = \begin{bmatrix} E(q_e)\omega \\ \dot{r}_e \end{bmatrix} \tag{4.138}$$

$$F_2(x) = [f_3^T(x) \ \ f_4^T(x)]^T \tag{4.139}$$

$$\begin{aligned} f_3(x) = {} & J_d^{-1} u_f - J_d^{-1} \left(\omega_e + \mathbb{C}(q_e)\omega_0\right)^{\times} J_d \left(\omega_e + \mathbb{C}(q_e)\omega_0\right) \\ & + \left((\omega_e)^{\times} \mathbb{C}(q_e)\omega_0 - \mathbb{C}(q_e)\dot{\omega}_0\right) \end{aligned} \tag{4.140}$$

$$\begin{aligned} f_4(x) = {} & (\mathbb{C}^T(q_e)\omega + \omega_o)^{\times} \left((\mathbb{C}^T(q_e)\omega + \omega_o)^{\times} (\mathbb{C}^T(q_e)p)\right) \\ & + (\mathbb{C}^T(q_e)\dot{\omega} + \omega_o)^{\times} (\mathbb{C}^T(q_e)p) - K(r_e - \mathbb{C}^T(q_e)p) \\ & - N\left(\dot{r}_e - (\mathbb{C}^T(q_e)\omega + \omega_o)^{\times} (\mathbb{C}^T(q_e)p)\right) + \frac{1}{m_d} u_T \end{aligned} \tag{4.141}$$

4.7.3 Implementation of UPF in formation flying control system

It is obtained that discretizing the formation flying system (4.136) has the form of (2.32)–(2.33). Then, the proposed UPF algorithm is applicable to estimate the states of the microsatellite formation flying system. In this subsection, the implementation of the UPF in this control system to estimate the relative position, the relative velocity, the relative attitude, and the relative angular velocity is clarified in detail.

It is known in engineering that four beacons are needed at least to measure the relative position and the relative attitude (Sun and Crassidis, 2002). Hence, it is assumed that four beacons are fixed on the chief microsatellite with the position of the ith beacon as $[X_i \ \ Y_i \ \ Z_i]$, $i = 1, 2, 3, 4$. Meanwhile, it follows from the formation flying control system that $p_i = 2$. It is seen in Algorithm 4.1 that the implementation of the UPF depends on the calculation of $\Lambda(\Delta t)$, $U(\hat{x}_k)$, and $\boldsymbol{Z}(\hat{\boldsymbol{x}}_k, \Delta t)$. Motivated by this, the detailed procedures of calculating them are given as

$$\boldsymbol{\Lambda}(\Delta t) = 0.5\Delta t^2 \boldsymbol{I}_{12} \tag{4.142}$$

$$\boldsymbol{Z}(\hat{\boldsymbol{x}}_k, \Delta t) = \Delta t \boldsymbol{L}_F^1 + 0.5(\Delta t)^2 \boldsymbol{L}_F^2 \tag{4.143}$$

$$\boldsymbol{U}(\hat{\boldsymbol{x}}_k) = \begin{bmatrix} \dfrac{\partial \boldsymbol{L}_F^1}{\partial \boldsymbol{\omega}} \boldsymbol{J}_d^{-1} & \dfrac{\partial \boldsymbol{L}_F^1}{\partial \dot{\boldsymbol{r}}_e} \end{bmatrix} \tag{4.144}$$

where Δt is the sampling time.

The term $\boldsymbol{L}_F^1$ in (4.143) is derived as

$$\begin{aligned} \boldsymbol{L}_F^1 &= \frac{\partial \boldsymbol{h}}{\partial \boldsymbol{x}} \boldsymbol{F}_1(\boldsymbol{x}) = \frac{\partial \boldsymbol{h}}{\partial \boldsymbol{q}_e} \boldsymbol{f}_1 + \frac{\partial \boldsymbol{h}}{\partial \boldsymbol{r}_e} \boldsymbol{f}_2 \\ &= \begin{bmatrix} \boldsymbol{E}_1^{\mathrm{T}}(\boldsymbol{q}_e)\boldsymbol{\psi}(\boldsymbol{r}_1) \\ \boldsymbol{E}_1^{\mathrm{T}}(\boldsymbol{q}_e)\boldsymbol{\psi}(\boldsymbol{r}_2) \\ \boldsymbol{E}_1^{\mathrm{T}}(\boldsymbol{q}_e)\boldsymbol{\psi}(\boldsymbol{r}_3) \\ \boldsymbol{E}_1^{\mathrm{T}}(\boldsymbol{q}_e)\boldsymbol{\psi}(\boldsymbol{r}_4) \end{bmatrix} \boldsymbol{E}_1(\boldsymbol{q})\boldsymbol{\omega} + \frac{A}{m^{\frac{3}{2}}} \begin{bmatrix} \frac{\partial \boldsymbol{r}_1}{\partial \boldsymbol{r}_e} \\ \frac{\partial \boldsymbol{r}_2}{\partial \boldsymbol{r}_e} \\ \frac{\partial \boldsymbol{r}_3}{\partial \boldsymbol{r}_e} \\ \frac{\partial \boldsymbol{r}_4}{\partial \boldsymbol{r}_e} \end{bmatrix} \dot{\boldsymbol{r}}_e \end{aligned} \tag{4.145}$$

where

$$\boldsymbol{\psi}(\boldsymbol{r}_i) = \begin{bmatrix} 0 & -\boldsymbol{r}_i^{\mathrm{T}} \\ \boldsymbol{r}_i & \boldsymbol{r}_i^{\times} \end{bmatrix} \tag{4.146}$$

$$\frac{\partial \boldsymbol{r}_i}{\partial \boldsymbol{r}_e} = \begin{bmatrix} (X_i - x)^2 - m & (X_i - x)(Y_i - y) & (X_i - x)(Z_i - z) \\ (X_i - x)(Y_i - y) & (Y_i - y)^2 - m & (Y_i - y)(Z_i - z) \\ (X_i - x)(Z_i - z) & (Y_i - y)(Z_i - z) & (Z_i - z)^2 - m \end{bmatrix} \tag{4.147}$$

$$m = (X_i - x)^2 + (Y_i - y)^2 + (Z_i - z)^2 \tag{4.148}$$

The term $\boldsymbol{L}_F^2$ in (4.143) can be calculated by using

$$\begin{aligned} \boldsymbol{L}_F^2 &= \frac{\partial \boldsymbol{L}_F^1}{\partial \boldsymbol{x}} \boldsymbol{F}_1(\boldsymbol{x}) + \frac{\partial \boldsymbol{L}_F^1}{\partial \dot{\boldsymbol{x}}} \boldsymbol{F}_2(\boldsymbol{x}) \\ &= \frac{\partial \boldsymbol{L}_F^1}{\partial \boldsymbol{q}_e} \boldsymbol{f}_1 + \frac{\partial \boldsymbol{L}_F^1}{\partial \boldsymbol{r}_e} \boldsymbol{f}_2 + \frac{\partial \boldsymbol{L}_F^1}{\partial \boldsymbol{\omega}} \boldsymbol{f}_3 + \frac{\partial \boldsymbol{L}_F^1}{\partial \dot{\boldsymbol{r}}_e} \boldsymbol{f}_4 \end{aligned} \tag{4.149}$$

where

$$\frac{\partial \boldsymbol{L}_F^1}{\partial \boldsymbol{q}_e} f_1 = \begin{bmatrix} \boldsymbol{E}^{\mathrm{T}}(\boldsymbol{q}_e)\boldsymbol{\Psi}^{\mathrm{T}}(\boldsymbol{\omega})\boldsymbol{E}_1(\boldsymbol{q}_e)\frac{\partial \boldsymbol{r}}{\partial \boldsymbol{r}_e} \\ \boldsymbol{E}^{\mathrm{T}}(\boldsymbol{q}_e)\boldsymbol{\Psi}^{\mathrm{T}}(\boldsymbol{\omega})\boldsymbol{E}_1(\boldsymbol{q}_e)\frac{\partial \boldsymbol{r}_2}{\partial \boldsymbol{r}_e} \\ \boldsymbol{E}^{\mathrm{T}}(\boldsymbol{q}_e)\boldsymbol{\Psi}^{\mathrm{T}}(\boldsymbol{\omega})\boldsymbol{E}_1(\boldsymbol{q}_e)\frac{\partial \boldsymbol{r}_3}{\partial \boldsymbol{r}_e} \\ \boldsymbol{E}^{\mathrm{T}}(\boldsymbol{q}_e)\boldsymbol{\Psi}^{\mathrm{T}}(\boldsymbol{\omega})\boldsymbol{E}_1(\boldsymbol{q}_e)\frac{\partial \boldsymbol{r}_4}{\partial \boldsymbol{r}_e} \end{bmatrix} \dot{\boldsymbol{r}}_e - \begin{bmatrix} \boldsymbol{\omega}^{\times}\boldsymbol{E}_1^{\mathrm{T}}(\boldsymbol{q}_e)\boldsymbol{\psi}(\boldsymbol{r}_1)\boldsymbol{E}_1(\boldsymbol{q}_e) \\ \boldsymbol{\omega}^{\times}\boldsymbol{E}_1^{\mathrm{T}}(\boldsymbol{q}_e)\boldsymbol{\psi}(\boldsymbol{r}_2)\boldsymbol{E}_1(\boldsymbol{q}_e) \\ \boldsymbol{\omega}^{\times}\boldsymbol{E}_1^{\mathrm{T}}(\boldsymbol{q}_e)\boldsymbol{\psi}(\boldsymbol{r}_3)\boldsymbol{E}_1(\boldsymbol{q}_e) \\ \boldsymbol{\omega}^{\times}\boldsymbol{E}_1^{\mathrm{T}}(\boldsymbol{q}_e)\boldsymbol{\psi}(\boldsymbol{r}_4)\boldsymbol{E}_1(\boldsymbol{q}_e) \end{bmatrix} \boldsymbol{\omega} \tag{4.150}$$

$$\frac{\partial \boldsymbol{L}_F^1}{\partial \boldsymbol{r}_e} f_2 = \begin{bmatrix} \boldsymbol{E}^{\mathrm{T}}(\boldsymbol{q}_e)\boldsymbol{\Psi}^{\mathrm{T}}(\boldsymbol{\omega})\boldsymbol{E}_1(\boldsymbol{q})\frac{\partial \boldsymbol{r}_1}{\partial \boldsymbol{r}_e} \\ \boldsymbol{E}^{\mathrm{T}}(\boldsymbol{q}_e)\boldsymbol{\Psi}^{\mathrm{T}}(\boldsymbol{\omega})\boldsymbol{E}_1(\boldsymbol{q}_e)\frac{\partial \boldsymbol{r}_2}{\partial \boldsymbol{r}_e} \\ \boldsymbol{E}^{\mathrm{T}}(\boldsymbol{q}_e)\boldsymbol{\Psi}^{\mathrm{T}}(\boldsymbol{\omega})\boldsymbol{E}_1(\boldsymbol{q})\frac{\partial \boldsymbol{r}_3}{\partial \boldsymbol{r}_e} \\ \boldsymbol{E}^{\mathrm{T}}(\boldsymbol{q}_e)\boldsymbol{\Psi}^{\mathrm{T}}(\boldsymbol{\omega})\boldsymbol{E}_1(\boldsymbol{q}_e)\frac{\partial \boldsymbol{r}_4}{\partial \boldsymbol{r}_e} \end{bmatrix} \dot{\boldsymbol{r}}_e + \frac{\boldsymbol{A}}{m^4} \begin{bmatrix} \boldsymbol{B}_1 \\ \boldsymbol{B}_2 \\ \boldsymbol{B}_3 \\ \boldsymbol{B}_4 \end{bmatrix} \dot{\boldsymbol{r}}_e \tag{4.151}$$

$$\frac{\partial \boldsymbol{L}_F^1}{\partial \boldsymbol{\omega}} f_3 = \begin{bmatrix} \boldsymbol{E}^{\mathrm{T}}(\boldsymbol{q})\boldsymbol{\psi}(\boldsymbol{r}_1)\boldsymbol{E}(\boldsymbol{q}) \\ \boldsymbol{E}^{\mathrm{T}}(\boldsymbol{q})\boldsymbol{\psi}(\boldsymbol{r}_2)\boldsymbol{E}(\boldsymbol{q}) \\ \boldsymbol{E}^{\mathrm{T}}(\boldsymbol{q})\boldsymbol{\psi}(\boldsymbol{r}_3)\boldsymbol{E}(\boldsymbol{q}) \\ \boldsymbol{E}^{\mathrm{T}}(\boldsymbol{q})\boldsymbol{\psi}(\boldsymbol{r}_4)\boldsymbol{E}(\boldsymbol{q}) \end{bmatrix} f_3 \tag{4.152}$$

$$\frac{\partial \boldsymbol{L}_F^1}{\partial \dot{\boldsymbol{r}}_e} f_4 = \frac{\boldsymbol{A}}{m^{\frac{3}{2}}} \begin{bmatrix} \frac{\partial \boldsymbol{r}_1}{\partial \boldsymbol{r}_e} \\ \frac{\partial \boldsymbol{r}_2}{\partial \boldsymbol{r}_e} \\ \frac{\partial \boldsymbol{r}_3}{\partial \boldsymbol{r}_e} \\ \frac{\partial \boldsymbol{r}_4}{\partial \boldsymbol{r}_e} \end{bmatrix} f_4 \tag{4.153}$$

$$\boldsymbol{E}_1(\boldsymbol{q}_e) = \begin{bmatrix} \boldsymbol{q}_{e24}^{\mathrm{T}} \\ -q_{e1}\boldsymbol{I}_3 + \boldsymbol{q}_{e24}^{\times} \end{bmatrix} \tag{4.154}$$

$$\boldsymbol{B}_i = \begin{bmatrix} b_{11} & b_{12} & b_{13} \\ b_{21} & b_{22} & b_{23} \\ b_{31} & b_{32} & b_{33} \end{bmatrix}_i \tag{4.155}$$

$$\begin{aligned} b_{11} = {} & 3(X_i - x)((X_i - x)^2 - m)v_x + (Y_i - y)(3(Y_i - y)^2 - m)v_y \\ & + (Z_i - z)(3(X_i - x)^2 - m)v_z \end{aligned} \tag{4.156}$$

$$\begin{aligned} b_{22} = (X_i - x)(3(Y_i - y)^2 - m)v_x + 3(Y_i - y)((Y_i - y)^2 - m)v_y \\ + (Z_i - z)(3(Y_i - y)^2 - m)v_z \end{aligned} \tag{4.157}$$

$$\begin{aligned} b_{33} = (X_i - x)(3(Z_i - z)^2 - m)v_x + (Y_i - y)(3(Z_i - z)^2 - m)v_y \\ + 3(Z_i - z)((Z_i - z)^2 - m)v_z \end{aligned} \tag{4.158}$$

$$\begin{aligned} b_{12} = b_{21} = (Y_i - y)(3(X_i - x)^2 - m)v_x \\ + (X_i - x)(3(Y_i - y)^2 - m)v_y + 3(X_i - x)(Y_i - y)(Z_i - z)v_z \end{aligned} \tag{4.159}$$

$$\begin{aligned} b_{13} = b_{31} = (Z_i - z)(3(X_i - x)^2 - m)v_x \\ + 3(X_i - x)(Y_i - y)(Z_i - z)v_y + (X_i - x)(3(Z_i - z)^2 - m)v_z \end{aligned} \tag{4.160}$$

$$\begin{aligned} b_{23} = b_{32} = 3(X_i - x)(Y_i - y)(Z_i - z)v_x \\ + (Z_i - z)(3(Y_i - y)^2 - m)v_y + (Y_i - y)(3(Z_i - z)^2 - m)v_z \end{aligned} \tag{4.161}$$

Based on the above results, the UPF and the classical PF in Algorithm 2.1 can be used for the relative position and the relative attitude estimation of the microsatellite formation flying control system.

4.7.4 Simulation results

In this subsection, simulations are carried out to evaluate the performance of the UPF for the relative position and attitude estimation. To demonstrate the superior advantages of the UPF, the PF in Algorithm 2.1 and the standard UKF are also examined for the purpose of comparison.

In simulations, the initial orbit elements of the chief microsatellite are listed in Table 4.1. The inertia of the deputy microsatellite is $\boldsymbol{J}_d = \text{diag}([49 \quad 55 \quad 63])$ kg·m^2. The position of the PSD sensor (feature point) at the deputy microsatellite is $\boldsymbol{p} = [0.2 \quad 0.2 \quad 0.2]^{\mathrm{T}}$ m in the frame $\mathcal{F}_{Bd}$. Four beacons are fixed in the chief microsatellite with $[X_1 \quad Y_1 \quad Z_1] = [0.5 \quad 0.5 \quad 0.5]$ m, $[X_2 \quad Y_2 \quad Z_2] = [-0.5 \quad -0.5 \quad 0.5]$ m, $[X_3 \quad Y_3 \quad Z_3] = [-0.5 \quad 0.5 \quad -0.5]$ m, and $[X_4 \quad Y_4 \quad Z_4] = [0.5 \quad -0.5 \quad -0.5]$ m.

The PF, the UKF, and the UPF are implemented with the sampling interval set as $\Delta t = 0.5$ seconds. The initial state estimation error covariance is designed as

$$\boldsymbol{P}_{\tilde{x},0} = \text{diag}([(\tilde{\boldsymbol{q}}_{e24}(t_0) - \boldsymbol{q}_{e24}(t_0))^2 \quad (\tilde{\boldsymbol{\omega}}(t_0) - \boldsymbol{\omega}(t_0))^2]) \tag{4.162}$$

The weighted matrix for the UPF and the PF is selected as

$$\boldsymbol{W}_E = \text{diag}([10^5 \quad 10^5 \quad 10^5 \quad 10^7 \quad 10^7 \quad 10^7]^{\mathrm{T}}) \tag{4.163}$$

TABLE 4.1 Initial elliptical orbital parameters of the chief microsatellite.

Parameters	Value	Unit
α	6978	km
ω	10	degree
M	0	degree
e	0.005	-
i	70	degree
Ω	320	degree

The parameters of sigma-points for the UPF and the UKF are chosen as $n = 12$ and $\kappa = 2$.

The initial relative position and relative velocity of the formation flying system are set as $\boldsymbol{r}_{e0}(t_0) = [200 \quad 200 \quad 100]^{\mathrm{T}}$ m and

$$\boldsymbol{V}_0(t_0) = \begin{bmatrix} 0.01 & \dfrac{-200\dot{\theta}(2+e)}{\sqrt{(1+e)(1-e)^3}} & 0.01 \end{bmatrix}^{\mathrm{T}} \text{ m} \tag{4.164}$$

respectively. The initial estimated relative position and relative velocity of the formation flying system are chosen as $\tilde{\boldsymbol{r}}_{e0}(t_0) = [205 \quad 205 \quad 105]^{\mathrm{T}}$ m and

$$\tilde{\boldsymbol{V}}_0(t_0) = \begin{bmatrix} 0.011 & \dfrac{-205\dot{\theta}(2+e)}{\sqrt{(1+e)(1-e)^3}} & 0.011 \end{bmatrix}^{\mathrm{T}} \text{ m/sec} \tag{4.165}$$

respectively. The initial relative attitude $\boldsymbol{q}_e(t_0) = [1 \quad 0 \quad 0 \quad 0]^{\mathrm{T}}$ and the initial relative angular velocity $\boldsymbol{\omega}(t_0) = [0 \quad 0 \quad 0.1]^{\mathrm{T}}$ deg/sec are selected. The corresponding initial estimation values are set as $\tilde{\boldsymbol{\omega}}(t_0) = [0.1 \quad 0.1 \quad 0.12]^{\mathrm{T}}$ deg/sec and

$$\tilde{\boldsymbol{q}}_e(t_0) = \begin{bmatrix} \cos\dfrac{\Theta_0}{2} & \dfrac{\sqrt{3}}{3}\sin\dfrac{\Theta_0}{2} & \dfrac{\sqrt{3}}{3}\sin\dfrac{\Theta_0}{2} & \dfrac{\sqrt{3}}{3}\sin\dfrac{\Theta_0}{2} \end{bmatrix}^{\mathrm{T}} \tag{4.166}$$

with $\Theta_0 = 5$ degrees. The modeling error is assumed to have a form of

$$\boldsymbol{u}_{df} = \Delta\boldsymbol{N}_{c1} + \Delta\boldsymbol{N}_{e1} \tag{4.167}$$

$$\boldsymbol{F} + \boldsymbol{d}_2 = \Delta\boldsymbol{N}_{c2} + \Delta\boldsymbol{N}_{e2} \tag{4.168}$$

where $\Delta\boldsymbol{N}_{c1} \in \mathbb{R}^3$ and $\Delta\boldsymbol{N}_{c2} \in \mathbb{R}^3$ are constant. $\Delta\boldsymbol{N}_{e1} \in \mathbb{R}^3$ and $\Delta\boldsymbol{N}_{e2} \in \mathbb{R}^3$ are time-varying. The following two scenarios of modeling error are considered.

1) Scenario #1: This is a nominal case. $\Delta\boldsymbol{N}_{c1} = [-3\omega_1 \quad -2\omega_1 \quad 3\omega_1]^{\mathrm{T}}$ Nm, $\Delta\boldsymbol{N}_{e1} = [3\omega_1\cos(\omega_T t) \quad \omega_1\cos(\omega_T t) \quad -2\omega_1\cos(\omega_T t)]^{\mathrm{T}}$ Nm,

$\Delta \boldsymbol{N}_{e2} = [-3\omega_1 \cos(\omega_T t) \quad 4\omega_1 \cos(\omega_T t) \quad 2\omega_1 \cos(\omega_T t)]^{\mathrm{T}}$ N, and $\Delta \boldsymbol{N}_{c2} = [5\omega_1 \quad -3\omega_1 \quad 2\omega_1]^{\mathrm{T}}$ N are assumed with $\omega_T = 0.66$ deg/sec and $\omega_1 = 0.0005$. The standard deviation of the measurement noise in (4.134) is assumed as 0.001 degrees.

2) Scenario #2: This case is with large modeling error and severe measurement noise. $\Delta \boldsymbol{N}_{e1} = [3\omega_1 \cos(\omega_T t) \quad \omega_1 \cos(\omega_T t) \quad -2\omega_1 \cos(\omega_T t)]^{\mathrm{T}}$ Nm, $\Delta \boldsymbol{N}_{e2} = [-3\omega_1 \cos(\omega_T t) \quad 4\omega_1 \cos(\omega_T t) \quad 2\omega_1 \cos(\omega_T t)]^{\mathrm{T}}$ N, $\Delta \boldsymbol{N}_{c1} = [-3\omega_1 \quad -2\omega_1 \quad 3\omega_1]^{\mathrm{T}}$ Nm, and $\Delta \boldsymbol{N}_{c2} = [5\omega_1 \quad -3\omega_1 \quad 2\omega_1]^{\mathrm{T}}$ N are assumed with $\omega_T = 0.66$ deg/sec and $\omega_1 = 0.005$. The standard deviation of the measurement noise in (4.134) is assumed as 0.01 degrees.

Remark 4.6. When applying state estimation approaches to the microsatellite formation flying control system, the corresponding state estimation error defined in (2.69) is specified as $\tilde{\boldsymbol{x}} = [\tilde{\boldsymbol{q}}_e^{\mathrm{T}} \quad \boldsymbol{\rho}_e \quad \tilde{\boldsymbol{\omega}}^{\mathrm{T}} \quad \dot{\boldsymbol{\rho}}_e]^{\mathrm{T}}$. As pointed out in Remark 4.5, the attitude Euler angles can explicitly describe the relative attitude between two microsatellites, the state estimation error described by the unit quaternion $\tilde{\boldsymbol{q}}_e$ can be transformed into the attitude Euler angles estimation error throughout the book, which is also denoted as $\boldsymbol{\Theta}_e = [\phi_e \quad \theta_e \quad \psi_e]^{\mathrm{T}}$. The estimation error $\tilde{\boldsymbol{\omega}}_{BI}$ of the relative angular velocity state is also denoted as $\boldsymbol{\omega}_E = [\omega_{Ex} \quad \omega_{Ey} \quad \omega_{Ez}]^{\mathrm{T}} = \tilde{\boldsymbol{\omega}}$. $\boldsymbol{\rho}_e$ and $\dot{\boldsymbol{\rho}}_e$ are re-represented by $\boldsymbol{\rho}_e = [x_e \quad y_e \quad z_e]^{\mathrm{T}}$ and $\dot{\boldsymbol{\rho}}_e = [v_{ex} \quad v_{ey} \quad v_{ez}]^{\mathrm{T}}$, respectively. The estimation error of the modeling error $[\boldsymbol{u}_{df}^{\mathrm{T}} \quad (\boldsymbol{F} + \boldsymbol{d}_2)^{\mathrm{T}}]^{\mathrm{T}}$ is re-denoted as $\boldsymbol{d}_e = [d_{e1} \quad d_{e2} \quad d_{e3} \quad d_{e4} \quad d_{e5} \quad d_{e6}]^{\mathrm{T}}$.

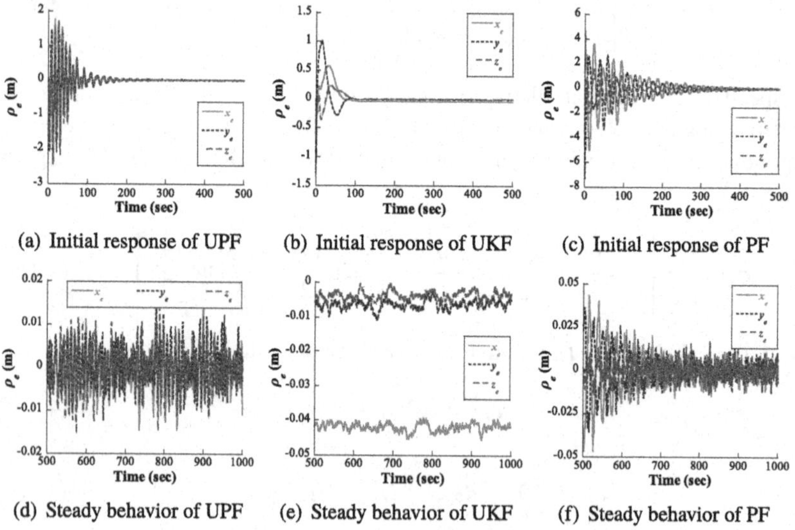

FIGURE 4.11 Relative position state estimation error using UPF, UKF, and PF in Scenario #1.

With the application of the UPF, the UKF, and the PF, the corresponding state estimation results in Scenario #1 were shown in Figs. 4.11–4.14. It was

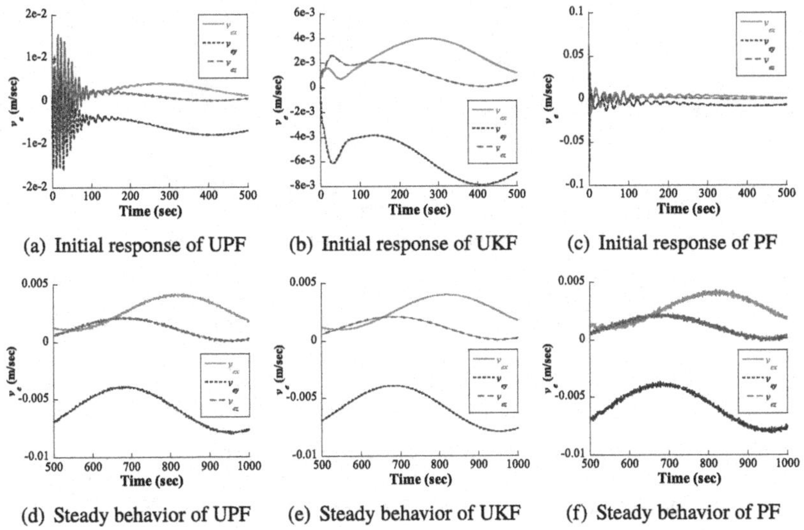

FIGURE 4.12 Relative velocity estimation error using UPF, UKF, and PF in Scenario #1.

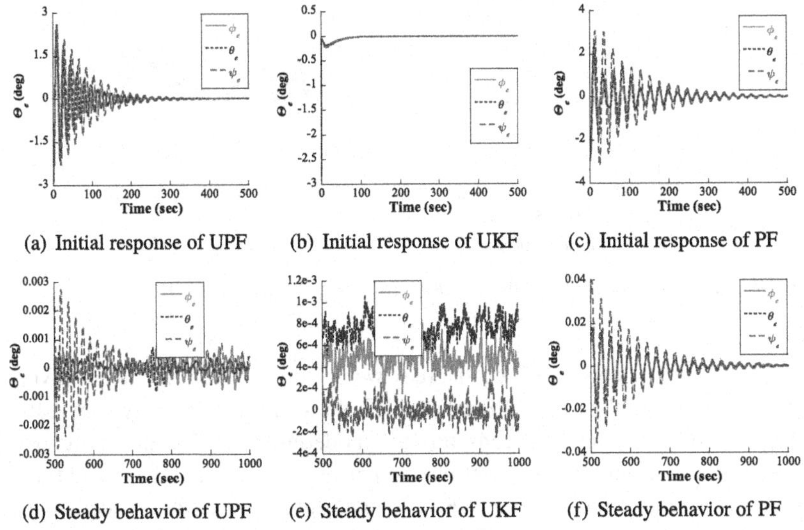

FIGURE 4.13 Relative attitude state estimation error using UPF, UKF, and PF in Scenario #1.

seen that the UPF ensured a faster estimation rate than the PF. The UPF required 200 seconds to estimate the actual states, while the PF needed 400 seconds. The conclusion that the UPF improves the estimation rate of the PF was validated. However, better estimation accuracy was achieved by the UPF and the PF than the UKF. This can be seen in Fig. 4.11 and Fig. 4.13 that a relative position estimation error $z_e = -0.041$ m and a relative attitude estimation error $\theta_e = 0.0008$

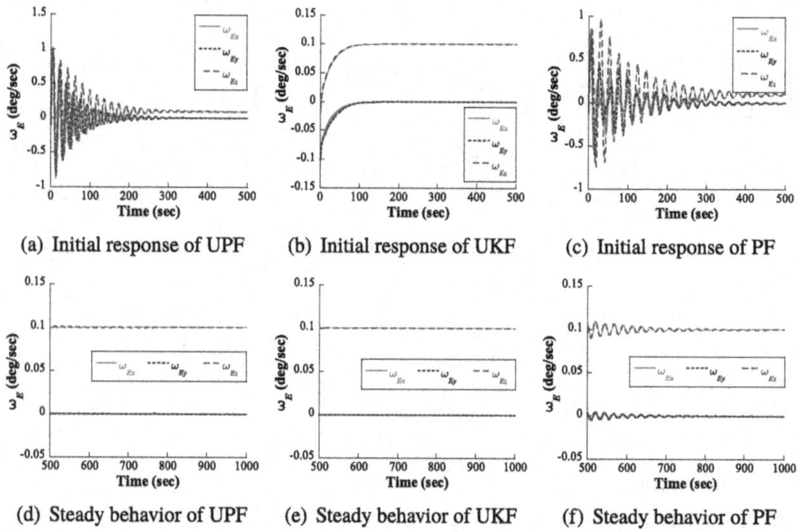

(a) Initial response of UPF (b) Initial response of UKF (c) Initial response of PF

(d) Steady behavior of UPF (e) Steady behavior of UKF (f) Steady behavior of PF

FIGURE 4.14 Relative angular velocity estimation error using UPF, UKF, and PF in Scenario #1.

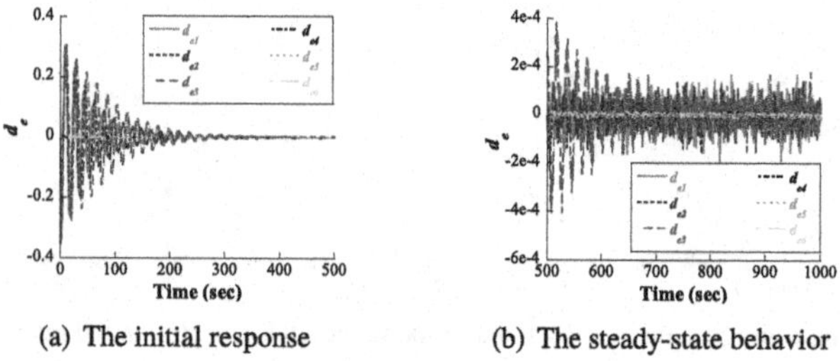

(a) The initial response (b) The steady-state behavior

FIGURE 4.15 Estimation error of the modeling error using UPF in Scenario #1.

degrees as well as $\phi_e = 0.0005$ degrees were seen for the UKF, respectively. That is because unlike the UPF and the PF, the UKF does not have any capability of estimating and compensating the modeling error. The modeling error in Scenario #1 deteriorated the estimation performance of the UKF. In addition, as shown in Figs. 4.15–4.16, the PF provided inferior estimation accuracy of the modeling error than the UPF. Hence, the modeling error had more negative effect in the PF than the UPF. This led to the inferior state estimation accuracy of the PF than the UPF.

When the microsatellite formation flying control system was subject to the large modeling error and the severe measurement noise assumed in Scenario #2, the estimation error of the relative position, the relative velocity, the relative attitude, and the relative angular velocity were shown in Figs. 4.17–4.20, respectively. Comparing with the estimation results in Scenario #1, worse estimation performance was seen for the UPF, the UKF, and the PF in Scenario #2. This was

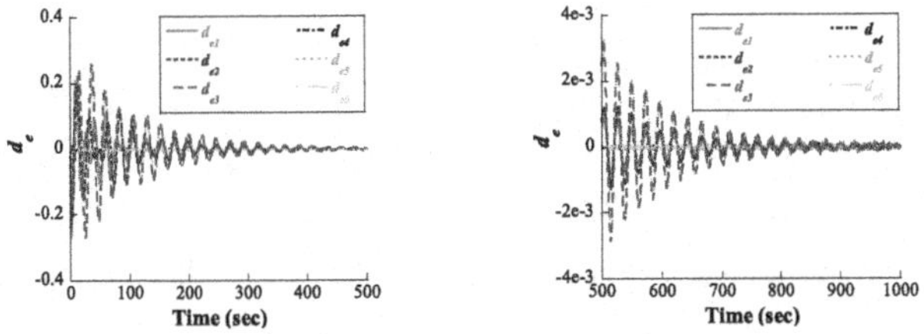

(a) The initial response (b) The steady-state behavior

FIGURE 4.16 Estimation error of the modeling error using PF in Scenario #1.

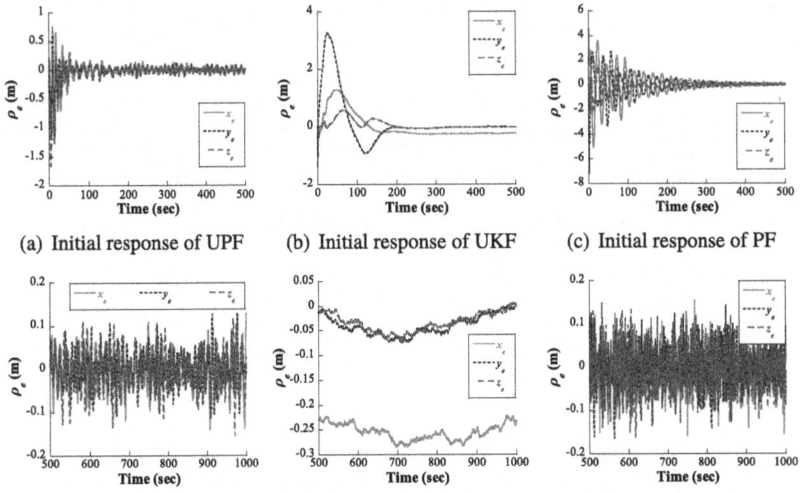

(a) Initial response of UPF (b) Initial response of UKF (c) Initial response of PF

(d) Steady behavior of UPF (e) Steady behavior of UKF (f) Steady behavior of PF

FIGURE 4.17 Relative position estimation error using UPF, UKF, and PF in Scenario #2.

due to the effect of the large modeling error and the severe measurement noise. However, the UPF still guaranteed better estimation performance than the UKF and the PF. More specifically, better relative position and relative attitude estimation accuracy was seen in Fig. 4.17 and Fig. 4.19 for the UPF than the other two filters. The advantage of the UPF that it has more robustness than the UKF and the PF to large modeling error and severe measurement noise was therefore verified. Due to the influence of the large modeling error, worse estimation of the modeling error was led in Scenario #2. This was found by comparing Fig. 4.15 with Fig. 4.20. Nevertheless, the UPF had a better estimation capability of the modeling error than the PF, as we can see in Figs. 4.21–4.22.

To this end, the above numerical results successfully demonstrated that the proposed UPF achieves better estimation performance than the UKF and the PF for the microsatellite formation flying control system even in the presence of large modeling error and severe measurement noise.

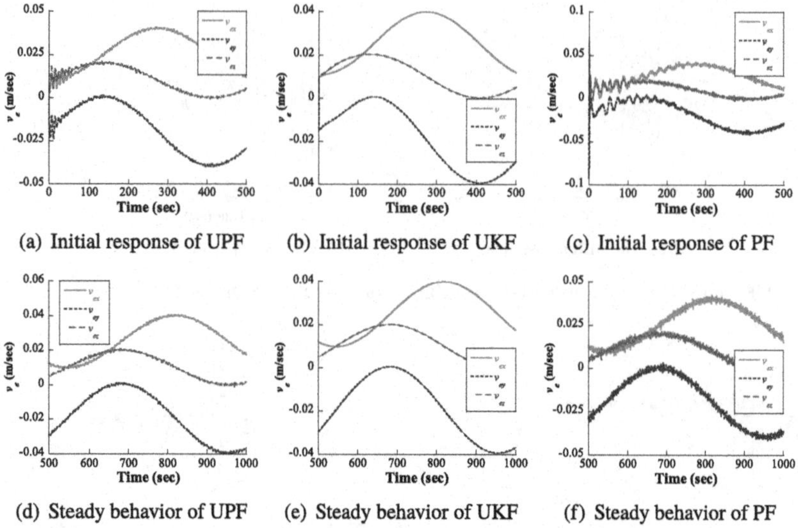
(a) Initial response of UPF
(b) Initial response of UKF
(c) Initial response of PF
(d) Steady behavior of UPF
(e) Steady behavior of UKF
(f) Steady behavior of PF
Time (sec)

FIGURE 4.18 Relative velocity estimation error using UPF, UKF, and PF in Scenario #2.

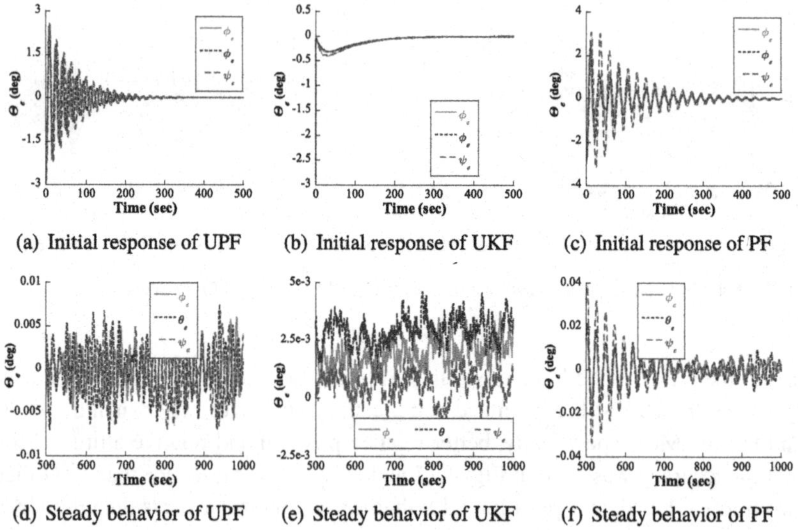
(a) Initial response of UPF
(b) Initial response of UKF
(c) Initial response of PF
(d) Steady behavior of UPF
(e) Steady behavior of UKF
(f) Steady behavior of PF
Time (sec)

FIGURE 4.19 Relative attitude state estimation error using UPF, UKF, and PF in Scenario #2.

4.8 Summary

In this chapter, a new filtering approach, namely the unscented predictive filter (UPF), was derived for a class of nonlinear systems with modeling error and non-Gaussian measurement noise. This was achieved by using the unscented transformation-based sigma point technique. The drawbacks of the classical PF

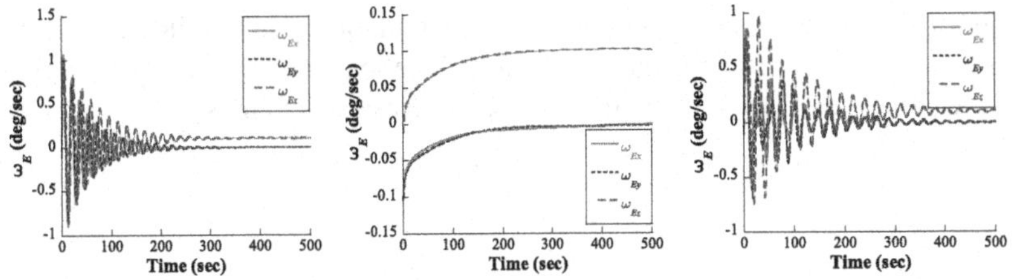

(a) Initial response of UPF (b) Initial response of UKF (c) Initial response of PF

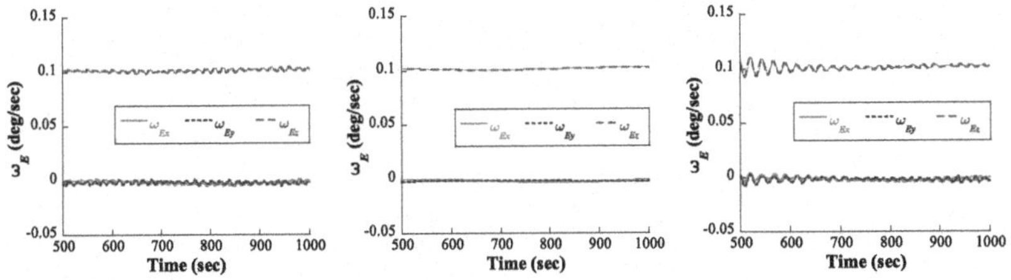

(d) Steady behavior of UPF (e) Steady behavior of UKF (f) Steady behavior of PF

FIGURE 4.20 Relative angular velocity estimation error using UPF, UKF, and PF in Scenario #2.

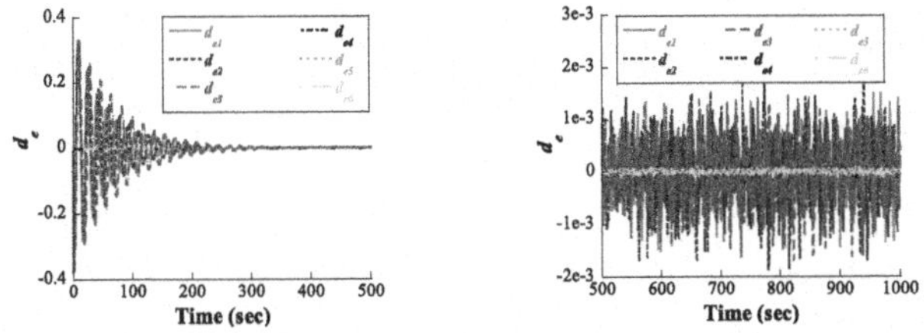

(a) The initial response (b) The steady-state behavior

FIGURE 4.21 Estimation error of the modeling error using UPF in Scenario #2.

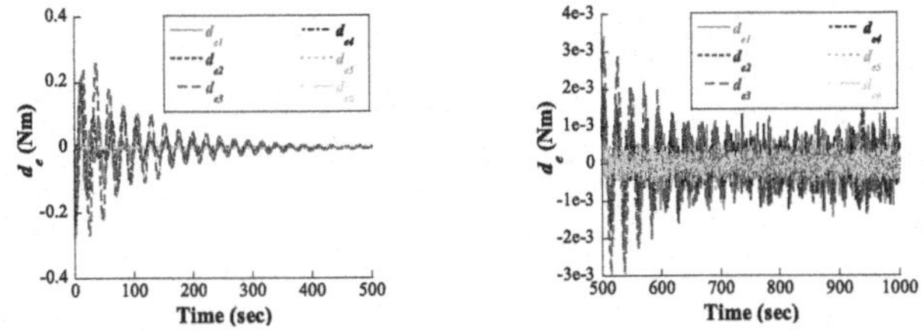

(a) The initial response (b) The steady-state behavior

FIGURE 4.22 Estimation error of the modeling error using PF in Scenario #2.

were solved by the UPF. It was rigorously proved that the UPF achieved better state estimation performance than the classical PF. Moreover, the stability of the estimation error was conditionally ensured. The application of the UPF to the microsatellite attitude control system and the microsatellite formation flying control system were verified in this chapter. Simulation results validated that the UPF ensures better state estimation performance than the UKF and the PF even in the presence of severe modeling error and measurement noise.

Chapter 5

Central difference predictive filter

5.1 Introduction

Due to limitations of cost and mass, the attitude control system of a microsatellite is usually designed with low-complexity, low-cost, and low-precision sensors. Gyros to measure angular velocity would not be used. This leads the attitude control system to be gyroless (Carmi and Oshman, 2006; Tsao and Liu, 2009). Owing to finite modeling technique, any microsatellite is also subject to severe modeling error. Hence, the problem of state estimation or attitude determination for microsatellite with low precision sensors and modeling error should be addressed. Although many filtering approaches such as the extended Kalman filter (Lefferts et al., 1982), the unscented Kalman filter (Crassidis and Markley, 2003), and even the classical predictive filter have been presented to this problem, there is not a unified state estimation framework. That is because the preceding filters each have its own disadvantages, as reviewed in Section 1.5.

Motivated by solving the above-mentioned challenges, an alternative solution namely the central difference predictive filter is presented in this chapter. This is achieved by using the Stirling polynomial interpolation technique (Froberg, 1969) to calculate the mean and covariance of the modeling error and the system state. This derivative-less method has the classical predictive filter's advantages of high-accuracy state estimation, robustness to modeling error, and low-computation implementation. Moreover, the estimated modeling error provided by this filter is able to capture the posterior mean accurately up to the 3rd order for nonlinear Gaussian system, with the estimation errors only introduced in the 4th and high orders. The estimated model error can capture the posterior mean accurately to the 2nd order for any nonlinearity. It is further proved that the estimation accuracy of the system states is better than the classical PF. With the application of this filtering approach to the problem of state estimation of the microsatellite attitude system using low precision sensors and in the presence of severe modeling error, simulation results are presented to verify this superiority.

5.2 Central difference predictive filter

It is known from Nørgaard et al. (2000) that the multi-dimensional Stirling interpolation of (2.48) can be used to approximate the posterior mean and covariance

Predictive Filtering for Microsatellite Control System. https://doi.org/10.1016/B978-0-12-821865-5.00019-X

of the modeling error $\boldsymbol{d}_k$. This technique provides engineers a method to outperform the PF not only in estimation accuracy of the modeling error but also in estimation accuracy of the system states. Hence, using the second-order Stirling interpolation technique (Froberg, 1969) to calculate the mean and the covariance of the system state distribution. A center difference predictive filter (CDPF) is presented in Algorithm 5.1 for estimating the states of the nonlinear system (2.32)–(2.33) subject to modeling error $\boldsymbol{d}_k$ and measurement noise $\boldsymbol{v}_k$, where the constant $h \in \mathbb{R}_+$ is the central difference step. Moreover, the numerical integral method is adopted to update the state estimation in Algorithm 5.1.

Algorithm 5.1: Central difference predictive filter.

Input Data: Initial estimation $\hat{\boldsymbol{x}}_0$, $\hat{\boldsymbol{y}}_0$, and measurement $\boldsymbol{y}_k$
Result: State estimation $\hat{\boldsymbol{x}}_k$
begin
Initializing the state and the covariance matrix as

$$\hat{\boldsymbol{x}}_0 = \mathbb{E}(\boldsymbol{x}_0)$$
$$\boldsymbol{P}_{\tilde{x},0} = \mathrm{Var}(\boldsymbol{x}_0) = \mathbb{E}\left((\boldsymbol{x}_0 - \hat{\boldsymbol{x}}_0)(\boldsymbol{x}_0 - \hat{\boldsymbol{x}}_0)^{\mathrm{T}}\right)$$

for $k = 1, 2, \cdots$ **do**

Updating the covariance square-root using the Cholesky factor

$$\boldsymbol{P}_{\tilde{x},k-1} = \boldsymbol{S}_{k-1}\boldsymbol{S}_{k-1}^{\mathrm{T}} = \sum_{i=1}^{n} \boldsymbol{\sigma}_{i,k-1}(\boldsymbol{\sigma}_{i,k-1})^{\mathrm{T}}$$

Updating the modeling error's estimation $\hat{\boldsymbol{d}}_{k-1}$ as

$$\hat{\boldsymbol{x}}_{k-1,i}^{+} = \hat{\boldsymbol{x}}_{k-1} + h\boldsymbol{\sigma}_{i,k-1},\ i = 1, 2, \cdots, n$$
$$\hat{\boldsymbol{x}}_{k-1,i}^{-} = \hat{\boldsymbol{x}}_{k-1} - h\boldsymbol{\sigma}_{i,k-1},\ i = 1, 2, \cdots, n$$
$$\boldsymbol{d}_{k-1}(\hat{\boldsymbol{x}}_{k-1,i}^{+}) = -\underline{\boldsymbol{Z}}_{k-1}(\hat{\boldsymbol{x}}_{k-1,i}^{+}, \Delta t) + \underline{\boldsymbol{h}}_{k-1}(\hat{\boldsymbol{x}}_{k-1,i}^{+})$$
$$\boldsymbol{d}_{k-1}(\hat{\boldsymbol{x}}_{k-1,i}^{-}) = -\underline{\boldsymbol{Z}}_{k-1}(\hat{\boldsymbol{x}}_{k-1,i}^{-}, \Delta t) + \underline{\boldsymbol{h}}_{k-1}(\hat{\boldsymbol{x}}_{k-1,i}^{-})$$
$$\boldsymbol{d}_{k-1}(\hat{\boldsymbol{x}}_{k-1}) = -\underline{\boldsymbol{Z}}_{k-1}(\hat{\boldsymbol{x}}_{k-1}, \Delta t) + \underline{\boldsymbol{h}}_{k-1}(\hat{\boldsymbol{x}}_{k-1})$$
$$\hat{\boldsymbol{d}}_{k-1} = \frac{h^2-n}{h^2}\boldsymbol{d}_{k-1}(\hat{\boldsymbol{x}}_{k-1}) + \frac{\sum_{i=1}^{n}(\boldsymbol{d}_{k-1}(\hat{\boldsymbol{x}}_{k-1,i}^{+}) + \boldsymbol{d}_{k-1}(\hat{\boldsymbol{x}}_{k-1,i}^{-}))}{2h^2}$$

Updating the state estimation

$$\hat{\boldsymbol{x}}_{k|k-1} = \frac{h^2-n}{h^2}\boldsymbol{f}(\hat{\boldsymbol{x}}_{k-1}) + \frac{1}{2h^2}\sum_{i=1}^{n}(\boldsymbol{f}(\hat{\boldsymbol{x}}_{k-1,i}^{+}) + \boldsymbol{f}(\hat{\boldsymbol{x}}_{k-1,i}^{-}))$$
$$\hat{\boldsymbol{x}}_k = \hat{\boldsymbol{x}}_{k|k-1} + \boldsymbol{g}(\hat{\boldsymbol{x}}_{k|k-1})\hat{\boldsymbol{d}}_{k-1}$$

Updating the state estimation error covariance

$$\boldsymbol{P}_{\tilde{x},k} = \frac{1}{4h^2}\sum_{i=1}^{n}\begin{pmatrix} \boldsymbol{f}(\hat{\boldsymbol{x}}_{k-1,i}^{+}) + \boldsymbol{d}_{k-1}(\hat{\boldsymbol{x}}_{k-1,i}^{+}) \\ -\boldsymbol{f}(\hat{\boldsymbol{x}}_{k-1,i}^{-}) - \boldsymbol{d}_{k-1}(\hat{\boldsymbol{x}}_{k-1,i}^{-}) \end{pmatrix}^2$$
$$+\frac{h^2-1}{4h^2}\sum_{i=1}^{n}\begin{pmatrix} \boldsymbol{f}(\hat{\boldsymbol{x}}_{k-1,i}^{+}) + \boldsymbol{f}(\hat{\boldsymbol{x}}_{k-1,i}^{-}) + \boldsymbol{d}_{k-1}(\hat{\boldsymbol{x}}_{k-1,i}^{+}) \\ -2\boldsymbol{f}(\hat{\boldsymbol{x}}_{k-1}) + \boldsymbol{d}_{k-1}(\hat{\boldsymbol{x}}_{k-1,i}^{-}) - 2\boldsymbol{d}_{k-1}(\hat{\boldsymbol{x}}_{k-1}) \end{pmatrix}^2$$

end

end

Remark 5.1. It is seen in Algorithm 5.1 that the CDPF is a straightforward application of Stirling's interpolation for the posterior statistics approximation to form a recursive filter. The formulas of the sigma-points as given in UPF are not explicitly presented for CDPF. Instead, the general expression of Stirling interpolation is established. In practical applications, following the expression established for UPF, the CDPF in Algorithm 5.1 can be re-summarized into the presentation of using the sigma-points. Hence, the CDPF can also be called as a sigma point predictive filtering approach.

5.3 Estimation performance of CDPF

In this section, the estimation performance of the modeling errors and the covariance guaranteed by the CDPF is rigorously analyzed in mathematics.

5.3.1 Estimation accuracy of modeling error

Let $(\boldsymbol{S}_k)_{\mathrm{CDPF}}$ be the Cholesky decomposition of the state estimation error covariance matrix $\boldsymbol{P}_{\tilde{x},k}$, *i.e.*,

$$\boldsymbol{P}_{\tilde{x},k} = \boldsymbol{S}_k \boldsymbol{S}_k^{\mathrm{T}} = \sum_{i=1}^{n} \boldsymbol{\sigma}_{i,k} \boldsymbol{\sigma}_{i,k}^{\mathrm{T}} \tag{5.1}$$

where $\boldsymbol{\sigma}_{i,k}$, $i = 1, 2, \cdots, n$, is the ith column of the matrix $\boldsymbol{S}_k$.

Using the Cholesky factor $(\boldsymbol{s}_k)_{\mathrm{CDPF}}$, the following linear transformation is first introduced to stochastically decouple the prior random variable $\boldsymbol{x}_k$.

$$\boldsymbol{c} = \boldsymbol{S}_k^{-1} \boldsymbol{x}_k \tag{5.2}$$

For this transformation, it follows that

$$\hat{\boldsymbol{c}} = \mathbb{E}(\boldsymbol{c}) = \boldsymbol{S}_k^{-1} \hat{\boldsymbol{x}}_k \tag{5.3}$$

$$\mathbb{E}((\boldsymbol{c} - \hat{\boldsymbol{c}})(\boldsymbol{c} - \hat{\boldsymbol{c}})^{\mathrm{T}}) = \boldsymbol{I}_n \tag{5.4}$$

It is seen in (5.3)–(5.4) that the transformation (5.2) stochastically decouples the variables in $\boldsymbol{x}_k$. The individual components of $\boldsymbol{c}$ become mutually uncorrelated (with unity variance). In correspondence, the random variable $\Delta \boldsymbol{c} = \boldsymbol{c} - \hat{\boldsymbol{c}}$ has a distribution of zero mean and covariance $\boldsymbol{I}_n$.

Let a nonlinear function $\tilde{\boldsymbol{Z}}(\boldsymbol{c})$ be defined as $\tilde{\boldsymbol{Z}}(\boldsymbol{c}) = \underline{\boldsymbol{Z}}(\boldsymbol{S}_k \boldsymbol{c}, \Delta t)$, we have

$$\tilde{\boldsymbol{Z}}(\boldsymbol{c}) = \underline{\boldsymbol{Z}}(\boldsymbol{S}_k \boldsymbol{c}, \Delta t) = \underline{\boldsymbol{Z}}(\boldsymbol{x}_k, \Delta t) \tag{5.5}$$

Define $\boldsymbol{\delta}_i$, $\boldsymbol{\mu}_i$, and $\boldsymbol{\delta}_i^2$ as the "partial" first-order difference and mean operators and "partial" second-order difference operator, respectively, then the following equations are valid.

$$\begin{aligned}\delta_i \tilde{\mathbf{Z}}(\hat{\mathbf{c}}) &= \tilde{\mathbf{Z}}\left(\hat{\mathbf{c}} + \frac{h}{2}\mathbf{e}_i\right) - \tilde{\mathbf{Z}}\left(\hat{\mathbf{c}} - \frac{h}{2}\mathbf{e}_i\right) \\ &= \underline{\mathbf{Z}}\left(\hat{\mathbf{x}}_k + \frac{h\boldsymbol{\sigma}_{i,k}}{2}, \Delta t\right) - \underline{\mathbf{Z}}\left(\hat{\mathbf{x}}_k - \frac{h\boldsymbol{\sigma}_{i,k}}{2}, \Delta t\right)\end{aligned} \tag{5.6}$$

$$\begin{aligned}\mu_i \tilde{\mathbf{Z}}(\hat{\mathbf{c}}) &= \frac{1}{2}\tilde{\mathbf{Z}}\left(\hat{\mathbf{c}} + \frac{h}{2}\mathbf{e}_i\right) + \frac{1}{2}\tilde{\mathbf{Z}}\left(\hat{\mathbf{c}} - \frac{h}{2}\mathbf{e}_i\right) \\ &= \frac{1}{2}\underline{\mathbf{Z}}\left(\hat{\mathbf{x}}_k + \frac{h\boldsymbol{\sigma}_{i,k}}{2}, \Delta t\right) + \frac{1}{2}\underline{\mathbf{Z}}\left(\hat{\mathbf{x}}_k - \frac{h\boldsymbol{\sigma}_{i,k}}{2}, \Delta t\right)\end{aligned} \tag{5.7}$$

$$\begin{aligned}\mu_i \delta_i \tilde{\mathbf{Z}}(\hat{\mathbf{c}}) &= \frac{\tilde{\mathbf{Z}}\left(\hat{\mathbf{c}} + h\mathbf{e}_i\right) + \tilde{\mathbf{Z}}\left(\hat{\mathbf{c}} - h\mathbf{e}_i\right)}{2} \\ &= \frac{1}{2}\underline{\mathbf{Z}}\left(\hat{\mathbf{x}}_k + h\boldsymbol{\sigma}_{i,k}, \Delta t\right) - \frac{1}{2}\underline{\mathbf{Z}}\left(\hat{\mathbf{x}}_k - h\boldsymbol{\sigma}_{i,k}, \Delta t\right)\end{aligned} \tag{5.8}$$

$$\begin{aligned}\delta_i^2 \tilde{\mathbf{Z}}(\hat{\mathbf{c}}) &= \tilde{\mathbf{Z}}\left(\hat{\mathbf{c}} + h\mathbf{e}_i\right) - \tilde{\mathbf{Z}}\left(\hat{\mathbf{c}} - h\mathbf{e}_i\right) - 2\tilde{\mathbf{Z}}(\hat{\mathbf{c}}) \\ &= \underline{\mathbf{Z}}\left(\hat{\mathbf{x}}_k + h\boldsymbol{\sigma}_{i,k}, \Delta t\right) + \underline{\mathbf{Z}}\left(\hat{\mathbf{x}}_k - h\boldsymbol{\sigma}_{i,k}, \Delta t\right) - 2\underline{\mathbf{Z}}(\hat{\mathbf{x}}_k, \Delta t)\end{aligned} \tag{5.9}$$

where $\mathbf{e}_i$ is the ith unit vector.

Approximating $\underline{\mathbf{Z}}(\mathbf{x}_k, \Delta t)$ with a second-order polynomial derived via the interpolation formula leads to

$$\underline{\mathbf{Z}}(\mathbf{x}_k, \Delta t) = \underline{\mathbf{Z}}(\mathbf{S}_k\mathbf{c}, \Delta t) = \tilde{\mathbf{Z}}(\mathbf{c}) \approx \tilde{\mathbf{Z}}(\hat{\mathbf{c}}) + \tilde{\mathbf{D}}_{\Delta c}\tilde{\mathbf{Z}} + \frac{\tilde{\mathbf{D}}^2_{\Delta c}\tilde{\mathbf{Z}}}{2!} \tag{5.10}$$

where $\tilde{\mathbf{D}}_{\Delta c}\tilde{\mathbf{Z}}$ and $\tilde{\mathbf{D}}^2_{\Delta c}\tilde{\mathbf{Z}}$ are the first and the second order central divided difference operators acting on $\underline{\mathbf{Z}}(\mathbf{x}_k, \Delta t)$, respectively; and they are given by

$$\tilde{\mathbf{D}}_{\Delta c}\tilde{\mathbf{Z}} = \frac{1}{h}\sum_{i=1}^{n} \Delta c_i \mu_i \delta_i \tilde{\mathbf{Z}}(\hat{\mathbf{c}}) \tag{5.11}$$

$$\tilde{\mathbf{D}}^2_{\Delta c}\tilde{\mathbf{Z}} = \frac{1}{h^2}\left(\sum_{i=1}^{n}(\Delta c_i)^2\delta_i^2 + \sum_{i=1}^{n}\sum_{j=1, i\neq j}^{n} \Delta c_i \Delta c_j (\mu_i\delta_i)(\mu_j\delta_j)\right)\tilde{\mathbf{Z}}(\hat{\mathbf{c}}) \tag{5.12}$$

Furthermore, similar Stirling interpolation formula can be applied to $\boldsymbol{h}(\mathbf{x}_k)$. Consequently, more accurate estimation of the mean and covariance of $\boldsymbol{d}_k$ can be obtained by approximating the function with a second-order polynomial derived with the interpolation formula. Then, (2.48) can be specified by

$$
\begin{aligned}
-\boldsymbol{d}_k &= \underline{\boldsymbol{Z}}(\boldsymbol{x}_k, \Delta t) + \underline{\boldsymbol{h}}(\boldsymbol{x}_k) \\
&= \tilde{\boldsymbol{Z}}(\boldsymbol{c}) + \tilde{\boldsymbol{h}}(\boldsymbol{c}) \\
&= \tilde{\boldsymbol{Z}}(\hat{\boldsymbol{c}} + \Delta \boldsymbol{c}) + \tilde{\boldsymbol{h}}(\hat{\boldsymbol{c}} + \Delta \boldsymbol{c}) \\
&\approx \tilde{\boldsymbol{Z}}(\hat{\boldsymbol{c}}) + \tilde{\boldsymbol{D}}_{\Delta c}\tilde{\boldsymbol{Z}} + \frac{\tilde{\boldsymbol{D}}^2_{\Delta c}\tilde{\boldsymbol{Z}}}{2!} + \tilde{\boldsymbol{h}}(\hat{\boldsymbol{c}}) + \tilde{\boldsymbol{D}}_{\Delta c}\tilde{\boldsymbol{h}} + \frac{\tilde{\boldsymbol{D}}^2_{\Delta c}\tilde{\boldsymbol{h}}}{2!} \\
&= \tilde{\boldsymbol{Z}}(\hat{\boldsymbol{c}}) + \tilde{\boldsymbol{h}}(\hat{\boldsymbol{c}}) + \frac{1}{h}\left(\sum_{i=1}^{n} \Delta c_i \mu_i \delta_i\right)(\tilde{\boldsymbol{Z}}(\hat{\boldsymbol{c}}) + \tilde{\boldsymbol{h}}(\hat{\boldsymbol{c}})) \\
&\quad + \frac{1}{2h^2}\sum_{i=1}^{n}(\Delta c_i)^2 \delta_i^2 (\tilde{\boldsymbol{Z}}(\hat{\boldsymbol{c}}) + \tilde{\boldsymbol{h}}(\hat{\boldsymbol{c}})) \\
&\quad + \frac{1}{2h^2}\sum_{i=1}^{n}\sum_{j=1, i\neq j}^{n} \Delta c_i \Delta c_j (\mu_i \delta_i)(\mu_j \delta_j)(\tilde{\boldsymbol{Z}}(\hat{\boldsymbol{c}}) + \tilde{\boldsymbol{h}}(\hat{\boldsymbol{c}}))
\end{aligned} \tag{5.13}
$$

where $\tilde{\boldsymbol{h}}(\boldsymbol{c}) = \underline{\boldsymbol{h}}(\boldsymbol{S}_k \boldsymbol{c}) = \underline{\boldsymbol{h}}(\boldsymbol{x}_k)$.

Based on the result in Nørgaard et al. (2000), the estimation of the modeling error provided by the CDPF algorithm can be written as

$$
\begin{aligned}
-(\hat{\boldsymbol{d}}_k)_{\mathrm{CDPF}} &= \tilde{\boldsymbol{Z}}(\boldsymbol{c}) + \tilde{\boldsymbol{h}}(\boldsymbol{c}) + \frac{1}{2h^2}\sum_{i=1}^{n}\delta_i^2 \tilde{\boldsymbol{Z}}(\hat{\boldsymbol{c}}) + \frac{1}{2h^2}\sum_{i=1}^{n}\delta_i^2 \tilde{\boldsymbol{h}}(\hat{\boldsymbol{c}}) \\
&= \tilde{\boldsymbol{Z}}(\boldsymbol{c}) + \frac{1}{2h^2}\sum_{i=1}^{n}(\tilde{\boldsymbol{Z}}(\hat{\boldsymbol{c}} + h\boldsymbol{e}_i) + \tilde{\boldsymbol{Z}}(\hat{\boldsymbol{c}} - h\boldsymbol{e}_i) - 2\tilde{\boldsymbol{Z}}(\hat{\boldsymbol{c}})) \\
&\quad + \tilde{\boldsymbol{h}}(\boldsymbol{c}) + \frac{1}{2h^2}\sum_{i=1}^{n}(\tilde{\boldsymbol{h}}(\hat{\boldsymbol{c}} + h\boldsymbol{e}_i) + \tilde{\boldsymbol{h}}(\hat{\boldsymbol{c}} - h\boldsymbol{e}_i) - 2\tilde{\boldsymbol{h}}(\hat{\boldsymbol{c}}))
\end{aligned} \tag{5.14}
$$

Substituting (5.6)–(5.9) into (5.14), the posterior mean approximation in terms of the prior statistics of $\boldsymbol{x}_k$ can be simplified as

$$
\begin{aligned}
-(\hat{\boldsymbol{d}}_k)_{\mathrm{CDPF}} &= \frac{1}{2h^2}\sum_{i=1}^{n}(\underline{\boldsymbol{Z}}(\hat{\boldsymbol{x}}_k + h\boldsymbol{\sigma}_{i,k}, \Delta t) + \underline{\boldsymbol{Z}}(\hat{\boldsymbol{x}}_k - h\boldsymbol{\sigma}_{i,k}, \Delta t)) \\
&\quad + \frac{1}{2h^2}\sum_{i=1}^{n}(\underline{\boldsymbol{h}}(\hat{\boldsymbol{x}}_k + h\boldsymbol{\sigma}_{i,k}) + \underline{\boldsymbol{h}}(\hat{\boldsymbol{x}}_k - h\boldsymbol{\sigma}_{i,k})) \\
&\quad + \frac{h^2 - n}{h^2}(\underline{\boldsymbol{Z}}(\hat{\boldsymbol{x}}_k, \Delta t) + \underline{\boldsymbol{h}}(\hat{\boldsymbol{x}}_k))
\end{aligned} \tag{5.15}
$$

The Taylor series expansion of the second and the third terms in the right-hand side of (5.15) are given by

$$\underline{Z}(\hat{x}_k + h\sigma_{i,k}, \Delta t) = \underline{Z}(\hat{x}_k, \Delta t) + \sum_{j=1}^{\infty} \frac{D_{\sigma_{i,k}}^{j} \underline{Z}(\hat{x}_k, \Delta t) h^j}{j!} \tag{5.16}$$

$$\underline{Z}(\hat{x}_k - h\sigma_{i,k}, \Delta t) = \underline{Z}(\hat{x}_k, \Delta t) + \sum_{j=1}^{\infty} \frac{(-1)^j D_{\sigma_{i,k}}^{j} \underline{Z}(\hat{x}_k, \Delta t) h^j}{j!} \tag{5.17}$$

Moreover, the Taylor series expansion can also be applied to $\underline{h}(\hat{x}_k + h\sigma_{i,k})$ and $\underline{h}(\hat{x}_k - h\sigma_{i,k})$ to obtain

$$\underline{h}(\hat{x}_k + h\sigma_{i,k}, \Delta t) = \underline{h}(\hat{x}_k) + \sum_{j=1}^{\infty} \frac{D_{\sigma_{i,k}}^{j} \underline{h}(\hat{x}_k) h^j}{j!} \tag{5.18}$$

$$\underline{h}(\hat{x}_k - h\sigma_{i,k}, \Delta t) = \underline{h}(\hat{x}_k) + \sum_{j=1}^{\infty} \frac{(-1)^j D_{\sigma_{i,k}}^{j} \underline{h}(\hat{x}_k) h^j}{j!} \tag{5.19}$$

Then, inserting (5.16)–(5.19) into (5.15) and simplifying it, it is ready to analyze the estimation accuracy of the modeling error $(-d_k)_{\mathrm{CDPF}}$ as

$$\begin{aligned} -(\hat{d}_k)_{\mathrm{CDPF}} = {} & \underline{Z}(\hat{x}_k, \Delta t) + \underline{h}(\hat{x}_k) + \frac{(\nabla^{\mathrm{T}} P_{\tilde{x},k} \nabla) \underline{Z}(x, \Delta t)|_{x=\hat{x}_k}}{2!} \\ & + \frac{(\nabla^{\mathrm{T}} P_{\tilde{x},k} \nabla) \underline{h}(x)|_{x=\hat{x}_k}}{2!} + \sum_{i=1}^{n} \sum_{j=1}^{\infty} \frac{D_{\sigma_{i,k}}^{2(j+1)} \underline{Z}(\hat{x}_k) h^{2(j+1)}}{(2(j+1))!} \\ & + \sum_{i=1}^{n} \sum_{j=1}^{\infty} \frac{D_{\sigma_{i,k}}^{2(j+1)} \underline{h}(\hat{x}_k) h^{2(j+1)}}{(2(j+1))!} \end{aligned} \tag{5.20}$$

According to Stirling's interpolation, the covariance matrix of the modeling error obtained from the CDPF can be calculated as

$$\begin{aligned} (P_{d,k})_{\mathrm{CDPF}} &= \mathbb{E}(d_{k,e} d_{k,e}^{\mathrm{T}}) \\ &= \mathbb{E}(((d_k)_{\mathrm{CDPF}} - \tilde{d}_k(\hat{c}))((d_k)_{\mathrm{CDPF}} - \tilde{d}_k(\hat{c}))^{\mathrm{T}}) \\ &\quad - \mathbb{E}((d_k)_{\mathrm{CDPF}} - \tilde{d}_k(\hat{c}))\mathbb{E}((d_k)_{\mathrm{CDPF}} - \tilde{d}_k(\hat{c}))^{\mathrm{T}} \end{aligned} \tag{5.21}$$

where $d_{k,e} = (d_k)_{\mathrm{CDPF}} - \mathbb{E}((d_k)_{\mathrm{CDPF}})$ and $\tilde{d}_k(\hat{c}) = d_k(S_k^{-1}\hat{c}) = d_k(\hat{x}_k)$.

Inserting (5.10) into (5.21), it leaves (5.21) as

$$
\begin{aligned}
(\boldsymbol{P}_{d,k})_{\text{CDPF}} &= \mathbb{E}\left(\left(\tilde{\boldsymbol{D}}_{\Delta c}\tilde{\boldsymbol{d}}_k + \frac{\tilde{\boldsymbol{D}}^2_{\Delta c}\tilde{\boldsymbol{d}}_k}{2!}\right)\left(\tilde{\boldsymbol{D}}_{\Delta c}\tilde{\boldsymbol{d}}_k + \frac{\tilde{\boldsymbol{D}}^2_{\Delta c}\tilde{\boldsymbol{d}}_k}{2!}\right)^{\mathrm{T}}\right) \\
&\quad - \mathbb{E}\left(\tilde{\boldsymbol{D}}_{\Delta c}\tilde{\boldsymbol{d}}_k + \frac{\tilde{\boldsymbol{D}}^2_{\Delta c}\tilde{\boldsymbol{d}}_k}{2!}\right)\mathbb{E}\left(\tilde{\boldsymbol{D}}_{\Delta c}\tilde{\boldsymbol{d}}_k + \frac{\tilde{\boldsymbol{D}}^2_{\Delta c}\tilde{\boldsymbol{d}}_k}{2!}\right)^{\mathrm{T}} \\
&= \mathbb{E}((\tilde{\boldsymbol{D}}_{\Delta c}\tilde{\boldsymbol{d}}_k)(\tilde{\boldsymbol{D}}_{\Delta c}\tilde{\boldsymbol{d}}_k)^{\mathrm{T}}) + \frac{1}{4}\mathbb{E}((\tilde{\boldsymbol{D}}^2_{\Delta c}\tilde{\boldsymbol{d}}_k)(\tilde{\boldsymbol{D}}^2_{\Delta c}\tilde{\boldsymbol{d}}_k)^{\mathrm{T}}) \\
&\quad - \frac{1}{4}\mathbb{E}(\tilde{\boldsymbol{D}}^2_{\Delta c}\tilde{\boldsymbol{d}}_k)\mathbb{E}(\tilde{\boldsymbol{D}}^2_{\Delta c}\tilde{\boldsymbol{d}}_k)^{\mathrm{T}}
\end{aligned}
\tag{5.22}
$$

Applying the result in Ito and Xiong (2000) and Simandl and Dunik (2009), we can further simplify $(\boldsymbol{P}_{d,k})_{\text{CDPF}}$ as

$$
\begin{aligned}
(\boldsymbol{P}_{d,k})_{\text{CDPF}} &= \frac{1}{h^2}\sum_{i=1}^{n}(\mu_i\delta_i\tilde{\boldsymbol{d}}_k(\hat{\boldsymbol{c}}))(\mu_i\delta_i\tilde{\boldsymbol{d}}_k(\hat{\boldsymbol{c}}))^{\mathrm{T}} \\
&\quad + \frac{h^2-1}{4h^4}\sum_{i=1}^{n}(\delta_i{}^2\tilde{\boldsymbol{d}}_k(\hat{\boldsymbol{c}}))(\delta_i{}^2\tilde{\boldsymbol{d}}_k(\hat{\boldsymbol{c}}))^{\mathrm{T}} \\
&= \frac{h^2-1}{4h^4}\sum_{i=1}^{n}(\tilde{\boldsymbol{d}}_{kh} - 2\tilde{\boldsymbol{d}}_k(\hat{\boldsymbol{c}}))(\tilde{\boldsymbol{d}}_{kh} - 2\tilde{\boldsymbol{d}}_k(\hat{\boldsymbol{c}}))^{\mathrm{T}} \\
&\quad + \frac{1}{4h^2}\sum_{i=1}^{n}\tilde{\boldsymbol{d}}_{kh}\tilde{\boldsymbol{d}}_{kh}^{\mathrm{T}} \\
&= \frac{1}{4h^2}\sum_{i=1}^{n}(\boldsymbol{d}_k(\hat{\boldsymbol{x}}^+_{k,i}) - \boldsymbol{d}_k(\hat{\boldsymbol{x}}^-_{k,i}))(\boldsymbol{d}_k(\hat{\boldsymbol{x}}^+_{k,i}) - \boldsymbol{d}_k(\hat{\boldsymbol{x}}^-_{k,i}))^{\mathrm{T}} \\
&\quad + \frac{h^2-1}{4h^4}\sum_{i=1}^{n}\boldsymbol{d}^+_k(\boldsymbol{d}^+_k)^{\mathrm{T}}
\end{aligned}
\tag{5.23}
$$

where $\tilde{\boldsymbol{d}}_{kh} = \tilde{\boldsymbol{d}}_k(\hat{\boldsymbol{c}} + h\boldsymbol{e}_i) - \tilde{\boldsymbol{d}}_k(\hat{\boldsymbol{c}} - h\boldsymbol{e}_i)$ and $\boldsymbol{d}^+_k = \boldsymbol{d}_k(\hat{\boldsymbol{x}}^+_{k,i}) + \boldsymbol{d}_k(\hat{\boldsymbol{x}}^-_{k,i}) - 2\boldsymbol{d}_k(\hat{\boldsymbol{x}}_k)$.

Based on (5.16)–(5.19), one has

$$
\boldsymbol{d}_k(\hat{\boldsymbol{x}}^+_{k,i}) - \boldsymbol{d}_k(\hat{\boldsymbol{x}}^-_{k,i}) = -2\sum_{j=1}^{\infty}\frac{\boldsymbol{D}^{2j-1}_{\sigma_{i,k}}(\underline{\boldsymbol{Z}} + \underline{\boldsymbol{h}})}{(2j-1)!}h^{2j-1} \tag{5.24}
$$

$$
\boldsymbol{d}^+_k = -2\sum_{j=1}^{\infty}\frac{\boldsymbol{D}^{2j}_{\sigma_{i,k}}(\underline{\boldsymbol{Z}} + \underline{\boldsymbol{h}})}{(2j-1)!}h^{2j} \tag{5.25}
$$

Then, substituting (5.24)–(5.25) into (5.23) results in

$$
\begin{aligned}
(\boldsymbol{P}_{d,k})_{\text{CDPF}} = {} & \boldsymbol{G}_{\underline{\boldsymbol{h}}} \boldsymbol{P}_{\tilde{x},k} \boldsymbol{G}_{\underline{\boldsymbol{h}}}^{\mathrm{T}} + \boldsymbol{G}_{\underline{\boldsymbol{Z}}} \boldsymbol{P}_{\tilde{x},k} \boldsymbol{G}_{\underline{\boldsymbol{Z}}}^{\mathrm{T}} + \boldsymbol{G}_{\underline{\boldsymbol{h}}} \boldsymbol{P}_{\tilde{x},k} \boldsymbol{G}_{\underline{\boldsymbol{Z}}}^{\mathrm{T}} \\
& + \boldsymbol{G}_{\underline{\boldsymbol{Z}}} \boldsymbol{P}_{\tilde{x},k} \boldsymbol{G}_{\underline{\boldsymbol{h}}}^{\mathrm{T}} + \boldsymbol{\Sigma}_1 - \boldsymbol{\Sigma}_2 \\
& - \mathbb{E}\left(\frac{\boldsymbol{D}_{\Delta x}^2 (\underline{\boldsymbol{Z}} + \underline{\boldsymbol{h}})}{i!} \right) \mathbb{E}\left(\left(\frac{\boldsymbol{D}_{\Delta x}^2 (\underline{\boldsymbol{Z}} + \underline{\boldsymbol{h}})}{2!} \right)^{\mathrm{T}} \right)
\end{aligned}
\tag{5.26}
$$

where

$$
\boldsymbol{\Sigma}_1 = \underbrace{\sum_{m=1}^{n} \sum_{i=1}^{\infty} \sum_{j=1}^{\infty} \frac{h^{i+j-2}}{i!j!} (\boldsymbol{D}_{\sigma_{m,k}}^{i} (\underline{\boldsymbol{Z}} + \underline{\boldsymbol{h}}))(\boldsymbol{D}_{\sigma_{m,k}}^{j} (\underline{\boldsymbol{Z}} + \underline{\boldsymbol{h}}))^{\mathrm{T}}}_{\text{condition\_1}} \tag{5.27}
$$

$$
\boldsymbol{\Sigma}_2 = \underbrace{\sum_{m=1}^{n} \sum_{i=1}^{\infty} \sum_{j=1}^{\infty} \frac{h^{2(i+j-2)}}{(2i)!(2j)!} (\boldsymbol{D}_{\sigma_{m,k}}^{2i} (\underline{\boldsymbol{Z}} + \underline{\boldsymbol{h}}))(\boldsymbol{D}_{\sigma_{m,k}}^{2j} (\underline{\boldsymbol{Z}} + \underline{\boldsymbol{h}}))^{\mathrm{T}}}_{\text{condition\_2}} \tag{5.28}
$$

Theorem 5.1. *Consider the nonlinear discrete system* (2.32)–(2.33) *with the modeling error* $\boldsymbol{d}_k$ *and the measurement noise* $\boldsymbol{v}_k$, *let its states be estimated by using the CDPF in Algorithm 5.1. If system* (2.32)–(2.33) *is Gaussian, then the CDPF can capture the posterior mean of the modeling error accurately to* 3rd *order with the estimation errors only introduced in the* 4th *and higher orders. Moreover, the CDPF completely capture the posterior mean of the modeling error accurately to* 2nd *order for any nonlinear system* (2.32)–(2.33). *The CDPF has higher state estimation accuracy than the classical PF in Algorithm 2.1.*

Proof. Comparing (2.63)–(2.64) with (5.20)–(5.25), the conclusion in Theorem 5.1 can be directly proved. □

Remark 5.2. The UPF and the CDPF guarantee almost the same estimation accuracy for the modeling error $\boldsymbol{d}_k$. The only difference is seen in the terms having more than four order by comparing (5.20) with (4.25).

Remark 5.3. The choice of h affects the approximation accuracy of the mean and the higher-order terms only. Therefore, tuning h appropriately can govern the posterior covariance of the estimation to approximate the real modeling error accurately.

5.3.2 Estimation accuracy of system states

In accordance with (2.67)–(2.68), the states estimation of the CDPF can be established as

$$
\begin{aligned}
(\hat{\boldsymbol{x}}_{k+1})_{\mathrm{CDPF}} =& (\hat{\boldsymbol{x}}_k)_{\mathrm{CDPF}} + \Delta t (\boldsymbol{f}(\hat{\boldsymbol{x}}_k))_{\mathrm{CDPF}} + \Delta t (\boldsymbol{\eta}(\hat{\boldsymbol{x}}_k))_{\mathrm{CDPF}} \\
&+ \boldsymbol{\mu}_{k,\mathrm{CDPF}}
\end{aligned}
\tag{5.29}
$$

where $\boldsymbol{\mu}_{k,\mathrm{CDPF}} = \boldsymbol{\mu}((\hat{\boldsymbol{x}}_k)_{\mathrm{CDPF}}, (\hat{\boldsymbol{d}}_k)_{\mathrm{CDPF}})$.

Using (5.20), the terms $(\boldsymbol{f}(\hat{\boldsymbol{x}}_k))_{\mathrm{CDPF}}$ and $(\boldsymbol{\eta}(\hat{\boldsymbol{x}}_k))_{\mathrm{CDPF}}$ are determined as

$$
\begin{aligned}
(\boldsymbol{f}(\hat{\boldsymbol{x}}_k))_{\mathrm{CDPF}} =& \boldsymbol{f}((\hat{\boldsymbol{x}}_k)_{\mathrm{CDPF}}) + \frac{(\nabla^{\mathrm{T}} \boldsymbol{P}_{\tilde{x},k} \nabla) \boldsymbol{f}(\boldsymbol{x})|_{\boldsymbol{x}=(\hat{\boldsymbol{x}}_k)_{\mathrm{CDPF}}}}{2!} \\
&+ \sum_{j=1}^{\infty} \frac{\boldsymbol{D}_{\sigma i,k}^{2(j+1)} \boldsymbol{f}}{(2(j+1))!} h^{2j}
\end{aligned}
\tag{5.30}
$$

$$
\begin{aligned}
(\boldsymbol{\eta}(\hat{\boldsymbol{x}}_k))_{\mathrm{CDPF}} =& \boldsymbol{\eta}((\hat{\boldsymbol{x}}_k)_{\mathrm{CDPF}}) + \frac{(\nabla^{\mathrm{T}} \boldsymbol{P}_{\tilde{x},k} \nabla) \boldsymbol{\eta}(\boldsymbol{x})|_{\boldsymbol{x}=(\hat{\boldsymbol{x}}_k)_{\mathrm{CDPF}}}}{2!} \\
&+ \sum_{j=1}^{\infty} \frac{\boldsymbol{D}_{\sigma i,k}^{2(j+1)} \boldsymbol{\eta}}{(2(j+1))!} h^{2j}
\end{aligned}
\tag{5.31}
$$

Then, it leaves the state estimation error of the CDPF as

$$
\begin{aligned}
(\tilde{\boldsymbol{x}}_{k+1})_{\mathrm{CDPF}} &= \boldsymbol{x}_{k+1} - (\hat{\boldsymbol{x}}_{k+1})_{\mathrm{CDPF}} \\
&= (\tilde{\boldsymbol{x}}_k)_{\mathrm{CDPF}} + \Delta t \left(\boldsymbol{G}_f (\tilde{\boldsymbol{x}}_k)_{\mathrm{CDPF}} + \boldsymbol{G}_\eta (\tilde{\boldsymbol{x}}_k)_{\mathrm{CDPF}} \right) \\
&\quad + \bar{\boldsymbol{\mu}}_k (\boldsymbol{x}_k, \boldsymbol{d}_k, (\hat{\boldsymbol{x}}_k)_{\mathrm{CDPF}}, (\hat{\boldsymbol{d}}_k)_{\mathrm{CDPF}}) + (\boldsymbol{s}_k)_{\mathrm{CDPF}} \\
&= \boldsymbol{A}_k (\tilde{\boldsymbol{x}}_k)_{\mathrm{CDPF}} + (\bar{\boldsymbol{\mu}}_k)_{\mathrm{CDPF}} + (\boldsymbol{s}_k)_{\mathrm{CDPF}}
\end{aligned}
\tag{5.32}
$$

where

$$
\boldsymbol{A}_k = \boldsymbol{I}_n + \Delta t \boldsymbol{G}_f + \Delta t \boldsymbol{G}_\eta
\tag{5.33}
$$

$$
\begin{aligned}
(\bar{\boldsymbol{\mu}}_k)_{\mathrm{CDPF}} &= \bar{\boldsymbol{\mu}}_k (\boldsymbol{x}_k, \boldsymbol{d}_k, (\hat{\boldsymbol{x}}_k)_{\mathrm{CDPF}}, (\hat{\boldsymbol{d}}_k)_{\mathrm{CDPF}}) \\
&= \boldsymbol{\mu}(\boldsymbol{x}_k, \boldsymbol{d}_k) \boldsymbol{v}_k - \boldsymbol{\mu}(\hat{\boldsymbol{x}}_k, \hat{\boldsymbol{d}}_k) - \boldsymbol{\varepsilon}(\bar{\boldsymbol{B}}((\hat{\boldsymbol{x}}_k)_{\mathrm{CDPF}})) \boldsymbol{v}_k \\
&\quad + \Delta t (\Delta \boldsymbol{f}(\boldsymbol{x}_k, (\hat{\boldsymbol{x}}_k)_{\mathrm{CDPF}}) \\
&\quad + \Delta \boldsymbol{\eta}(\boldsymbol{x}_k, \boldsymbol{d}_k, (\hat{\boldsymbol{x}}_k)_{\mathrm{CDPF}}, (\hat{\boldsymbol{d}}_k)_{\mathrm{CDPF}}))
\end{aligned}
\tag{5.34}
$$

$$
\Delta \boldsymbol{f}(\boldsymbol{x}_k, (\hat{\boldsymbol{x}}_k)_{\mathrm{CDPF}}) = \sum_{i=2}^{\infty} \frac{1}{i!} \boldsymbol{D}_{\hat{x}_k}^{i} \boldsymbol{f} - \sum_{j=1}^{n} \sum_{i=1}^{\infty} \frac{\boldsymbol{D}_{\sigma j,k}^{2i} \boldsymbol{f}}{(2i)!} h^i
\tag{5.35}
$$

$$
\Delta \boldsymbol{\eta}(\boldsymbol{x}_k, \boldsymbol{d}_k, (\hat{\boldsymbol{x}}_k)_{\mathrm{CDPF}}, (\hat{\boldsymbol{d}}_k)_{\mathrm{CDPF}}) = \sum_{i=2}^{\infty} \frac{1}{i!} \boldsymbol{D}_{\hat{x}_k}^{i} \boldsymbol{\eta} - \sum_{j=1}^{n} \sum_{i=1}^{\infty} \frac{\boldsymbol{D}_{\sigma j,k}^{2i} \boldsymbol{\eta}}{(2i)!} h^i
\tag{5.36}
$$

$$(\boldsymbol{s}_k)_{\mathrm{CDPF}} = -\Delta t \bar{\boldsymbol{B}}((\hat{\boldsymbol{x}}_k)_{\mathrm{CDPF}})\boldsymbol{\nu}_k = -(\boldsymbol{B}_k)_{\mathrm{CDPF}}\boldsymbol{\nu}_k \tag{5.37}$$

According to the definition of the covariance, the covariance of the state estimation error can be computed as

$$(\boldsymbol{P}_{\tilde{x},k+1})_{\mathrm{CDPF}} = \frac{1}{4h^2}\sum_{i=1}^{n}\boldsymbol{\Omega}_1\boldsymbol{\Omega}_1^{\mathrm{T}} + \frac{h^2-1}{4h^4}\sum_{i=1}^{n}\boldsymbol{\Omega}_2\boldsymbol{\Omega}_2^{\mathrm{T}} \tag{5.38}$$

where $\boldsymbol{\Omega}_1 = (\hat{\boldsymbol{x}}_{k+1})_{\mathrm{CDPF}}((\hat{\boldsymbol{x}}_k)_{\mathrm{CDPF}} + h\boldsymbol{\sigma}_{i,k}) - (\hat{\boldsymbol{x}}_{k+1})_{\mathrm{CDPF}}((\hat{\boldsymbol{x}}_k)_{\mathrm{CDPF}} - h\boldsymbol{\sigma}_{i,k})$ and $\boldsymbol{\Omega}_2 = (\hat{\boldsymbol{x}}_{k+1})_{\mathrm{CDPF}}((\hat{\boldsymbol{x}}_k)_{\mathrm{CDPF}} + h\boldsymbol{\sigma}_{i,k}) + (\hat{\boldsymbol{x}}_{k+1})_{\mathrm{CDPF}}((\hat{\boldsymbol{x}}_k)_{\mathrm{CDPF}} - h\boldsymbol{\sigma}_{i,k}) - 2(\hat{\boldsymbol{x}}_{k+1})_{\mathrm{CDPF}}(\hat{\boldsymbol{x}}_k)_{\mathrm{CDPF}}$. Since the following equations always hold

$$\boldsymbol{\Omega}_1 = 2h\boldsymbol{\sigma}_{i,k} + 2\boldsymbol{D}_{\boldsymbol{\sigma}_{i,k}}(\boldsymbol{f}+\boldsymbol{\eta})h + \frac{2\boldsymbol{D}^3_{\boldsymbol{\sigma}_{j,k}}(\boldsymbol{f}+\boldsymbol{\eta})}{3!}h^3 + \cdots + \boldsymbol{\mu}_{k1} \tag{5.39}$$

$$\boldsymbol{\Omega}_2 = \frac{2\boldsymbol{D}^2_{\boldsymbol{\sigma}_{j,k}}(\boldsymbol{f}+\boldsymbol{\eta})}{2!}h^2 + \frac{2\boldsymbol{D}^4_{\boldsymbol{\sigma}_{j,k}}(\boldsymbol{f}+\boldsymbol{\eta})}{4!}h^4 + \cdots + \boldsymbol{\mu}_{k2} \tag{5.40}$$

$$\boldsymbol{\mu}_{k1} = \boldsymbol{\mu}_k((\hat{\boldsymbol{x}}_k)_{\mathrm{CDPF}} + h\boldsymbol{\sigma}_{i,k}) - \boldsymbol{\mu}_k((\hat{\boldsymbol{x}}_k)_{\mathrm{CDPF}} - h\boldsymbol{\sigma}_{i,k}) \tag{5.41}$$

$$\boldsymbol{\mu}_{k2} = \boldsymbol{\mu}_k((\hat{\boldsymbol{x}}_k)_{\mathrm{CDPF}} + h\boldsymbol{\sigma}_{i,k}) + \boldsymbol{\mu}_k((\hat{\boldsymbol{x}}_k)_{\mathrm{CDPF}} - h\boldsymbol{\sigma}_{i,k}) - 2\boldsymbol{\mu}_k(\hat{\boldsymbol{x}}_k)_{\mathrm{CDPF}} \tag{5.42}$$

From (5.39)–(5.42), we can compute (5.38) as

$$\begin{aligned}(\boldsymbol{P}_{\tilde{x},k+1})_{\mathrm{CDPF}} = &\sum_{i=1}^{n}\boldsymbol{\Omega}_3\boldsymbol{\Omega}_3^{\mathrm{T}} + (\Delta t)^2\sum_{i=1}^{n}\boldsymbol{\Omega}_4\boldsymbol{\Omega}_4^{\mathrm{T}} - (\Delta t)^2\sum_{i=1}^{n}\boldsymbol{\Omega}_5\boldsymbol{\Omega}_5^{\mathrm{T}} \\ &+ \bar{\boldsymbol{\psi}}(\boldsymbol{\nu}_k) + \bar{\boldsymbol{o}}(\boldsymbol{\mu}_{k,1}, \boldsymbol{\mu}_{k,2})\end{aligned} \tag{5.43}$$

where $\bar{\boldsymbol{o}}(\boldsymbol{\mu}_{k1}, \boldsymbol{\mu}_{k2})$ is the sum of the polynomial of $\boldsymbol{\mu}_{k1}, \boldsymbol{\mu}_{k2}$, as well as other polynomials, $\bar{\boldsymbol{\psi}}(\boldsymbol{\nu}_k)$ is the sum of the polynomial of $\boldsymbol{\nu}_k$, $\boldsymbol{\Omega}_3 = \boldsymbol{\sigma}_{j,k} + \Delta t\boldsymbol{\Omega}_6$, and

$$\boldsymbol{\Omega}_4 = \frac{\boldsymbol{D}^2_{\boldsymbol{\sigma}_{j,k}}(\boldsymbol{f}+\boldsymbol{\eta})}{2!}h + \frac{\boldsymbol{D}^4_{\boldsymbol{\sigma}_{j,k}}(\boldsymbol{f}+\boldsymbol{\eta})}{4!}h^3 + \cdots \tag{5.44}$$

$$\boldsymbol{\Omega}_5 = \frac{\boldsymbol{D}^2_{\boldsymbol{\sigma}_{j,k}}(\boldsymbol{f}+\boldsymbol{\eta})}{2!}h + \frac{\boldsymbol{D}^4_{\boldsymbol{\sigma}_{j,k}}(\boldsymbol{f}+\boldsymbol{\eta})}{4!}h^2 + \cdots \tag{5.45}$$

$$\boldsymbol{\Omega}_6 = \boldsymbol{D}_{\boldsymbol{\sigma}_{i,k}}(\boldsymbol{f}+\boldsymbol{\eta}) + \frac{\boldsymbol{D}^3_{\boldsymbol{\sigma}_{j,k}}(\boldsymbol{f}+\boldsymbol{\eta})}{3!}h^2 + \cdots \tag{5.46}$$

It is further obtained from the above analysis that

$$\sum_{i=1}^{n}\sigma_{j,k}\Omega_6^{\mathrm{T}} = P_{\tilde{x},k}G_f^{\mathrm{T}} + P_{\tilde{x},k}G_\eta^{\mathrm{T}} + \sum_{i=1}^{n}\sum_{j=1}^{\infty}\frac{h^{2j}\sigma_{j,k}D_{\sigma_{j,k}}^{2j+1}(f+\eta)}{(2j+1)!} \tag{5.47}$$

$$\begin{aligned}&\sum_{i=1}^{n}\Omega_6\Omega_6^{\mathrm{T}} + \sum_{i=1}^{n}\Omega_4\Omega_4^{\mathrm{T}}\\ &= G_fP_{\tilde{x},k}G_f^{\mathrm{T}} + G_fP_{\tilde{x},k}G_\eta^{\mathrm{T}} + G_\eta P_{\tilde{x},k}G_\eta^{\mathrm{T}} + G_\eta P_{\tilde{x},k}G_f^{\mathrm{T}}\\ &\quad + \underbrace{\sum_{i=1}^{n}\sum_{l=1}^{\infty}\sum_{j=1}^{\infty}\frac{h^{l+j-2}}{l!j!}(D_{\sigma_{j,k}}^{l}(f+\eta))(D_{\sigma_{j,k}}^{j}(f+\eta))^{\mathrm{T}}}_{\text{condition\_1}}\end{aligned} \tag{5.48}$$

$$\begin{aligned}\sum_{i=1}^{n}\Omega_5\Omega_5^{\mathrm{T}} &= \mathbb{E}\left(\frac{D_{\Delta x}^2(f+\eta)}{2!}\right)\mathbb{E}\left(\frac{D_{\Delta x}^2(f+\eta)}{2!}\right)^{\mathrm{T}}\\ &\quad + \underbrace{\sum_{i=1}^{n}\sum_{l=1}^{\infty}\sum_{j=1}^{\infty}\frac{h^{2(l+j-2)}}{(2l)!(2j)!}(D_{\sigma_{j,k}}^{2l}(f+\eta))(D_{\sigma_{j,k}}^{2j}(f+\eta))^{\mathrm{T}}}_{\text{condition\_2}}\end{aligned} \tag{5.49}$$

Therefore, inserting (5.47)–(5.49) into (5.43) leads to

$$(P_{\tilde{x},k+1})_{\text{CDPF}} = A_k(P_{\tilde{x},k})_{\text{CDPF}}A_k^{\mathrm{T}} - K_kK_k^{\mathrm{T}} + Q_k \tag{5.50}$$

where $Q_k = \tilde{\Sigma}_1 + \tilde{\Sigma}_2 + \bar{o}_k + \bar{\psi}_k$ and

$$K_k = \Delta t\mathbb{E}\left(\frac{D_{\Delta x}^2 f}{2!}\right) + \Delta t\mathbb{E}\left(\frac{D_{\Delta x}^2\eta}{2!}\right) \tag{5.51}$$

$$\tilde{\Sigma}_1 = \Delta t\sum_{i=1}^{n}\sum_{j=1}^{\infty}\frac{h^{2j}}{(2j+1)!}(\sigma_{j,k}D_{\sigma_{i,k}}^{2j+1}(f+\eta) + D_{\sigma_{i,k}}^{2j+1}(f+\eta)\sigma_{i,k}) \tag{5.52}$$

$$\begin{aligned}\tilde{\Sigma}_2 &= \underbrace{(\Delta t)^2\sum_{i=1}^{n}\sum_{l=1}^{\infty}\sum_{j=1}^{\infty}\frac{h^{(l+j-2)}}{l!j!}D_{\sigma_{i,k}}^{l}(f+\eta)(D_{\sigma_{i,k}}^{j}(f+\eta))^{\mathrm{T}}}_{\text{condition\_1}}\\ &\quad \underbrace{-(\Delta t)^2\sum_{i=1}^{n}\sum_{l=1}^{\infty}\sum_{j=1}^{\infty}\frac{h^{2(l+j-2)}}{(2l)!(2j)!}(D_{\sigma_{i,k}}^{2l}(f+\eta))(D_{\sigma_{i,k}}^{2j}(f+\eta))^{\mathrm{T}}}_{\text{condition\_2}}\end{aligned} \tag{5.53}$$

In (5.50), a special matrix $\boldsymbol{Q}_k$ is introduced. If the initial value satisfies $(\boldsymbol{P}_{\tilde{x},0})_{\text{CDPF}} = \mathbb{E}(\tilde{\boldsymbol{x}}_0\tilde{\boldsymbol{x}}_0^{\text{T}})$, then $(\boldsymbol{P}_{\tilde{x},k})_{\text{CDPF}}$ calculated from (5.50) will approximate the actual value. That is $(\boldsymbol{P}_{\tilde{x},k})_{\text{CDPF}} \approx \mathbb{E}(\tilde{\boldsymbol{x}}_k\tilde{\boldsymbol{x}}_k^{\text{T}})$. Hence, (5.50) can be used to estimate the covariance of the state estimation error. Moreover, the covariance of the traditional PF has been proved to be stable in Li and Zhang (2006). Therefore, the proposed CDPF is also stable.

Theorem 5.2. *For the nonlinear discrete system* (2.32)–(2.33), *if the CDPF is applied to estimate its states with the initial value satisfying* $(\boldsymbol{P}_{\tilde{x},0})_{CDPF} = (\boldsymbol{P}_{\tilde{x},0})$, *then* $(\boldsymbol{P}_{\tilde{x},k})_{CDPF} \leq (\boldsymbol{P}_{\tilde{x},k})_{PF}$ *is guaranteed for every* $k \geq 0$. *The covariance of the CDPF has a faster convergence rate than the classical PF. Better filtering performance is achieved by the CDPF than the classical PF.*

Proof. Comparing (2.74) with (5.50), one can easily prove the conclusion in this theorem. □

5.4 Stochastic stability of CDPF

In this section, the stability of achieved by the CDPF is rigorously proved.

5.4.1 Boundedness analysis of state estimation covariance

Theorem 5.3. *Let the CDPF presented in Algorithm 5.1 be applied to the state estimation problem of the nonlinear stochastic system described by* (2.32) *and* (2.33), *for every* $k \geq 0$, *if there exist positive real numbers* $0 < z \leq 1$, $\bar{k} > 0$, $\bar{r} \geq \underline{r} \geq 0$, *and* $\bar{q} \geq \underline{q} > 0$ *satisfying*

$$\mathbf{0} \leq \boldsymbol{A}_k\boldsymbol{A}_k^{\text{T}} \leq (1-z)\boldsymbol{I}_n \tag{5.54}$$

$$\mathbf{0} \leq \frac{1}{4}(\Delta t)^2\boldsymbol{\Upsilon}\boldsymbol{\Upsilon}^{\text{T}} \leq \bar{k}\boldsymbol{I}_n \tag{5.55}$$

$$\underline{r}\boldsymbol{I}_n \leq \boldsymbol{R}_k \leq \bar{r}\boldsymbol{I}_n \tag{5.56}$$

$$\underline{q}\boldsymbol{I}_n \leq \boldsymbol{Q}_k \leq \bar{q}\boldsymbol{I}_n \tag{5.57}$$

and the initial $(\boldsymbol{P}_{\tilde{x},0})_{CDPF}$ *satisfies* $\underline{p}\boldsymbol{I}_n \leq (\boldsymbol{P}_{\tilde{x},0})_{CDPF} \leq \bar{p}\boldsymbol{I}_n$, *where* $\boldsymbol{\Upsilon} = (\nabla^{\text{T}}\nabla)\boldsymbol{f}(\boldsymbol{x})|_{\boldsymbol{x}=\hat{\boldsymbol{x}}_k} + (\nabla^{\text{T}}\nabla)\boldsymbol{\eta}(\boldsymbol{x})|_{\boldsymbol{x}=\hat{\boldsymbol{x}}_k}$, *then the recursion* (5.50) *is bounded as*

$$\underline{p}\boldsymbol{I}_n \leq (\boldsymbol{P}_{\tilde{x},k})_{CDPF} \leq \bar{p}\boldsymbol{I}_n,\ k \geq 0 \tag{5.58}$$

Proof. Following the proof of Theorem 4.3 and using (4.67)–(4.68), if $\underline{p}\boldsymbol{I}_n \leq (\boldsymbol{P}_{\tilde{x},0})_{\text{CDPF}} \leq \bar{p}\boldsymbol{I}_n$ is satisfied, then it can be proved that

$$(\boldsymbol{P}_{\tilde{x},1})_{\text{CDPF}} = \boldsymbol{A}_0(\boldsymbol{P}_{\tilde{x},0})_{\text{CDPF}}\boldsymbol{A}_0^{\text{T}} - \boldsymbol{K}_0\boldsymbol{K}_0^{\text{T}} + \boldsymbol{Q}_0 \leq \bar{p}\boldsymbol{I}_n \tag{5.59}$$

$$(\boldsymbol{P}_{\tilde{x},1})_{\mathrm{CDPF}} = \boldsymbol{A}_0(\boldsymbol{P}_{\tilde{x},0})_{\mathrm{CDPF}}\boldsymbol{A}_0^{\mathrm{T}} - \boldsymbol{K}_0\boldsymbol{K}_0^{\mathrm{T}} + \boldsymbol{Q}_0 \geq \underline{p}\boldsymbol{I}_n \tag{5.60}$$

which yields $\underline{p}\boldsymbol{I}_n \leq (\boldsymbol{P}_{\tilde{x},1})_{\mathrm{CDPF}} \leq \bar{p}\boldsymbol{I}_n$. Repeating this step, $\underline{p}\boldsymbol{I}_n \leq (\boldsymbol{P}_{\tilde{x},k})_{\mathrm{CDPF}} \leq \bar{p}\boldsymbol{I}_n$ can be proved for $k \geq 0$. The proof is thereby completed here. □

5.4.2 Boundedness analysis of state estimation error

The stochastic boundedness of the state estimation error obtained from the CDPF in Algorithm 5.1 is analyzed in this subsection. To carry out this work, the following three assumptions are preliminarily made.

Assumption 5.1. For every $k \geq 0$, there exist positive real constants $\bar{a} \in \mathbb{R}_+$, $\underline{a} \in \mathbb{R}_+$, $\bar{p} \in \mathbb{R}_+$, $\underline{p} \in \mathbb{R}_+$, $\bar{q} \in \mathbb{R}_+$, $\underline{q} \in \mathbb{R}_+$, $\underline{r} \in \mathbb{R}_+$, $\bar{r} \in \mathbb{R}_+$, $\underline{\lambda} \in \mathbb{R}_+$, $\bar{\lambda} \in \mathbb{R}_+$, $\underline{s} \in \mathbb{R}_+$, $\bar{s} \in \mathbb{R}_+$, $\bar{k} \in \mathbb{R}_+$, $\delta \in \mathbb{R}_+$ satisfying

$$\underline{a} \leq ||\boldsymbol{A}_k|| \leq \bar{a} \tag{5.61}$$

$$\frac{1}{2}\|\Delta t\boldsymbol{\Upsilon}\| \leq \sqrt{\bar{k}} \tag{5.62}$$

$$\underline{q}\boldsymbol{I}_n \leq \boldsymbol{Q}_k \leq \bar{q}\boldsymbol{I}_n \tag{5.63}$$

$$\underline{r}\boldsymbol{I}_n \leq \boldsymbol{R}_k \tag{5.64}$$

$$\boldsymbol{0} \leq \boldsymbol{W}_E \leq \bar{w}\boldsymbol{I}_n \tag{5.65}$$

$$\underline{\lambda} \leq ||\boldsymbol{\Lambda}(\Delta t)|| \leq \bar{\lambda} \tag{5.66}$$

$$\underline{s} \leq \|(\boldsymbol{U}(\hat{\boldsymbol{x}}_k))_{\mathrm{CDPF}}\| \leq \bar{s} \tag{5.67}$$

$$\underline{p}\boldsymbol{I}_n \leq (\boldsymbol{P}_{\tilde{x},k})_{\mathrm{CDPF}} \leq \bar{p}\boldsymbol{I}_n \tag{5.68}$$

Assumption 5.2. The matrix $\boldsymbol{A}_k$, $k \geq 0$, in (5.33) is nonsingular.

Assumption 5.3. Two positive real numbers $\kappa_{\bar{\mu}} > 0$ and $\varepsilon' > 0$ exist such that the nonlinear function given by (5.32) is bounded, *i.e.*,

$$\left\|\bar{\boldsymbol{\mu}}_k(\boldsymbol{x}_k, \boldsymbol{d}_k, (\tilde{\boldsymbol{x}}_k)_{\mathrm{CDPF}}, (\hat{\boldsymbol{d}}_k)_{\mathrm{CDPF}})\right\| \leq \kappa_{\bar{\mu}}\left\|\boldsymbol{x}_k - (\hat{\boldsymbol{x}}_k)_{\mathrm{CDPF}}\right\|^2 \tag{5.69}$$

$$||\boldsymbol{x}_k - (\hat{\boldsymbol{x}}_k)_{\mathrm{CDPF}}|| \leq \varepsilon' \tag{5.70}$$

Lemma 5.1. *If Assumptions 5.1–5.3 are satisfied, then a real and positive scalar $0<\alpha<1$ exists with the following inequality be satisfied.*

$$(\boldsymbol{A}_k-\boldsymbol{K}_k\boldsymbol{C}_k)^{\mathrm{T}}\boldsymbol{\Pi}_{k+1}(\boldsymbol{A}_k-\boldsymbol{K}_k\boldsymbol{C}_k)\le(1-\alpha)\boldsymbol{\Pi}_k,\ k\ge 0 \tag{5.71}$$

where $\boldsymbol{\Pi}_k=(\boldsymbol{P}_{\tilde{x},k}^{-1})_{CDPF}$, $\boldsymbol{C}_k=((\boldsymbol{P}_{\tilde{x},k}^{-1})_{CDPF}\boldsymbol{A}_k^{-1}\boldsymbol{K}_k)^{\mathrm{T}}$, *and*

$$\alpha=1-\frac{1}{1+\frac{q}{\bar{p}(\bar{a}+\bar{p}^2\bar{k}/\underline{ap})^2}} \tag{5.72}$$

Proof. From (5.50) and the definition of $\boldsymbol{C}_k$, it follows that

$$\begin{aligned}(\boldsymbol{P}_{\tilde{x},k+1})_{\mathrm{CDPF}}&=\boldsymbol{A}_k(\boldsymbol{P}_{\tilde{x},k})_{\mathrm{CDPF}}\boldsymbol{A}_k^{\mathrm{T}}+\boldsymbol{Q}_k-\boldsymbol{A}_k(\boldsymbol{P}_{\tilde{x},k})_{\mathrm{CDPF}}\boldsymbol{C}_k^{\mathrm{T}}\boldsymbol{K}_k^{\mathrm{T}}\\&=(\boldsymbol{A}_k-\boldsymbol{K}_k\boldsymbol{C}_k)(\boldsymbol{P}_{\tilde{x},k})_{\mathrm{CDPF}}(\boldsymbol{A}_k-\boldsymbol{K}_k\boldsymbol{C}_k)^{\mathrm{T}}+\boldsymbol{Q}_k\\&+\boldsymbol{K}_k\boldsymbol{C}_k(\boldsymbol{P}_{\tilde{x},k})_{\mathrm{CDPF}}(\boldsymbol{A}_k-\boldsymbol{K}_k\boldsymbol{C}_k)^{\mathrm{T}}\end{aligned} \tag{5.73}$$

and

$$\begin{aligned}\boldsymbol{A}_k^{-1}(\boldsymbol{A}_k-\boldsymbol{K}_k\boldsymbol{C}_k)(\boldsymbol{P}_{\tilde{x},k})_{\mathrm{CDPF}}&=(\boldsymbol{P}_{\tilde{x},k})_{\mathrm{CDPF}}\\&-(\boldsymbol{P}_{\tilde{x},k})_{\mathrm{CDPF}}\boldsymbol{C}_k^{\mathrm{T}}\boldsymbol{C}_k(\boldsymbol{P}_{\tilde{x},k})_{\mathrm{CDPF}}\end{aligned} \tag{5.74}$$

which is a symmetric matrix.

From (5.32)–(5.37) and (5.50), the linear term and the first-order term of the Taylor Series expansion of $(\boldsymbol{x}_{k+1})_{\mathrm{CDPF}}$ is $\boldsymbol{A}_k(\tilde{\boldsymbol{x}}_k)_{\mathrm{CDPF}}$. $\boldsymbol{K}_k$ is the expectation of the second-order term of the Taylor Series expansion of $(\boldsymbol{x}_{k+1})_{\mathrm{CDPF}}$. It is seen that the zeroth and the first-order terms of any nonlinear function account for the major than that of the residual terms. Then, the following inequalities are valid.

$$|\boldsymbol{K}_k|<|\boldsymbol{A}_k(\tilde{\boldsymbol{x}}_k)_{\mathrm{CDPF}}| \tag{5.75}$$

$$\frac{|\boldsymbol{K}_k|}{\left|\boldsymbol{A}_k(\tilde{\boldsymbol{x}}_k)_{\mathrm{CDPF}}\right|}\left|(\tilde{\boldsymbol{x}}_k)_{\mathrm{CDPF}}\right|<\left|(\tilde{\boldsymbol{x}}_k)_{\mathrm{CDPF}}\right| \tag{5.76}$$

Based on (5.76), it can be obtained that

$$\begin{aligned}&\boldsymbol{A}_k^{-1}\boldsymbol{K}_k((\boldsymbol{A}_k^{-1})\boldsymbol{K}_k)^{\mathrm{T}}\\&<(\boldsymbol{P}_{\tilde{x},k})_{\mathrm{CDPF}}(\boldsymbol{P}_{\tilde{x},k}^{-1})_{\mathrm{CDPF}}((\boldsymbol{A}_k^{-1})\boldsymbol{K}_k)((\boldsymbol{A}_k^{-1})\boldsymbol{K}_k)^{\mathrm{T}}(\boldsymbol{P}_{\tilde{x},k}^{-1})_{\mathrm{CDPF}}\\&<(\boldsymbol{P}_{\tilde{x},k}^{-1})_{\mathrm{CDPF}}\end{aligned} \tag{5.77}$$

and

$$\boldsymbol{C}_k\boldsymbol{C}_k^{\mathrm{T}}<(\boldsymbol{P}_{\tilde{x},k}^{-1})_{\mathrm{CDPF}} \tag{5.78}$$

where $(\boldsymbol{P}_{\tilde{x},k}^{-1})_{\mathrm{CDPF}}=((\boldsymbol{P}_{\tilde{x},k}^{-1})_{\mathrm{CDPF}})^{\mathrm{T}}$ is used.

Applying $(\boldsymbol{P}_{\tilde{\boldsymbol{x}},k}^{-1})_{\text{CDPF}} > \mathbf{0}$ and $(\boldsymbol{P}_{\tilde{\boldsymbol{x}},k})_{\text{CDPF}} > \mathbf{0}$, substituting (5.78) into (5.70) yields

$$\begin{aligned} &\boldsymbol{A}_k^{-1}(\boldsymbol{A}_k - \boldsymbol{K}_k\boldsymbol{C}_k)(\boldsymbol{P}_{\tilde{\boldsymbol{x}},k})_{\text{CDPF}} \\ &= (\boldsymbol{P}_{\tilde{\boldsymbol{x}},k})_{\text{CDPF}} - (\boldsymbol{P}_{\tilde{\boldsymbol{x}},k})_{\text{CDPF}}\boldsymbol{C}_k^{\text{T}}\boldsymbol{C}_k(\boldsymbol{P}_{\tilde{\boldsymbol{x}},k})_{\text{CDPF}} > \mathbf{0} \end{aligned} \tag{5.79}$$

$$\boldsymbol{A}_k^{-1}\boldsymbol{K}_k\boldsymbol{C}_k = (\boldsymbol{A}_k^{-1}\boldsymbol{K}_k)(\boldsymbol{A}_k^{-1}\boldsymbol{K}_k)^{\text{T}}((\boldsymbol{P}_{\tilde{\boldsymbol{x}},k}^{-1})_{\text{CDPF}})^{\text{T}} \geq \mathbf{0} \tag{5.80}$$

Then, it can be got from (5.79) and (5.80) that

$$\begin{aligned} &\boldsymbol{K}_k\boldsymbol{C}_k(\boldsymbol{P}_{\tilde{\boldsymbol{x}},k})_{\text{CDPF}}(\boldsymbol{A}_k - \boldsymbol{K}_k\boldsymbol{C}_k)^{\text{T}} \\ &= \boldsymbol{A}_k(\boldsymbol{A}_k^{-1}\boldsymbol{K}_k\boldsymbol{C}_k)(\boldsymbol{A}_k^{-1}(\boldsymbol{A}_k - \boldsymbol{K}_k\boldsymbol{C}_k)(\boldsymbol{P}_{\tilde{\boldsymbol{x}},k})_{\text{CDPF}})^{\text{T}}\boldsymbol{A}_k^{\text{T}} \geq \mathbf{0} \end{aligned} \tag{5.81}$$

Moreover, inserting (5.81) into (5.73) can prove

$$\begin{aligned} (\boldsymbol{P}_{\tilde{\boldsymbol{x}},k+1})_{\text{CDPF}} &\geq (\boldsymbol{A}_k - \boldsymbol{K}_k\boldsymbol{C}_k)(\boldsymbol{P}_{\tilde{\boldsymbol{x}},k})_{\text{CDPF}}(\boldsymbol{A}_k - \boldsymbol{K}_k\boldsymbol{C}_k)^{\text{T}} + \boldsymbol{Q}_k \\ &= (\boldsymbol{A}_k - \boldsymbol{K}_k\boldsymbol{C}_k)\Big((\boldsymbol{P}_{\tilde{\boldsymbol{x}},k})_{\text{CDPF}}(\boldsymbol{A}_k - \boldsymbol{K}_k\boldsymbol{C}_k)^{\text{T}} \\ &\quad + (\boldsymbol{A}_k - \boldsymbol{K}_k\boldsymbol{C}_k)^{-1}\boldsymbol{Q}_k\Big) \end{aligned} \tag{5.82}$$

On the other hand, Assumption 5.1 guarantees that $||\boldsymbol{K}_k|| \leq \bar{p}\sqrt{\bar{k}}$ and $\|\boldsymbol{C}_k\| \leq \frac{\bar{p}\sqrt{\bar{k}}}{\underline{ap}}$. Then, (5.82) can be simplified as

$$(\boldsymbol{P}_{\tilde{\boldsymbol{x}},k+1})_{\text{CDPF}} \geq (\boldsymbol{A}_k - \boldsymbol{K}_k\boldsymbol{C}_k)\left((\boldsymbol{P}_{\tilde{\boldsymbol{x}},k})_{\text{CDPF}} + \frac{q}{\left(\bar{a} + \frac{\bar{p}^2\bar{k}}{\underline{ap}}\right)^2}\right)(\boldsymbol{A}_k - \boldsymbol{K}_k\boldsymbol{C}_k)^{\text{T}} \tag{5.83}$$

Because $(\boldsymbol{P}_{\tilde{\boldsymbol{x}},k})_{\text{CDPF}} > \underline{p}\boldsymbol{I}_n$ and $\boldsymbol{A}_k - \boldsymbol{K}_k\boldsymbol{C}_k$ are nonsingular, it can be proved based on Assumption 5.1 that

$$(\boldsymbol{A}_k - \boldsymbol{K}_k\boldsymbol{C}_k)^{\text{T}}\boldsymbol{\Pi}_{k+1}(\boldsymbol{A}_k - \boldsymbol{K}_k\boldsymbol{C}_k) \leq \left(1 + \frac{q}{\bar{p}(\bar{a} + \frac{\bar{p}^2\bar{k}}{\underline{ap}})^2}\right)^{-1}\boldsymbol{\Pi}_k \tag{5.84}$$

To this end, the conclusion in Lemma 5.1 is proved. □

Lemma 5.2. *Suppose that Assumptions 5.1–5.3 are satisfied, then*

$$\begin{aligned} &\left(\boldsymbol{K}_k\boldsymbol{C}_k\left(\boldsymbol{x}_k - (\hat{\boldsymbol{x}}_k)_{CDPF}\right)\right)^{\text{T}}\boldsymbol{\Pi}_k\begin{pmatrix} 2(\boldsymbol{A}_k - \boldsymbol{K}_k\boldsymbol{C}_k)(\boldsymbol{x}_k - (\hat{\boldsymbol{x}}_k)_{CDPF}) + \\ \boldsymbol{K}_k\boldsymbol{C}_k(\boldsymbol{x}_k - (\hat{\boldsymbol{x}}_k)_{CDPF}) + (\boldsymbol{B}_k)_{CDPF} \end{pmatrix} \\ &\leq \kappa_1\left\|\boldsymbol{x}_k - (\hat{\boldsymbol{x}}_k)_{CDPF}\right\|^2 \end{aligned} \tag{5.85}$$

holds for $||\boldsymbol{x}_k - (\hat{\boldsymbol{x}}_k)_{CDPF}|| \leq \varepsilon'$, *where* $\kappa_1 = \frac{k_1}{\underline{p}}((2\bar{a} + k_1) + \kappa_{\hat{\mu}}\varepsilon')$ *and* $\varepsilon' \in \mathbb{R}_+$.

Proof. Following the proof of Lemma 5.1, one has

$$\begin{aligned}\left\| \boldsymbol{K}_k \boldsymbol{C}_k \left(\boldsymbol{x}_k - (\hat{\boldsymbol{x}}_k)_{\text{CDPF}}\right) \right\| &\le \frac{\bar{p}^2 \bar{k}}{\underline{a}\underline{p}} \|\boldsymbol{x}_k - (\hat{\boldsymbol{x}}_k)_{\text{CDPF}}\| \\ &= k_1 \|\boldsymbol{x}_k - (\hat{\boldsymbol{x}}_k)_{\text{CDPF}}\|\end{aligned} \tag{5.86}$$

and

$$\begin{aligned}&\left(\boldsymbol{K}_k \boldsymbol{C}_k \left(\boldsymbol{x}_k - (\hat{\boldsymbol{x}}_k)_{\text{CDPF}}\right)\right)^{\text{T}} \boldsymbol{\Pi}_k \begin{pmatrix} 2(\boldsymbol{A}_k - \boldsymbol{K}_k \boldsymbol{C}_k)(\boldsymbol{x}_k - (\hat{\boldsymbol{x}}_k)_{\text{CDPF}}) + \\ \boldsymbol{K}_k \boldsymbol{C}_k (\boldsymbol{x}_k - (\hat{\boldsymbol{x}}_k)_{\text{CDPF}}) + (\boldsymbol{B}_k)_{\text{CDPF}} \end{pmatrix} \\ &\le \frac{k_1}{\underline{p}} \|\boldsymbol{x}_k - (\hat{\boldsymbol{x}}_k)_{\text{CDPF}}\| \left((2\bar{a} + k_1 + \kappa_{\bar{\mu}} \varepsilon') \|\boldsymbol{x}_k - (\hat{\boldsymbol{x}}_k)_{\text{CDPF}}\|\right) \\ &= \kappa_1 \left\|\boldsymbol{x}_k - (\hat{\boldsymbol{x}}_k)_{\text{CDPF}}\right\|^2\end{aligned} \tag{5.87}$$

where $k_1 = \frac{\bar{p}^2 \bar{k}}{\underline{a}\underline{p}}$. The proof is hence completed here. □

Lemma 5.3. *If Assumptions 5.1–5.3 are satisfied and $\boldsymbol{\Pi}_k = (\boldsymbol{P}^{-1}_{\tilde{x},k})_{CDPF}$ exists, then it follows that*

$$\begin{aligned}&(\boldsymbol{B}_k)^{\text{T}}_{CDPF} \boldsymbol{\Pi}_k \begin{pmatrix} 2(\boldsymbol{A}_k - \boldsymbol{K}_k \boldsymbol{C}_k)(\boldsymbol{x}_k - (\hat{\boldsymbol{x}}_k)_{CDPF}) + \\ \boldsymbol{K}_k \boldsymbol{C}_k (\boldsymbol{x}_k - (\hat{\boldsymbol{x}}_k)_{CDPF}) + (\boldsymbol{B}_k)_{CDPF} \end{pmatrix} \\ &\le \kappa_2 \left\|\boldsymbol{x}_k - (\hat{\boldsymbol{x}}_k)_{CDPF}\right\|^3\end{aligned} \tag{5.88}$$

where $\kappa_2 = \frac{\kappa_{\bar{\mu}}}{\underline{p}}((2\bar{a} + k_1) + \kappa_{\bar{\mu}} \varepsilon')$.

Proof. Since Assumptions 5.1–5.3 are satisfied, it can be obtained from (5.69), (5.70), and (5.87) in the proof of Lemma 5.2 that

$$\begin{aligned}&(\boldsymbol{B}_k)^{\text{T}}_{\text{CDPF}} \boldsymbol{\Pi}_k \begin{pmatrix} 2(\boldsymbol{A}_k - \boldsymbol{K}_k \boldsymbol{C}_k)(\boldsymbol{x}_k - (\hat{\boldsymbol{x}}_k)_{\text{CDPF}}) + \\ \boldsymbol{K}_k \boldsymbol{C}_k (\boldsymbol{x}_k - (\hat{\boldsymbol{x}}_k)_{\text{CDPF}}) + (\boldsymbol{B}_k)_{\text{CDPF}} \end{pmatrix} \\ &\le \frac{k_1}{\underline{p}} \|\boldsymbol{x}_k - (\hat{\boldsymbol{x}}_k)_{\text{CDPF}}\| \left((2\bar{a} + k_1 + \kappa_{\bar{\mu}} \varepsilon') \|\boldsymbol{x}_k - (\hat{\boldsymbol{x}}_k)_{\text{CDPF}}\|\right) \\ &= \kappa_2 \left\|\boldsymbol{x}_k - (\hat{\boldsymbol{x}}_k)_{\text{CDPF}}\right\|^3\end{aligned} \tag{5.89}$$

which proves that Lemma 5.3 is valid. □

Lemma 5.4. *If Assumptions 5.1–5.3 are valid, then*

$$\mathbb{E}\left((\boldsymbol{s}_k)^{\text{T}}_{CDPF} \boldsymbol{\Pi}_k (\boldsymbol{s}_k)_{CDPF}\right) \le \kappa_3 \delta \tag{5.90}$$

where $\boldsymbol{\kappa}_3 = \frac{\bar{\lambda}^2 \bar{s}^2}{\underline{p} \underline{r}^2 \underline{\lambda}^4 \underline{s}^4} n$ *and* $\delta = \bar{r}^3$.

Proof. From (5.37), it is ready to establish

$$\begin{aligned}(\boldsymbol{s}_k)_{\mathrm{CDPF}}^{\mathrm{T}}\boldsymbol{\Pi}_k(\boldsymbol{s}_k)_{\mathrm{CDPF}} &= ((\boldsymbol{B}_k)_{\mathrm{CDPF}}\boldsymbol{\nu}_k)^{\mathrm{T}}\boldsymbol{\Pi}_k\left((\boldsymbol{B}_k)_{\mathrm{CDPF}}\boldsymbol{\nu}_k\right)\\ &\le \frac{1}{\underline{p}}\boldsymbol{\nu}_k^{\mathrm{T}}(\boldsymbol{B}_k)_{\mathrm{CDPF}}^{\mathrm{T}}(\boldsymbol{B}_k)_{\mathrm{CDPF}}\boldsymbol{\nu}_k\\ &= \frac{1}{\underline{p}}tr(\boldsymbol{\nu}_k^{\mathrm{T}}(\boldsymbol{B}_k)_{\mathrm{CDPF}}^{\mathrm{T}}(\boldsymbol{B}_k)_{\mathrm{CDPF}}\boldsymbol{\nu}_k)\\ &= \frac{1}{\underline{p}}tr((\boldsymbol{B}_k)_{\mathrm{CDPF}}\boldsymbol{\nu}_k\boldsymbol{\nu}_k^{\mathrm{T}}(\boldsymbol{B}_k)_{\mathrm{CDPF}}^{\mathrm{T}})\end{aligned} \tag{5.91}$$

Applying $\mathbb{E}(\boldsymbol{\nu}_k\boldsymbol{\nu}_k{}^{\mathrm{T}}) = \boldsymbol{R}_k$ and (5.61)–(5.70), one can compute

$$\begin{aligned}\mathbb{E}((\boldsymbol{s}_k)_{\mathrm{CDPF}}^{\mathrm{T}}\boldsymbol{\Pi}_k(\boldsymbol{s}_k)_{\mathrm{CDPF}}) &\le \frac{\bar{r}}{\underline{p}}tr((\boldsymbol{B}_k)_{\mathrm{CDPF}}(\boldsymbol{B}_k)_{\mathrm{CDPF}}^{\mathrm{T}})\\ &\le \frac{\bar{r}^2\bar{\lambda}^2\bar{s}^2}{\underline{r}^2\underline{\lambda}^4\underline{s}^4}n = \kappa_3\delta\end{aligned} \tag{5.92}$$

The proof is hence ended here. □

Theorem 5.4. *Consider a nonlinear stochastic system* (2.32) *and* (2.33), *applying the CDPF in Algorithm 5.1, if Assumptions 5.1–5.3,* $||(\tilde{\boldsymbol{x}}_0)_{CDPF}|| \le \varepsilon$ *are satisfied for a constant* $\varepsilon > 0$, *and* $\boldsymbol{R}_k \le \bar{r}\boldsymbol{I}_n$ *are satisfied for a constant* $\varepsilon \in \mathbb{R}_+$, *where* $\bar{r}^3 = \delta$, *then the state estimation error* $(\tilde{\boldsymbol{x}}_{k+1})_{CDPF}$ *given in* (5.32) *is exponentially bounded in mean square and bounded with probability one.*

Proof. Based on Assumptions 5.1–5.3, choose a Lyapunov candidate function as

$$\boldsymbol{V}_k\left((\tilde{\boldsymbol{x}}_k)_{\mathrm{CDPF}}\right) = ((\tilde{\boldsymbol{x}}_k)_{\mathrm{CDPF}})^{\mathrm{T}}\boldsymbol{\Pi}_k(\tilde{\boldsymbol{x}}_k)_{\mathrm{CDPF}} \tag{5.93}$$

which is bounded by

$$\frac{1}{\bar{p}}\|(\tilde{\boldsymbol{x}}_k)_{\mathrm{CDPF}}\|^2 \le \boldsymbol{V}_k((\tilde{\boldsymbol{x}}_k)_{\mathrm{CDPF}}) \le \frac{1}{\underline{p}}\|(\tilde{\boldsymbol{x}}_k)_{\mathrm{CDPF}}\|^2 \tag{5.94}$$

From (5.32), the following can be obtained.

$$\boldsymbol{V}_{k+1}\left((\tilde{\boldsymbol{x}}_{k+1})_{\mathrm{CDPF}}\right) = \boldsymbol{\Upsilon}_3^{\mathrm{T}}\boldsymbol{\Pi}_{k+1}\boldsymbol{\Upsilon}_3 = \boldsymbol{\Upsilon}_4^{\mathrm{T}}\boldsymbol{\Pi}_{k+1}\boldsymbol{\Upsilon}_4 \tag{5.95}$$

where $\boldsymbol{\Upsilon}_3 = (\tilde{\boldsymbol{x}}_{k+1})_{\mathrm{CDPF}}\boldsymbol{A}_k + (\bar{\boldsymbol{\mu}}_k)_{\mathrm{CDPF}} + (\boldsymbol{s}_k)_{\mathrm{CDPF}}$ and $\boldsymbol{\Upsilon}_4 = (\tilde{\boldsymbol{x}}_{k+1})_{\mathrm{CDPF}}(\boldsymbol{A}_k - \boldsymbol{K}_k\boldsymbol{C}_k) + (\bar{\boldsymbol{\mu}}_k)_{\mathrm{CDPF}} + (\tilde{\boldsymbol{x}}_{k+1})_{\mathrm{CDPF}}(\boldsymbol{K}_k\boldsymbol{C}_k)^{\mathrm{T}} + (\boldsymbol{s}_k)_{\mathrm{CDPF}}$. Applying Lemma 5.1, it follows that $V_{k+1}((\tilde{x}_{k+1})_{\text{CDPF}})$ is bounded as

$$\begin{aligned} V_{k+1}\left((\tilde{x}_{k+1})_{\text{CDPF}}\right) \leq & (1-\alpha) V_k((\tilde{x}_k)_{\text{CDPF}}) \\ & + (K_k C_k (\tilde{x}_k)_{\text{CDPF}})^{\mathrm{T}} \Pi_k \big(K_k C_k (\tilde{x}_k)_{\text{CDPF}} \\ & + (\bar{\mu}_k)_{\text{CDPF}} + 2(A_k - K_k C_k)(\tilde{x}_k)_{\text{CDPF}}\big) \\ & + (\bar{\mu}_k)_{\text{CDPF}}^{\mathrm{T}} \Pi_k \big(2(A_k - K_k C_k)(\tilde{x}_k)_{\text{CDPF}} \\ & + K_k C_k (\tilde{x}_k)_{\text{CDPF}} + (B_k)_{\text{CDPF}}\big) \\ & + (s_k)_{\text{CDPF}}^{\mathrm{T}} \Pi_k ((s_k)_{\text{CDPF}})_{\text{CDPF}} \\ & + (s_k)_{\text{CDPF}}^{\mathrm{T}} \Pi_k \big(2(A_k - K_k C_k)(\tilde{x}_k)_{\text{CDPF}} \\ & + 2K_k C_k (\tilde{x}_k)_{\text{CDPF}} + 2(B_k)_{\text{CDPF}}\big) \end{aligned} \tag{5.96}$$

Using Lemma 5.2 and Lemma 5.3, it can be obtained that the following will hold for $||(\tilde{x}_k)_{\text{CDPF}}|| \leq \varepsilon'$.

$$\begin{aligned} & \mathbb{E}\left(V_{k+1}\left((\tilde{x}_{k+1})_{\text{CDPF}}\right) \middle| (\tilde{x}_{k+1})_{\text{CDPF}}\right) - V_k((\tilde{x}_k)_{\text{CDPF}}) \\ & \leq -\alpha V_k((\tilde{x}_k)_{\text{CDPF}}) + \kappa_1 \left\|(\tilde{x}_k)_{\text{CDPF}}\right\|^2 + \kappa_2 \left\|(\tilde{x}_k)_{\text{CDPF}}\right\|^3 + \kappa_3 \delta \end{aligned} \tag{5.97}$$

Defining $\varepsilon = \min\left(\varepsilon', \frac{\alpha}{2\bar{p}\kappa_2}\right)$, the following inequality

$$\kappa_2 \|(\tilde{x}_k)_{\text{CDPF}}\| \|(\tilde{x}_k)_{\text{CDPF}}\|^2 \leq \frac{\alpha}{2\bar{p}} \|(\tilde{x}_k)_{\text{CDPF}}\|^2 \leq \frac{\alpha}{2} V_k((\tilde{x}_k)_{\text{CDPF}}) \tag{5.98}$$

is seen for $||(\tilde{x}_k)_{\text{CDPF}}|| \leq \varepsilon$. Moreover, it is proved from (5.97) that the following is valid for $||(\tilde{x}_k)_{\text{CDPF}}|| \leq \varepsilon$.

$$\begin{aligned} & \mathbb{E}\left(V_{k+1}\left((\tilde{x}_{k+1})_{\text{CDPF}}\right) \middle| (\tilde{x}_{k+1})_{\text{CDPF}}\right) - V_k((\tilde{x}_k)_{\text{CDPF}}) \\ & \leq -\frac{\alpha}{2} V_k((\tilde{x}_k)_{\text{CDPF}}) + \tilde{\kappa}_3 \delta \end{aligned} \tag{5.99}$$

where $\tilde{\kappa}_3 = \kappa_3 + \frac{\kappa_1}{\delta} \|(\tilde{x}_k)_{\text{CDPF}}\|^2$.

Choosing $\delta = \frac{\alpha \tilde{\varepsilon}^2}{2\bar{p}\tilde{\kappa}_3}$ with some $\tilde{\varepsilon} < \varepsilon$, one has $||(\tilde{x}_k)_{\text{CDPF}}|| \geq \tilde{\varepsilon}$ and

$$\tilde{\kappa}_3 \delta = \frac{\alpha \tilde{\varepsilon}^2}{2\bar{p}} \leq \frac{\alpha}{2\bar{p}} \|(\tilde{x}_k)_{\text{CDPF}}\|^2 \leq \frac{\alpha}{2} V_k((\tilde{x}_k)_{\text{CDPF}}) \tag{5.100}$$

Then, applying $\tilde{\varepsilon} \leq ||(\tilde{x}_k)_{\text{CDPF}} \leq \varepsilon$ and Lemma 2.1 with $||(\tilde{x}_k)_{\text{CDPF}}|| \leq \varepsilon'$, $\underline{\upsilon} = \frac{1}{\bar{p}}$, $\bar{\upsilon} = \frac{1}{\underline{p}}$, and $\tilde{\mu} = \tilde{\kappa}_3 \delta$, it can be proved that

$$\begin{aligned} & \mathbb{E}\left(V_{k+1}\left((\tilde{x}_{k+1})_{\text{CDPF}}\right) \middle| (\tilde{x}_{k+1})_{\text{CDPF}}\right) - V_k((\tilde{x}_k)_{\text{CDPF}}) \\ & \leq -\frac{\alpha}{2} V_k((\tilde{x}_k)_{\text{CDPF}}) + \tilde{\kappa}_3 \delta \leq \mathbf{0} \end{aligned} \tag{5.101}$$

Hence, invoking Definition 2.1 and Definition 2.2, it can be proved from (5.101) that the state estimation error $\tilde{x}_k$ is bounded with probability one if the initial error and the measurement noise are bounded. The conclusion in Theorem 5.2 is proved. □

Remark 5.4. In the process of proving convergence for the covariance $(P_{\tilde{x},k})_{\mathrm{CDPF}}$, the high-order error is included into the term of Q_k in (5.50). Because the covariance $(P_{\tilde{x},k})_{\mathrm{CDPF}}$ is verified to be convergent and bounded, $\hat{d}(x_k)$ is the estimation of the model error in the UPF. If the CDPF is convergent, the estimation is bounded. Then, it can be concluded that if the CDPF is convergent, the expected value of the actual modeling error must be bounded, so that the actual modeling error also requires to be bounded.

5.5 Application to microsatellite attitude system with low precision sensors

The CDPF developed in Algorithm 5.1 is applied in this section to solve the state estimation problem (or called as the attitude determination problem in aerospace engineering) of the microsatellite attitude control system with a low-precision three-axis magnetometer and a low-precision sun sensor. The effectiveness and performance of the CDPF for the microsatellite control system is validated through numerical simulations.

5.5.1 Measurement model of magnetometer and sun sensor

Suppose that a microsatellite is fixed with a three-axis magnetometer and a sun sensor in the $\mathcal{F}_B$ to measure its attitude, then the measurements provided by this two types of sensors can be mathematically modeled as

$$y = \begin{bmatrix} \mathbb{C}(q)\,S_{Sun} \\ \mathbb{C}(q)\,S_M \end{bmatrix} + \begin{bmatrix} \Delta S_{Sun} \\ \Delta S_M \end{bmatrix} \tag{5.102}$$

where $y \in \mathbb{R}^6$ is the attitude measurement value obtained from those two sensors. $S_{Sun} \in \mathbb{R}^3$ and $S_M \in \mathbb{R}^3$ are the sun vector and the magnetic field expressed in $\mathcal{F}_I$. $\Delta S_{Sun} \in \mathbb{R}^3$ and $\Delta S_M \in \mathbb{R}^3$ are the sensor measurement noises.

5.5.2 Implementation of CDPF for attitude determination

Based on the attitude dynamic model (3.1)–(3.2), the microsatellite control system with its measurement given by (5.102) can be modeled as

$$\begin{cases} \dot{x} = f(x) + g(x)d \\ y = h(x) + v \end{cases} \tag{5.103}$$

where $x = [q^{\mathrm{T}} \quad \omega_{BI}^{\mathrm{T}}]^{\mathrm{T}}$, $d = [0 \quad u_d^{\mathrm{T}}]^{\mathrm{T}}$, $v = [(\Delta S_{Sun})^{\mathrm{T}} \quad (\Delta S_M)^{\mathrm{T}}]^{\mathrm{T}}$, $g(x) = J^{-1}$, and

$$h(x) = \begin{bmatrix} \mathbb{C}(q)\, S_{Sun} \\ \mathbb{C}(q)\, S_M \end{bmatrix} \tag{5.104}$$

$$f(x) = \begin{bmatrix} E(q)\omega_{BI} \\ -J^{-1}(\omega_{BI})^{\times} J\omega_{BI} + J^{-1}u_c \end{bmatrix} \tag{5.105}$$

From (5.103), it is seen that its discretized version has a form of (2.32)–(2.33). Hence, the proposed CDPF in Algorithm 5.1 is applicable for this microsatellite attitude control system to solve the attitude determination problem with low-precision sensors.

5.5.3 Simulation results

In this subsection, numerical simulations are conducted with the CDPF applied to a microsatellite attitude control system with $J = \mathrm{diag}([20 \quad 25 \quad 33]^{\mathrm{T}})$ kg·m^2 to demonstrate the effectiveness of the CDPF. The modeling error given by (4.119) is considered. The initial states of the microsatellite control system and the initial estimation of the attitude is chosen as the same as Section 4.6.2 with $\Theta_0 = 10$ degrees. The initial estimation of angular velocity is set as $[0.2 \quad 0.2 \quad 0.2]^{\mathrm{T}}$ deg/sec. The weighting matrix $W_E = 10^4 I_3$ is chosen for the CDPF and the PF. Moreover, the interpolation step of the CDPF is chosen as $h = \sqrt{3}$.

Because the microsatellite considered use a three-axis magnetometer to measure its attitude, the magnetic field of the Earth should be considered in simulation. Note that it is important to consider three aspects of the Earth's internal abnormalities, exogenous magnetic field, and orbital altitude error. They can seriously influence the magnetic field. The Earth's magnetic field error caused by its internal abnormalities is about 50 nT at a 600 km altitude orbit. The exogenous magnetic field is always influenced by magnetic storms with its fluctuation usually between 20 and 100 nT. For simplicity, we consider its average value to be 50 nT. Since most of microsatellites operate without orbit control, the orbit altitude error cannot be ignored. For example, if a microsatellite operates in a sun synchronous orbit with 600 km altitude with an altitude error of about 5 km, the corresponding error in the Earth's magnetic field is about 115 nT. The total error is about 215 nT when all the three factors mentioned above are considered. The error between the 13th order and the 7th order of the International Geomagnetic Reference Field (IGRF) is about 208 nT. Hence, it is assumed that the 13th order represents the Earth's real magnetic field, while the 7th order of the IGRF is used in the estimation model, when carrying out simulations. Moreover, the following four scenarios are considered.

1) Case #1: This is a nominal case, in which

$$\Delta \boldsymbol{N}_c = [-0.0015 \quad -0.002 \quad 0.0015]^{\mathrm{T}} \ \mathrm{Nm} \tag{5.106}$$

$$\Delta \boldsymbol{N}_e = [3\omega_1 \cos(\omega_T t) \quad \omega_1 \cos(\omega_T t) \quad -2\omega_1 \cos(\omega_T t)]^{\mathrm{T}} \ \mathrm{Nm} \tag{5.107}$$

are assumed with $\omega_T = 0.66$ deg/sec and $\omega_1 = 0.0005$. The standard deviation of the measurement noises of the sun sensor and the magnetometer in (5.102) are assumed as 0.05 degrees and 20 nT, respectively.

2) Case #2: This case is related to severe measurement noise with the standard deviation of the measurement noises of the sun sensor and the magnetometer in (5.102) assumed as 0.25 degrees and 100 nT, respectively. The modeling error is assumed as same as Case #1.

3) Case #3: Large modeling error is considered with

$$\Delta \boldsymbol{N}_c = [-0.0075 \quad -0.01 \quad 0.0075]^{\mathrm{T}} \ \mathrm{Nm} \tag{5.108}$$

$$\Delta \boldsymbol{N}_e = [3\omega_1 \cos(\omega_T t) \quad \omega_1 \cos(\omega_T t) \quad -2\omega_1 \cos(\omega_T t)]^{\mathrm{T}} \ \mathrm{Nm} \tag{5.109}$$

where $\omega_T = 0.66$ deg/sec and $\omega_1 = 0.0025$. The standard deviation of the measurement noises in (5.102) is as same as Case #1.

4) Case #4: This case is relative to severe measurement noise and modeling error. The standard deviation of the measurement noises assumed in Case #1 and the modeling error assumed in Case #3 are assumed in this case.

As same as Remark 4.5, $\boldsymbol{\Theta}_e = [\phi_e \quad \theta_e \quad \psi_e]^{\mathrm{T}}, \boldsymbol{\omega}_E = [\omega_{Ex} \quad \omega_{Ey} \quad \omega_{Ez}]^{\mathrm{T}} = \tilde{\boldsymbol{\omega}}_{BI}$, and $\boldsymbol{d}_e = [d_{e1} \quad d_{e2} \quad d_{e3}]^{\mathrm{T}}$ are used to denote the estimation error of the attitude, the angular velocity, and $\boldsymbol{u}_d$, respectively. If the CDPF, the UKF, and PF were applied in Case #1, the obtained state estimation were shown in Figs. 5.1–5.2. The CDPF and the UKF almost achieved the same estimation accuracy for the attitude state, which was better than the PF. However, the UKF led to lower angular velocity estimation accuracy than the CDPF. This is because the UKF has a weak capability of handling the modeling error. When severe measurement noise was met in Case #2, the attitude estimation and the angular velocity estimation in Figs. 5.3–5.4 illustrated that the CDPF ensured better estimation accuracy for the angular velocity state than the UKF and the PF, while the estimation provided by the PF was diverging. The superior state estimation performance of the CDPF over the UKF and the PF is owing to its high-accuracy estimation of the modeling error. Comparing Figs. 5.5–5.6 with Figs. 5.7–5.8, a faster and higher-accuracy estimation of modeling error was achieved by the CDPF than the PF. The effect of the modeling error on the state estimation performance was reduced.

In the presence of the large modeling error and the severe measurement noise in Case #3 and Case #4, the estimation performance achieved by those three filters was deteriorated by comparing Figs. 5.9–5.12 with the results in

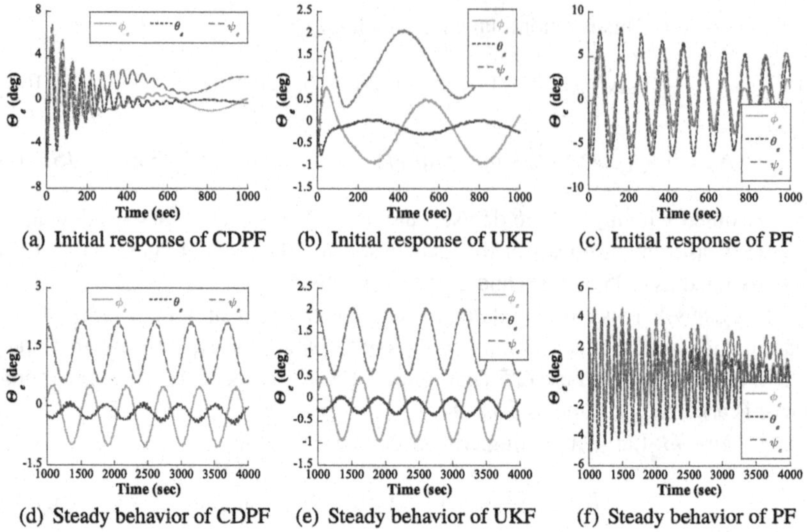

FIGURE 5.1 Attitude estimation error using CDPF, UKF, and PF in Case #1.

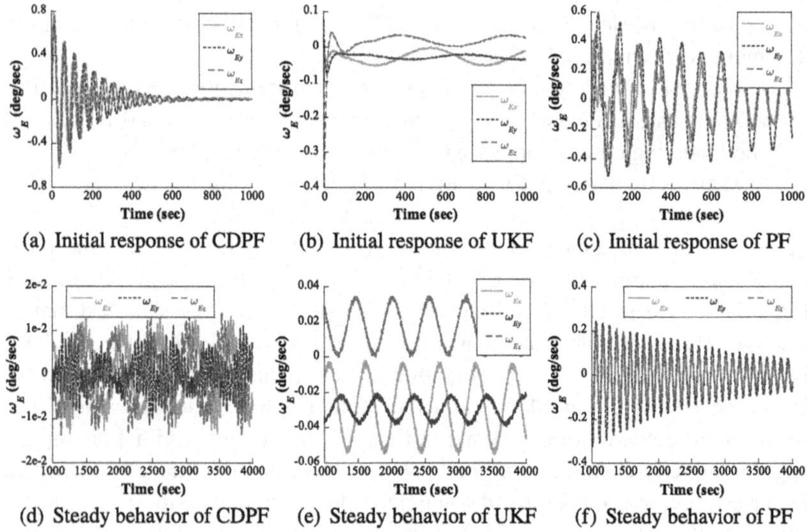

FIGURE 5.2 Angular velocity estimation error using CDPF, UKF, and PF in Case #1.

Case #1 and Case #2. However, the state estimation accuracy obtained from the CDPF was higher than the UKF and the PF. More specifically, it was observed in Fig. 5.12(d) that the CDPF can still provide the angular velocity state with high estimation accuracy even when the microsatellite was subject to the severe noise and large modeling error in Case #4. This good estimation results

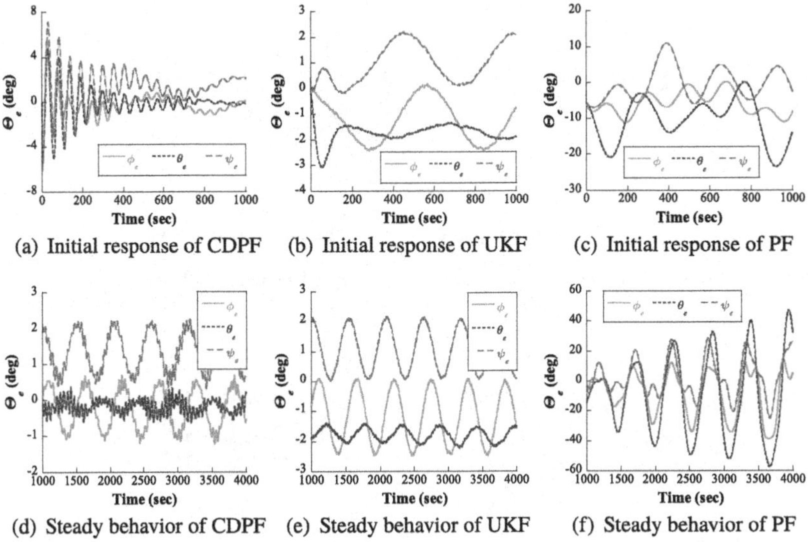

FIGURE 5.3 Attitude estimation error using CDPF, UKF, and PF in Case #2.

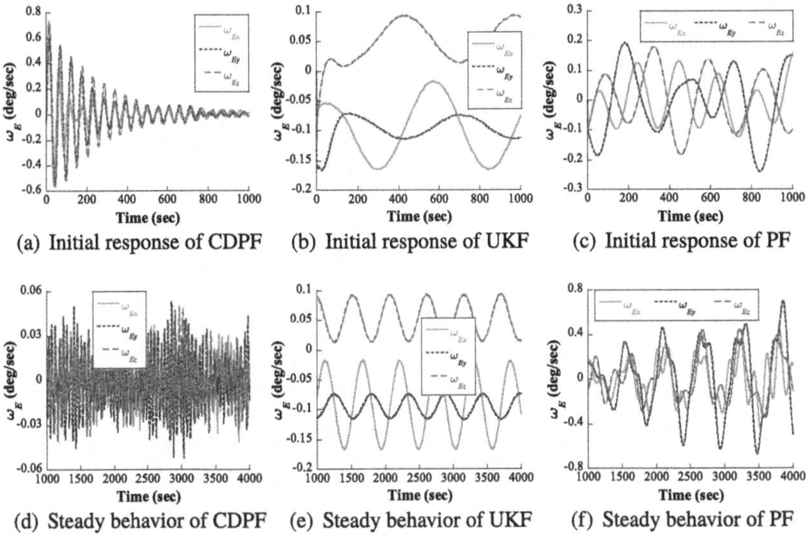

FIGURE 5.4 Angular velocity estimation error using CDPF, UKF, and PF in Case #2.

are owing to the fact that the CDPF has more robustness capability of handling modeling error and measurement noise than the UKF and the PF. Note that periodic state estimation errors were seen for the CDPF in Figs. 5.9–5.12. This is because the periodic modeling error can not be completely reconstructed and then compensated by the CDPF.

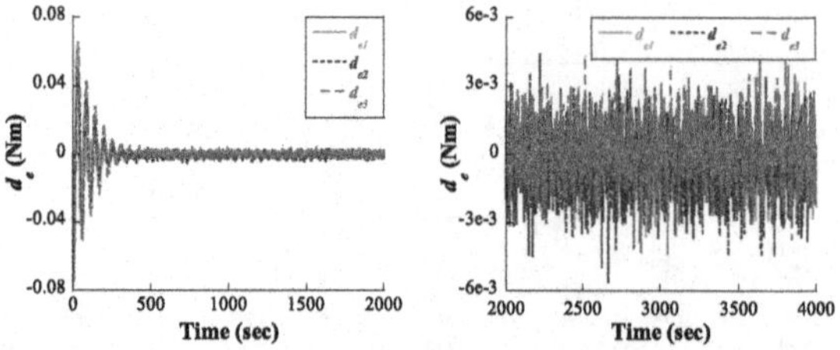

(a) The initial response (b) The steady-state behavior

FIGURE 5.5 Estimation error of the modeling error using CDPF in Case #1.

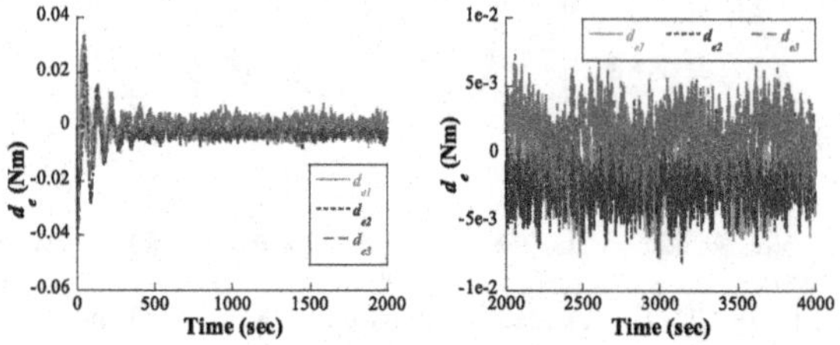

(a) The initial response (b) The steady-state behavior

FIGURE 5.6 Estimation error of the modeling error using CDPF in Case #2.

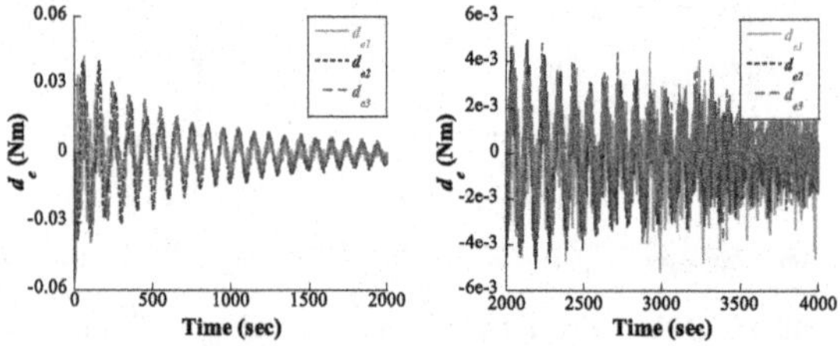

(a) The initial response (b) The steady-state behavior

FIGURE 5.7 Estimation error of the modeling error using PF in Case #1.

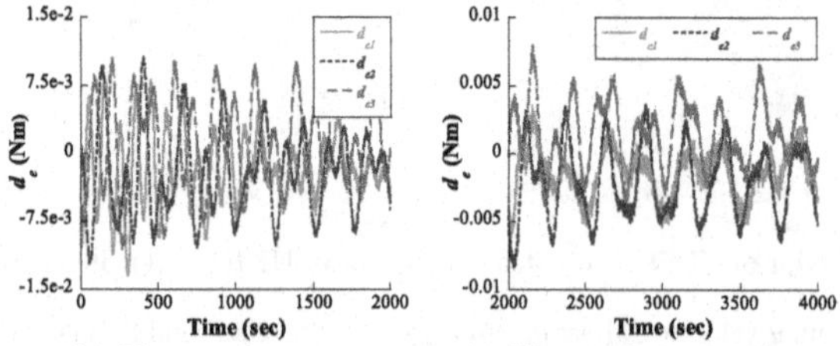

(a) The initial response (b) The steady-state behavior

FIGURE 5.8 Estimation error of the modeling error using PF in Case #2.

The above numerical simulations for each case and each filtering approach were conducted on the same computer. The computation environment was ensured to be same. When setting the simulation time as 4000 seconds, the compu-

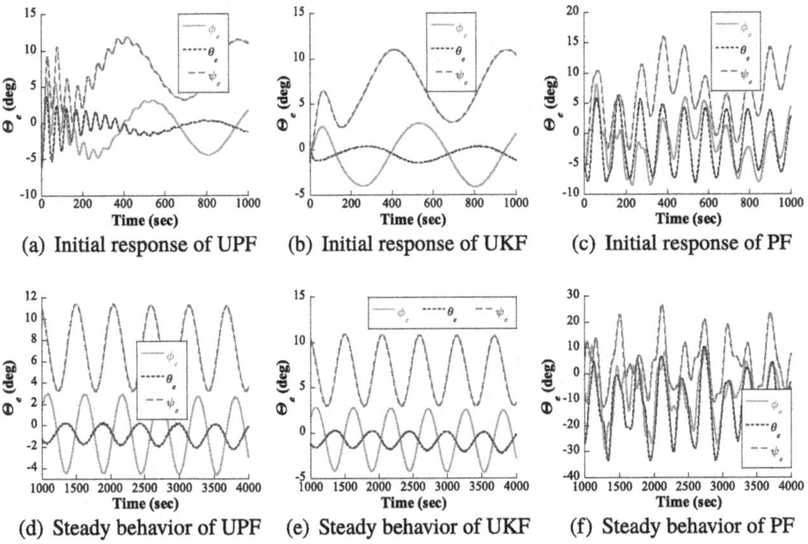

FIGURE 5.9 Attitude estimation error from CDPF, UKF, and PF in Case #3.

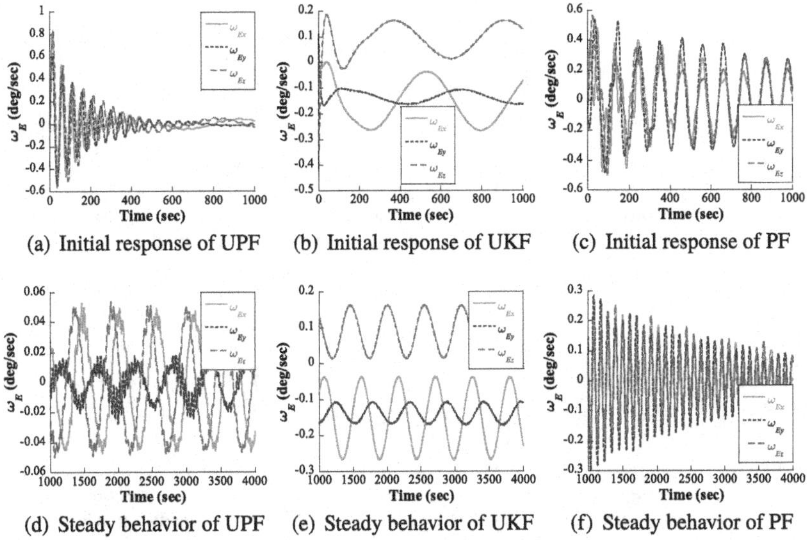

FIGURE 5.10 Angular velocity estimation error using CDPF, UKF, and PF in Case #3.

tation time of the PF, the UKF, and the CDPF for each case was 23.367 seconds, 30.865 seconds, and the CDPF 31.217 seconds, respectively. Additionally, the stability ensured by the CDPF was also numerically analyzed. It was found that the stability of the CDPF is conditionally. When the modeling error or the measurement noise is quite severe, the state estimation provided by the CDPF will

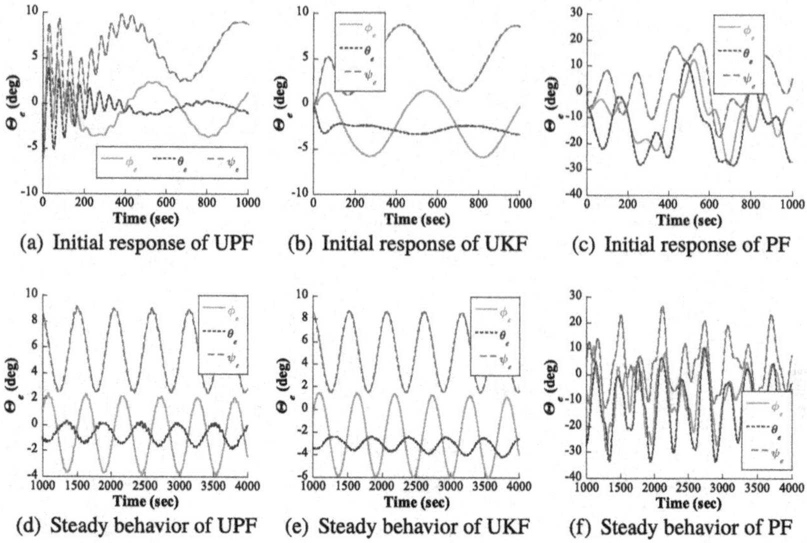

FIGURE 5.11 Attitude estimation error using CDPF, UKF, and PF in Case #4.

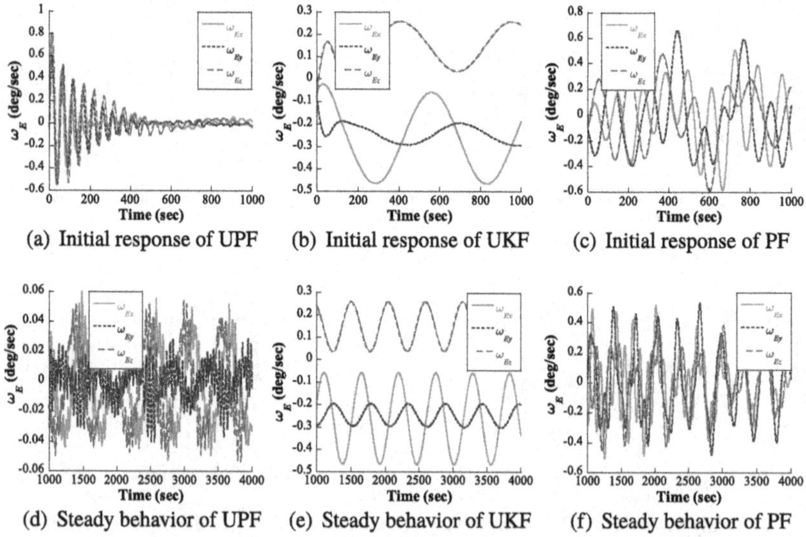

FIGURE 5.12 Angular velocity estimation error using CDPF, UKF, and PF in Case #4.

diverge. If the following Case #5 is met, then simulation result shows that divergence of the state estimation is seen for the CDPF, as shown in Fig. 5.13.

5) Case #5: $\Delta N_e = [3\omega_1 \cos(\omega_T t) \quad \omega_1 \cos(\omega_T t) \quad -2\omega_1 \cos(\omega_T t)]^{\mathrm{T}}$ Nm and $\Delta N_c = [-3 \quad -0.4 \quad 3]^{\mathrm{T}}$ Nm, where $\omega_T = 0.66$ deg/sec and $\omega_1 = 0.0001$. The initial estimation of the attitude and the angular velocity are $\Theta_0 = 10$

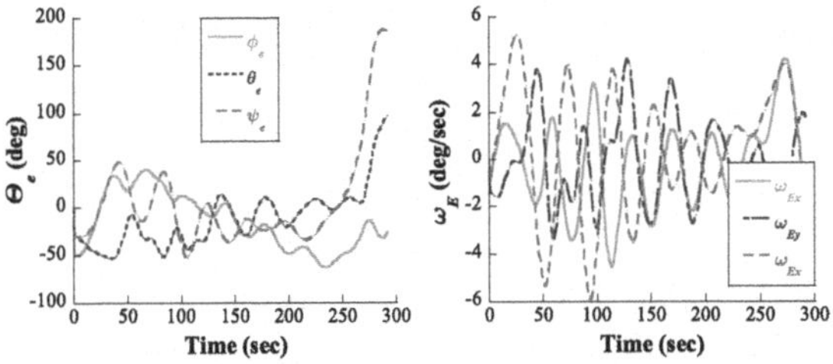

(a) Estimation error Θ_e (b) Estimation error ω_E

FIGURE 5.13 State estimation error ensured by the CDPF in Case #5.

degrees and $[0 \quad 0 \quad 0]^T$ deg/sec, respectively. The standard deviation of the measurement noises of the sun sensor and the magnetometer in (5.102) are 10 degrees and 600 nT, respectively.

5.6 Summary

In this chapter, a central difference predictive filter was proposed based on the polynomial approximations obtained with Taylor's formula. The approximations were obtained with a multi-variable extension of Stirling interpolation formula. It was proved that the estimation accuracy of the proposed filtering approach is higher than the classical PF and has a better convergence rate and robustness. The stochastic boundedness of the state estimation error was proved. Moreover, the estimation of the modeling error was proved to capture the posterior mean accurately to the second-order for any nonlinearity. With the application of this filtering approach, the state estimation or attitude determination problem of the microsatellite attitude control system with low-precision sensors and severe modeling error was successfully addressed. Simulation results show that the proposed filter provided the microsatellite attitude control system with better state estimation performance than the PF and the UKF. It lets this filtering method have great application potentials in the engineering area. It is worth mentioning that the presented central difference predictive filter was a parallel result to the UPF presented in Chapter 4.

Chapter 6

Cubature predictive filter

6.1 Introduction

Modeling error in linear or nonlinear control systems is the main issue that should be addressed in designing model-based filtering approaches to achieve high-accuracy state estimation. Although the classical PF is capable of handling system's unmodelled dynamics or error, it is characterized by low-accuracy and slow rate of state-estimation. Those two drawbacks limit its application to the microsatellite control system. That is because the microsatellite control system poses stringent requirements on its real-time state estimation performance.

When designing nonlinear filters to estimate states of linear or nonlinear systems, one of its key parts is to compute a weighted integral of the form

$$I(f) = \int_{\mathcal{D}_1} f(\boldsymbol{x}) w(\boldsymbol{x}) d\boldsymbol{x} \tag{6.1}$$

where $\mathcal{D}_1 \subseteq \mathbb{R}^n$ is the region of integration, $\boldsymbol{x} \in \mathcal{D}_1$, $\boldsymbol{f}(\cdot)$ is an arbitrary nonlinear function, and $w(\boldsymbol{x})$ is the known weighting function. For example, nonlinear filtering in the Gaussian domain finally reduces to a problem of computing integrals with the weighting function $w(\boldsymbol{x})$ as a Gaussian density. It is usually difficult to obtain (6.1). Numerical integration is therefore sought to compute (6.1). Currently, product rules-base approximation and non-product rules-based approximation are two methods to obtain that numerical integration. However, the curse of dimensionality is seen in the product rules-based approximation method. To mitigate this issue, non-product rules-based approximation method can be applied. More specifically, the cubature rule is available (Bhaumik and Swati, 2013). Invoking the advantages of the cubature rule, many cubature filters have been reported such as the CKF in Arasaratnam and Haykin (2009) and the adaptive cubature filter in Qiu et al. (2018).

Knowing the superior approximation capability of the cubature rule, the third-degree spherical-radial Cubature rule (Arasaratnam and Haykin, 2009; Tarn and Rasis, 1976; Macagnano and Abreu, 2012) is applied in this chapter to design a cubature predictive filtering approach for the microsatellite control system with modeling error. The deterministically chosen Cubature-points, which match the mean and covariance of a (not necessarily Gaussian-distributed) probability distribution, are generated to develop this filter. When these cubature-points are propagated through the nonlinear function of the modeling error to enhance the accuracy of estimation, the estimation of the modeling error for the

Predictive Filtering for Microsatellite Control System. https://doi.org/10.1016/B978-0-12-821865-5.00020-6

filtering method can capture the posterior mean accurately up to the 3rd order for nonlinear Gaussian system, while the estimation errors are only introduced in the 4th and higher orders. The estimated modeling error can capture the posterior mean accurately to the 2nd order for any nonlinearity.

6.2 Third-degree spherical-radial cubature rule

Let us consider an integral of the form

$$I(f) = \int_{\mathbb{R}^n} f(\boldsymbol{x}) \exp(-\boldsymbol{x}^{\mathrm{T}}\boldsymbol{x}) d\boldsymbol{x} \tag{6.2}$$

defined in the Cartesian coordinate system. In the spherical-radial cubature transformation, a new variable $\boldsymbol{x} = r\boldsymbol{s}$ with $\boldsymbol{s}^{\mathrm{T}}\boldsymbol{s} = 1$ and $\boldsymbol{x}^{\mathrm{T}}\boldsymbol{x} = r^2$ is first introduced for $r \in [0, \infty)$. In accordance, the integral (6.1) can be transformed in the spherical-radial coordinate system as (Arasaratnam and Haykin, 2009)

$$I(f) = \int_0^{\infty} \int_{U_n} f(r\boldsymbol{s}) r^{n-1} \exp(-r^2) d\sigma(\boldsymbol{s}) dr \tag{6.3}$$

where $U_n = \{\boldsymbol{s} \in \mathbb{R}^n | \boldsymbol{s}^{\mathrm{T}}\boldsymbol{s} = 1\}$, $\boldsymbol{s} = [s_1 \quad s_2 \quad \cdots \quad s_n]^{\mathrm{T}}$ is the surface of the sphere, and $\boldsymbol{\sigma}(\cdot)$ is the spherical surface measure or the area element on U_n. Then, the integral (6.2) can be decomposed into the spherical integral $S(r)$ and the radial integral I_f, which are expressed as

$$S(r) = \int_{U_n} f(r\boldsymbol{s}) d\sigma(\boldsymbol{s}) \tag{6.4}$$

$$I_f = \int_0^{\infty} S(r) r^{n-1} \exp(-r^2) dr \tag{6.5}$$

Theorem 6.1. *(Arasaratnam and Haykin, 2009) Let the radial integral be calculated numerically by the N_r-point Gaussian quadrature rule, i.e.,*

$$\int_0^{\infty} S(r) r^{n-1} \exp(-r^2) dr \approx \sum_{i=1}^{N_r} \omega_{r,i} S(r_i) \tag{6.6}$$

Moreover, the spherical integral is numerically computed by the N_s-point spherical rule, i.e.,

$$\int_{U_n} f(r\boldsymbol{s}) d\sigma(\boldsymbol{s}) \approx \sum_{j=1}^{N_s} \omega_{s,j} f(r\boldsymbol{s}_j) \tag{6.7}$$

Then, the integral in (6.2) *can be approximated by an* $(N_s \times N_r)$*-point spherical-radial cubature rule, i.e.,*

$$I(f) = \int_{\mathbb{R}^n} f(\boldsymbol{x}) \exp(-\boldsymbol{x}^{\mathrm{T}}\boldsymbol{x}) d\boldsymbol{x} \approx \sum_{i=1}^{N_r} \sum_{j=1}^{N_s} \omega_{r,i} \omega_{s,j} f(r_i \boldsymbol{s}_j) \tag{6.8}$$

where r_i, $\boldsymbol{s}_j$ *and* $\omega_{r,i}$, $\omega_{s,j}$ *are the sampling points and weights for calculating the radial integral and spherical integral, respectively.*

For the third-degree spherical-radial cubature rule, it follows that $N_r = 1$ and $N_s = 2n$. Hence, $2n$ cubature sigma points are generated. More specifically, when the weighting function $w(\boldsymbol{x})$ is the standard Gaussian density, invoking the third-degree spherical-radial rule, then the integral (6.1) can be computed by using (6.8) with $N_r = 1$, $N_s = 2n$, $r_1 = \sqrt{n}$, $\omega_{s,j} = \frac{1}{2n}$, $j = 1, 2, \cdots, 2n$, and $\boldsymbol{s}_j$ is the jth element of the following set $\mathfrak{D}_2$, *i.e.*,

$$\mathfrak{D}_2 = \left\{ \begin{bmatrix} 1 \\ 0 \\ \vdots \\ 0 \end{bmatrix}, \begin{bmatrix} 0 \\ 1 \\ \vdots \\ 0 \end{bmatrix}, \cdots, \begin{bmatrix} 0 \\ 0 \\ \vdots \\ 1 \end{bmatrix}, \begin{bmatrix} -1 \\ 0 \\ \vdots \\ 0 \end{bmatrix}, \begin{bmatrix} 0 \\ -1 \\ \vdots \\ 0 \end{bmatrix}, \cdots, \begin{bmatrix} 0 \\ 0 \\ \vdots \\ -1 \end{bmatrix} \right\} \tag{6.9}$$

6.3 Cubature predictive filtering

Consider the nonlinear discrete systems described by (2.32)–(2.33), invoking the third-degree spherical-radial cubature rule, the cubature predictive filter (CPF) to estimate their states is presented in Algorithm 6.1. Moreover, the numerical integral method is applied to update the state estimation and its error covariance in Algorithm 6.1.

6.4 Estimation performance of CPF

The estimation performance of the modeling error and its covariance guaranteed by the CPF is rigorously analyzed in this section.

6.4.1 Estimation accuracy of modeling error

For the CPF presented in Algorithm 6.1, the third-degree spherical-radial cubature rule is used to approximate the posterior distribution of the modeling error. A detailed analysis of the posterior distribution of the modeling error after this cubature rule is provided in this subsection.

As same as Chapter 5, let S_k be the Cholesky decomposition of the state estimation error covariance matrix $P_{\tilde{x},k}$, then $P_{\tilde{x},k}$ can be decomposed as follows, where $\sigma_{i,k}$ is the ith column of the matrix S_k, $i = 1, 2, \cdots, n$.

$$P_{\tilde{x},k} = S_k S_k^{\mathrm{T}} = \sum_{i=1}^{n} \sigma_{i,k} \sigma_{i,k}^{\mathrm{T}} \tag{6.10}$$

Algorithm 6.1: Cubature predictive filter.

Input Data: Initial estimation $\hat{x}_0$, $\hat{y}_0$, and measurement y_k
Result: State estimation $\hat{x}_k$
begin
  Initializing the state and the covariance matrix as
  $\hat{x}_0 = \mathbb{E}(x_0)$
  $P_{\tilde{x},0} = \mathrm{Var}(x_0) = \mathbb{E}\left((x_0 - \hat{x}_0)(x_0 - \hat{x}_0)^{\mathrm{T}}\right)$
  for $k = 1, 2, \cdots$ **do**
    Generating the following cubature points using the third-degree spherical-radial cubature rule, $\hat{x}_{k-1}$, and $P_{\tilde{x},k-1}$
    $\xi_{i,k-1}, i = 1, 2, \cdots, 2n$
    Updating the modeling error's estimation $\hat{d}_{k-1}$ as
    $\beta_{i,k-1} = M(\xi_{i,k-1}), i = 0, 1, \cdots, 2n$
    $\gamma_{i,k-1} = \beta_{i,k-1}(h_{k-1}(\xi_{i,k-1}) - y_k), i = 0, 1, \cdots, 2n$
    $\alpha_{i,k-1} = \beta_{i,k-1}\underline{Z}(\xi_{i,k-1}, \Delta t), i = 0, 1, \cdots, 2n$
    $\overline{h}_{k-1} = \frac{1}{2n}\sum_{i=0}^{2n} \gamma_{i,k-1} = \frac{1}{2n}\sum_{i=0}^{2n} \underline{h}(\xi_{i,k-1})$
    $\overline{Z}_{k-1} = \frac{1}{2n}\sum_{i=0}^{2n} \alpha_{i,k} = \frac{1}{2n}\sum_{i=0}^{2n} \underline{Z}(\xi_{i,k-1}, \Delta t)$
    $\hat{d}_{i,k-1} = -(\alpha_{i,k-1} + \gamma_{i,k-1}), i = 0, 1, \cdots, 2n$
    $\hat{d}_{k-1} = -(\overline{Z}_{k-1} + \overline{h}_{k-1})$
    Updating the state estimation
    $\delta_{i,k|k-1} = f(\xi_{i,k-1}), i = 0, 1, \cdots, 2n$
    $\hat{x}_{k|k-1} = \frac{1}{2n}\sum_{i=0}^{2n} \delta_{i,k|k-1}$
    $\hat{x}_k = \hat{x}_{k|k-1} + g(\hat{x}_{k|k-1})\hat{d}_{k-1}$
    Updating the state estimation error covariance
    $\chi_{i,k} = \delta_{i,k|k-1} + g(\xi_{i,k-1})\hat{d}_{i,k-1}, i = 0, 1, \cdots, 2n$
    $P_{\tilde{x},k} = \frac{1}{2n}\sum_{i=0}^{2n} (\chi_{i,k} - \hat{x}_k)(\chi_{i,k} - \hat{x}_k)^{\mathrm{T}}$
  end
end

The set of cubature-points is deterministically chosen in order to ensure that these points can completely capture the true mean and covariance of $\boldsymbol{x}_k$. More specifically, the set of $2n$ cubature-points $\boldsymbol{\xi}_{i,k}$, $i = 1, 2, \cdots, 2n$, are given by

$$\boldsymbol{\xi}_{i,k} = \hat{\boldsymbol{x}}_k + \sqrt{n}\boldsymbol{\sigma}_{i,k} = \hat{\boldsymbol{x}}_k + \hat{\boldsymbol{\sigma}}_{i,k}, \; i = 1, 2, \cdots, n \tag{6.11}$$

$$\boldsymbol{\xi}_{i+n,k} = \hat{\boldsymbol{x}}_k - \sqrt{n}\boldsymbol{\sigma}_{i,k} = \hat{\boldsymbol{x}}_k - \hat{\boldsymbol{\sigma}}_{i,k}, \; i = 1, 2, \cdots, n \tag{6.12}$$

where $\hat{\boldsymbol{\sigma}}_{i,k} = \sqrt{n}\boldsymbol{\sigma}_{i,k}$.

Each cubature-point $\boldsymbol{\xi}_{i,k}$, $i = 1, 2, \cdots, 2n$, is now propagated through the nonlinear function to generate $\boldsymbol{\gamma}_{i,k}$.

$$\boldsymbol{\gamma}_{i,k} = \boldsymbol{h}(\boldsymbol{\xi}_{i,k}) = \boldsymbol{h}(\hat{\boldsymbol{x}}_k) + \boldsymbol{D}_{\hat{\sigma}_{i,k}}\boldsymbol{h} + \sum_{j=2}^{\infty} \frac{\boldsymbol{D}^j_{\hat{\sigma}_{i,k}}\boldsymbol{h}}{j!}, \; i = 1, 2, \cdots, n \tag{6.13}$$

$$\boldsymbol{\gamma}_{i+n,k} = \boldsymbol{h}(\boldsymbol{\xi}_{i+n,k}) = \boldsymbol{h}(\hat{\boldsymbol{x}}_k) + \boldsymbol{D}_{-\hat{\sigma}_{i,k}}\boldsymbol{h} + \sum_{j=2}^{\infty} \frac{\boldsymbol{D}^j_{-\hat{\sigma}_{i,k}}\boldsymbol{h}}{j!}, \; i = 1, 2, \cdots, n \tag{6.14}$$

$$\boldsymbol{\alpha}_{i,k} = \underline{\boldsymbol{Z}}(\boldsymbol{\xi}_{i,k}, \Delta t) = \underline{\boldsymbol{Z}}(\hat{\boldsymbol{x}}_k, \Delta t) + \boldsymbol{D}_{\hat{\sigma}_{i,k}}\underline{\boldsymbol{Z}} + \sum_{j=2}^{\infty} \frac{\boldsymbol{D}^j_{\hat{\sigma}_{i,k}}\underline{\boldsymbol{Z}}}{j!}, \; i = 1, 2, \cdots, n \tag{6.15}$$

$$\boldsymbol{\alpha}_{i+n,k} = \underline{\boldsymbol{Z}}(\boldsymbol{\xi}_{i+n,k}, \Delta t) = \underline{\boldsymbol{Z}}(\hat{\boldsymbol{x}}_k, \Delta t) + \sum_{j=1}^{\infty} \frac{\boldsymbol{D}^j_{-\hat{\sigma}_{i,k}}\underline{\boldsymbol{Z}}}{j!}, \; i = 1, 2, \cdots, n \tag{6.16}$$

Applying the vector differential operators (2.54) and (2.55), one has

$$\boldsymbol{D}^m_{-\hat{\sigma}_{i,k}}\boldsymbol{h} = ((-\hat{\boldsymbol{\sigma}}_{i,k})^{\mathrm{T}}\nabla)^m \boldsymbol{h}(\boldsymbol{x})|_{x=\hat{\boldsymbol{x}}_k} = -\boldsymbol{D}^m_{\hat{\sigma}_{i,k}}\boldsymbol{h} \tag{6.17}$$

$$\boldsymbol{D}^p_{-\hat{\sigma}_{i,k}}\boldsymbol{h} = ((-\hat{\boldsymbol{\sigma}}_{i,k})^{\mathrm{T}}\nabla)^m \boldsymbol{h}(\boldsymbol{x})|_{x=\hat{\boldsymbol{x}}_k} = \boldsymbol{D}^p_{\hat{\sigma}_{i,k}}\boldsymbol{h} \tag{6.18}$$

$$\boldsymbol{D}^m_{-\hat{\sigma}_{i,k}}\underline{\boldsymbol{Z}} = ((-\hat{\boldsymbol{\sigma}}_{i,k})^{\mathrm{T}}\nabla)^m \underline{\boldsymbol{Z}}(\boldsymbol{x})|_{x=\hat{\boldsymbol{x}}_k} = -\boldsymbol{D}^m_{\hat{\sigma}_{i,k}}\underline{\boldsymbol{Z}} \tag{6.19}$$

$$\boldsymbol{D}^p_{-\hat{\sigma}_{i,k}}\underline{\boldsymbol{Z}} = ((-\hat{\boldsymbol{\sigma}}_{i,k})^{\mathrm{T}}\nabla)^m \underline{\boldsymbol{Z}}(\boldsymbol{x})|_{x=\hat{\boldsymbol{x}}_k} = -\boldsymbol{D}^p_{\hat{\sigma}_{i,k}}\underline{\boldsymbol{Z}} \tag{6.20}$$

where $m \in \mathbb{N}$ is an odd number and $p \in \mathbb{N}$ is an even number.

Invoking

$$\frac{1}{2n}\sum_{i=1}^{2n}\frac{D_{\hat{\sigma}_{i,k}}^{2}h}{2!}=\frac{1}{2n}\mathbb{E}\left(\frac{D_{\hat{\sigma}_{i,k}}(D_{\hat{\sigma}_{i,k}}h)}{2!}\right)=\frac{(\nabla^{\mathrm{T}}P_{\tilde{x},k}\nabla)h(x)\big|_{x=\hat{x}_k}}{2} \tag{6.21}$$

and

$$\frac{1}{2n}\sum_{i=1}^{2n}\frac{D_{\hat{\sigma}_{i,k}}^{2}\underline{Z}}{2!}=\mathbb{E}\left(\frac{D_{\Delta x}^{2}\underline{Z}}{2!}\right) \tag{6.22}$$

leaves the posterior distribution of different nonlinear functions after the approximation be calculated as

$$\begin{aligned}(\bar{h}_k)_{\mathrm{CPF}}&=\frac{1}{2n}\sum_{i=1}^{n}(\gamma_{i,k}+\gamma_{i+n,k})=h(\hat{x}_k)+\frac{1}{n}\sum_{i=1}^{n}\sum_{j=1}^{\infty}\frac{D_{\hat{\sigma}_{i,k}}^{2j}h}{(2j)!}\\&=h(\hat{x}_k)+\mathbb{E}\left(\frac{D_{\Delta x}^{2}h}{2!}\right)+\frac{1}{n}\sum_{i=1}^{n}\sum_{j=2}^{\infty}\frac{D_{\hat{\sigma}_{i,k}}^{2j}h}{(2j)!}\end{aligned} \tag{6.23}$$

$$\begin{aligned}(\bar{\underline{Z}}_k)_{\mathrm{CPF}}&=\frac{1}{2n}\sum_{i=1}^{n}(\gamma_{i,k}+\gamma_{i+n,k})\\&=\underline{Z}\left(\hat{x}_k\right)+\frac{1}{n}\sum_{i=1}^{n}\sum_{j=1}^{\infty}\frac{D_{\hat{\sigma}_{i,k}}^{2j}\underline{Z}}{(2j)!}\\&=\underline{Z}\left(\hat{x}_k\right)+\mathbb{E}\left(\frac{D_{\Delta x}^{2}\underline{Z}}{2!}\right)+\frac{1}{n}\sum_{i=1}^{n}\sum_{j=2}^{\infty}\frac{D_{\hat{\sigma}_{i,k}}^{2j}\underline{Z}}{(2j)!}\end{aligned} \tag{6.24}$$

Then, we obtain from Algorithm 6.1 and (6.24)–(6.25) that the estimation of modeling error after the third-degree spherical-radial cubature rule is given by

$$\begin{aligned}(\hat{d}_k)_{\mathrm{CPF}}=&-\left(h\left(\hat{x}_k\right)+\mathbb{E}\left(\frac{D_{\Delta x}^{2}h}{2!}\right)+\frac{1}{n}\sum_{i=1}^{n}\sum_{j=2}^{\infty}\frac{D_{\hat{\sigma}_{i,k}}^{2j}h}{(2j)!}\right)\\&-\left(\underline{Z}\left(\hat{x}_k\right)+\mathbb{E}\left(\frac{D_{\Delta x}^{2}\underline{Z}}{2!}\right)+\frac{1}{n}\sum_{i=1}^{n}\sum_{j=2}^{\infty}\frac{D_{\hat{\sigma}_{i,k}}^{2j}\underline{Z}}{(2j)!}\right)\end{aligned} \tag{6.25}$$

Additionally, following almost the same calculation procedure given in Section 4.3.1, the covariance matrix of the estimated modeling error obtained from

CPF can be calculated as

$$
\begin{aligned}
(\boldsymbol{P}_{d,k})_{\mathrm{CPF}} &= \frac{1}{2n}\sum_{i=0}^{2n}\mathbb{E}(\boldsymbol{A}\boldsymbol{A}^{\mathrm{T}}) \\
&= \boldsymbol{G}_{\underline{h}}\boldsymbol{P}_{\tilde{x},k}\boldsymbol{G}_{\underline{h}}^{\mathrm{T}} + \boldsymbol{G}_{\underline{Z}}\boldsymbol{P}_{\tilde{x},k}\boldsymbol{G}_{\underline{Z}}^{\mathrm{T}} + \boldsymbol{G}_{\underline{h}}\boldsymbol{P}_{\tilde{x},k}\boldsymbol{G}_{\underline{Z}}^{\mathrm{T}} + \boldsymbol{G}_{\underline{Z}}\boldsymbol{P}_{\tilde{x},k}\boldsymbol{G}_{\underline{h}}^{\mathrm{T}} \\
&\quad - \mathbb{E}\left(\frac{\boldsymbol{D}_{\Delta x}^{2}(\underline{\boldsymbol{Z}}+\underline{\boldsymbol{h}})}{2!}\right)\mathbb{E}\left(\left(\frac{\boldsymbol{D}_{\Delta x}^{2}(\underline{\boldsymbol{Z}}+\underline{\boldsymbol{h}})}{2!}\right)^{\mathrm{T}}\right) + \frac{1}{n}\boldsymbol{\Sigma}_1 + \frac{1}{n^2}\boldsymbol{\Sigma}_2
\end{aligned} \tag{6.26}
$$

with $\boldsymbol{A} = \boldsymbol{\gamma}_{i,k} + \boldsymbol{\alpha}_{i,k} - (\bar{\boldsymbol{h}}_k)_{\mathrm{CPF}} - (\bar{\underline{\boldsymbol{Z}}}_k)_{\mathrm{CPF}}$ and

$$
\boldsymbol{\Sigma}_1 = \underbrace{\sum_{m=1}^{\infty}\sum_{i=1}^{\infty}\sum_{j=1}^{\infty}\frac{1}{i!j!}\mathbb{E}((\boldsymbol{D}_{\hat{\sigma}_{km}}^{i}(\underline{\boldsymbol{Z}}+\underline{\boldsymbol{h}}))(\boldsymbol{D}_{\hat{\sigma}_{km}}^{j}(\underline{\boldsymbol{Z}}+\underline{\boldsymbol{h}}))^{\mathrm{T}})}_{\text{condition\_1}} \tag{6.27}
$$

$$
\boldsymbol{\Sigma}_2 = \underbrace{\sum_{i=1}^{\infty}\sum_{j=1}^{\infty}\sum_{l=1}^{n}\sum_{m=1}^{n}\frac{1}{(2i)!(2j)!}\mathbb{E}(\boldsymbol{D}_{\hat{\sigma}_{kl}}^{2i}(\underline{\boldsymbol{Z}}+\underline{\boldsymbol{h}})\mathbb{E}(\boldsymbol{D}_{\hat{\sigma}_{km}}^{2j}(\underline{\boldsymbol{Z}}+\underline{\boldsymbol{h}}))^{\mathrm{T}}}_{\text{condition\_2}} \tag{6.28}
$$

Theorem 6.2. *Let the CPF presented in Algorithm 6.1 be applied to estimate the modeling error in the nonlinear discrete system* (2.32)–(2.33)*, if the system* (2.32)–(2.33) *is Gaussian, then the CPF can capture the posterior mean of the modeling error accurately to* 3rd *order with its estimation errors only introduced in the* 4th *and higher orders. Moreover, the CPF completely capture the posterior mean of the modeling error accurately to* 2nd *order at least for any nonlinear system* (2.32)–(2.33)*. The CPF has higher estimation accuracy for modeling error than the classical PF in Algorithm 2.1.*

Proof. Comparing (2.63)–(2.64) with (6.25)–(5.26), the conclusion in Theorem 6.2 can be directly proved. □

Remark 6.1. The CPF guarantees almost the same estimation accuracy as the UPF in Chapter 4 and the CDPF in Chapter 5 for the modeling error $\boldsymbol{d}_k$. The only difference is seen in the terms having more than four order by comparing (6.25) with (5.20) and (4.25).

6.4.2 Estimation accuracy of system state

The state and the covariance estimation performance achieved by the CPF are analyzed in this subsection.

From (6.11) and (6.12), the state estimation error covariance $\boldsymbol{P}_{\tilde{x},k}$ is to be calculated and decomposed by using the state $(\boldsymbol{x}_k)_{\mathrm{CPF}}$ for any time. At the same time, the set of cubature-points $\boldsymbol{\xi}_{i,k}$, $i = 1, 2, ..., 2n$, is symmetrically represented by

$$\boldsymbol{\xi}_{i,k} = (\hat{\boldsymbol{x}}_k)_{\text{CPF}} + \sqrt{n}\boldsymbol{\sigma}_{i,k} = \hat{\boldsymbol{x}}_k + \hat{\boldsymbol{\sigma}}_{i,k},\ i = 1, 2, \cdots, n \tag{6.29}$$

$$\boldsymbol{\xi}_{i+n,k} = (\hat{\boldsymbol{x}}_k)_{\text{CPF}} - \sqrt{n}\boldsymbol{\sigma}_{i,k} = \hat{\boldsymbol{x}}_k - \hat{\boldsymbol{\sigma}}_{i,k},\ i = 1, 2, \cdots, n \tag{6.30}$$

Then, using those points to propagate through the nonlinear transformation yields

$$\boldsymbol{\delta}_{i,k} = \boldsymbol{f}(\boldsymbol{\xi}_{i,k}) = \boldsymbol{f}((\hat{\boldsymbol{x}}_k)_{\text{CPF}}) + \boldsymbol{D}_{\hat{\boldsymbol{\sigma}}_{i,k}}\boldsymbol{f} + \sum_{j=2}^{\infty}\frac{\boldsymbol{D}^j_{\hat{\boldsymbol{\sigma}}_{i,k}}\boldsymbol{f}}{j!},\ i = 1, 2, \cdots, n \tag{6.31}$$

$$\boldsymbol{\delta}_{i+n,k} = \boldsymbol{f}(\boldsymbol{\xi}_{i+n,k}) = \boldsymbol{f}((\hat{\boldsymbol{x}}_k)_{\text{CPF}}) + \sum_{j=1}^{\infty}\frac{\boldsymbol{D}^j_{-\hat{\boldsymbol{\sigma}}_{i,k}}\boldsymbol{f}}{j!},\ i = 1, 2, \cdots, n \tag{6.32}$$

$$\boldsymbol{\eta}_{i,k} = \boldsymbol{\eta}(\boldsymbol{\xi}_{i,k}) = \boldsymbol{\eta}((\hat{\boldsymbol{x}}_k)_{\text{CPF}}) + \boldsymbol{D}_{\hat{\boldsymbol{\sigma}}_{i,k}}\boldsymbol{\eta} + \sum_{j=2}^{\infty}\frac{\boldsymbol{D}^j_{\hat{\boldsymbol{\sigma}}_{i,k}}\boldsymbol{\eta}}{j!},\ i = 1, 2, \cdots, n \tag{6.33}$$

$$\boldsymbol{\eta}_{i+n,k} = \boldsymbol{\eta}(\boldsymbol{\xi}_{i+n,k}) = \boldsymbol{\eta}((\hat{\boldsymbol{x}}_k)_{\text{CPF}}) + \sum_{j=1}^{\infty}\frac{\boldsymbol{D}^j_{-\hat{\boldsymbol{\sigma}}_{i,k}}\boldsymbol{\eta}}{j!},\ i = 1, 2, \cdots, n \tag{6.34}$$

In accordance with (2.67)–(2.68) and Algorithm 6.1, the states estimation provided by the CPF has a form of

$$\begin{aligned}
(\hat{\boldsymbol{x}}_{k+1})_{\text{CPF}} &= \frac{1}{2n}\sum_{i=1}^{2n}(\boldsymbol{\xi}_{i,k} + \Delta t\boldsymbol{\delta}_{i,k} + \Delta t\boldsymbol{\rho}_{i,k} + \boldsymbol{\mu}_{i,k}) \\
&= (\boldsymbol{x}_k)_{\text{CPF}} + \Delta t\boldsymbol{f}((\boldsymbol{x}_k)_{\text{CPF}}) + \Delta t\boldsymbol{\eta}((\boldsymbol{x}_k)_{\text{CPF}}) + \frac{1}{2n}\sum_{i=1}^{2n}\boldsymbol{\mu}_{i,k} \\
&\quad + \frac{\Delta t}{n}\sum_{i=1}^{n}\sum_{j=1}^{\infty}\frac{\boldsymbol{D}^{2j}_{\hat{\boldsymbol{\sigma}}_{i,k}}\boldsymbol{\eta}}{(2j)!} + \frac{\Delta t}{n}\sum_{i=1}^{n}\sum_{j=1}^{\infty}\frac{\boldsymbol{D}^{2j}_{\hat{\boldsymbol{\sigma}}_{i,k}}\boldsymbol{\eta}}{(2j)!}
\end{aligned} \tag{6.35}$$

Consequently, following the same procedures in Chapter 4 and Chapter 5, the state estimation error provided by the CPF is governed by

$$\begin{aligned}
(\tilde{\boldsymbol{x}}_{k+1})_{\text{CPF}} &= \boldsymbol{x}_{k+1} - (\hat{\boldsymbol{x}}_{k+1})_{\text{CPF}} \\
&= (\tilde{\boldsymbol{x}}_k)_{\text{CPF}} + \Delta t(\boldsymbol{G}_f(\tilde{\boldsymbol{x}}_k)_{\text{CPF}} + \boldsymbol{G}_\eta(\tilde{\boldsymbol{x}}_k)_{\text{CPF}}) \\
&\quad + \bar{\boldsymbol{\mu}}_k(\boldsymbol{x}_k, \boldsymbol{d}_k, (\hat{\boldsymbol{x}}_k)_{\text{CPF}}, (\hat{\boldsymbol{d}}_k)_{\text{CPF}}) + (\boldsymbol{s}_k)_{\text{CPF}} \\
&= \boldsymbol{A}_k(\tilde{\boldsymbol{x}}_k)_{\text{CPF}} + \bar{\boldsymbol{\mu}}_k(\boldsymbol{x}_k, \boldsymbol{d}_k, (\hat{\boldsymbol{x}}_k)_{\text{CPF}}, (\hat{\boldsymbol{d}}_k)_{\text{CPF}}) + (\boldsymbol{s}_k)_{\text{CPF}}
\end{aligned} \tag{6.36}$$

where $(\bar{\boldsymbol{\mu}}_k)_{\mathrm{CPF}} = \bar{\boldsymbol{\mu}}_k(\boldsymbol{x}_k, \boldsymbol{d}_k, (\hat{\boldsymbol{x}}_k)_{\mathrm{CPF}}, (\hat{\boldsymbol{d}}_k)_{\mathrm{CPF}}) = \boldsymbol{\mu}(\boldsymbol{x}_k, \boldsymbol{d}_k)\boldsymbol{v}_k - \boldsymbol{\mu}(\hat{\boldsymbol{x}}_k, \hat{\boldsymbol{d}}_k) - \boldsymbol{\varepsilon}(\bar{\boldsymbol{B}}(\hat{\boldsymbol{x}}_k)_{\mathrm{CPF}})\boldsymbol{v}_k + \Delta t(\Delta \boldsymbol{f}(\boldsymbol{x}_k, (\hat{\boldsymbol{x}}_k)_{\mathrm{CPF}}) + \Delta \boldsymbol{\eta}(\boldsymbol{x}_k, \boldsymbol{d}_k, (\hat{\boldsymbol{x}}_k)_{\mathrm{CPF}}, (\hat{\boldsymbol{d}}_k)_{\mathrm{CPF}}))$ and

$$\boldsymbol{A}_k = \boldsymbol{I} + \Delta t \boldsymbol{G}_f + \Delta t \boldsymbol{G}_\eta \tag{6.37}$$

$$\Delta \boldsymbol{f}(\boldsymbol{x}_k, (\hat{\boldsymbol{x}}_k)_{\mathrm{CPF}}) = \sum_{i=2}^{\infty} \frac{1}{i!} \boldsymbol{D}_{\hat{\boldsymbol{x}}_k}^i \boldsymbol{f} - \frac{\Delta t}{n} \sum_{i=1}^{n} \sum_{j=1}^{\infty} \frac{\boldsymbol{D}_{\boldsymbol{\sigma}_{i,k}}^{2j} \boldsymbol{f}}{(2j)!} h^j \tag{6.38}$$

$$\Delta \boldsymbol{\eta}(\boldsymbol{x}_k, \boldsymbol{d}_k, (\hat{\boldsymbol{x}}_k)_{\mathrm{CPF}}, (\hat{\boldsymbol{d}}_k)_{\mathrm{CPF}}) = \sum_{i=2}^{\infty} \frac{1}{i!} \boldsymbol{D}_{\hat{\boldsymbol{x}}_k}^i \boldsymbol{\eta} - \frac{\Delta t}{n} \sum_{i=1}^{n} \sum_{j=1}^{\infty} \frac{\boldsymbol{D}_{\boldsymbol{\sigma}_{i,k}}^{2j} \boldsymbol{\eta}}{(2j)!} h^j \tag{6.39}$$

$$(\boldsymbol{s}_k)_{\mathrm{CPF}} = -\Delta t \bar{\boldsymbol{B}}((\hat{\boldsymbol{x}}_k)_{\mathrm{CPF}})\boldsymbol{v}_k = -(\boldsymbol{B}_k)_{\mathrm{CPF}} \boldsymbol{v}_k \tag{6.40}$$

Based on the definition of the covariance, it can be calculated that the covariance of state estimation achieved by the CPF is

$$\begin{aligned}
(\boldsymbol{P}_{\tilde{x},k+1})_{\mathrm{CPF}} &= \frac{1}{2n} \sum_{t=1}^{2n} (\boldsymbol{\xi}_{i,k+1} - (\hat{\boldsymbol{x}}_{k+1})_{\mathrm{CPF}})(\boldsymbol{\xi}_{i,k+1} - (\hat{\boldsymbol{x}}_{k+1})_{\mathrm{CPF}})^{\mathrm{T}} \\
&= (\boldsymbol{P}_{\tilde{x},k})_{\mathrm{CPF}} + \bar{\boldsymbol{\psi}}(\boldsymbol{v}_k) + \bar{\boldsymbol{o}}(\boldsymbol{\delta}_{i,k}, \boldsymbol{\mu}_{i,k}) \\
&+ \frac{\Delta t}{2n} \sum_{i=1}^{2n} (\boldsymbol{\xi}_{i,k} - (\boldsymbol{x}_k)_{\mathrm{CPF}})(\boldsymbol{\delta}_{i,k} - \boldsymbol{f}((\hat{\boldsymbol{x}}_k)_{\mathrm{CPF}}))^{\mathrm{T}} \\
&+ \frac{\Delta t}{2n} \sum_{i=1}^{2n} (\boldsymbol{\xi}_{i,k} - (\boldsymbol{x}_k)_{\mathrm{CPF}})(\boldsymbol{\rho}_{i,k} - \boldsymbol{\eta}((\hat{\boldsymbol{x}}_k)_{\mathrm{CPF}}))^{\mathrm{T}} \\
&+ \frac{\Delta t}{2n} \sum_{i=1}^{2n} (\boldsymbol{\delta}_{i,k} - \boldsymbol{f}((\boldsymbol{x}_k)_{\mathrm{CPF}}))(\boldsymbol{\xi}_{i,k} - (\hat{\boldsymbol{x}}_k)_{\mathrm{CPF}})^{\mathrm{T}} \\
&+ \frac{\Delta t}{2n} \sum_{i=1}^{2n} (\boldsymbol{\rho}_{i,k} - \boldsymbol{\eta}((\boldsymbol{x}_k)_{\mathrm{CPF}}))(\boldsymbol{\xi}_{i,k} - (\hat{\boldsymbol{x}}_k)_{\mathrm{CPF}})^{\mathrm{T}} \\
&+ \frac{(\Delta t)^2}{2n} \sum_{i=0}^{2n} (\boldsymbol{\delta}_{i,k} - \boldsymbol{f}((\hat{\boldsymbol{x}}_k)_{\mathrm{CPF}}))(\boldsymbol{\delta}_{i,k} - \boldsymbol{f}((\hat{\boldsymbol{x}}_k)_{\mathrm{CPF}}))^{\mathrm{T}} \\
&+ \frac{(\Delta t)^2}{2n} \sum_{i=0}^{2n} (\boldsymbol{\rho}_{i,k} - \boldsymbol{\eta}((\hat{\boldsymbol{x}}_k)_{\mathrm{CPF}}))(\boldsymbol{\rho}_{i,k} - \boldsymbol{\eta}((\hat{\boldsymbol{x}}_k)_{\mathrm{CPF}}))^{\mathrm{T}} \\
&+ \frac{(\Delta t)^2}{2n} \sum_{i=0}^{2n} (\boldsymbol{\delta}_{i,k} - \boldsymbol{f}((\hat{\boldsymbol{x}}_k)_{\mathrm{CPF}}))(\boldsymbol{\rho}_{i,k} - \boldsymbol{\eta}((\hat{\boldsymbol{x}}_k)_{\mathrm{CPF}}))^{\mathrm{T}} \\
&+ \frac{(\Delta t)^2}{2n} \sum_{i=0}^{2n} (\boldsymbol{\rho}_{i,k} - \boldsymbol{\eta}((\hat{\boldsymbol{x}}_k)_{\mathrm{CPF}}))(\boldsymbol{\delta}_{i,k} - \boldsymbol{f}((\hat{\boldsymbol{x}}_k)_{\mathrm{CPF}}))^{\mathrm{T}}
\end{aligned} \tag{6.41}$$

where $\bar{o}(\boldsymbol{\delta}_{i,k}, \boldsymbol{\mu}_{i,k})$ is the sum of the polynomial of $\boldsymbol{\mu}_{i,k} - \frac{1}{2n}\sum_{i=1}^{2n}\boldsymbol{\mu}_{k,i}$ and $\bar{\boldsymbol{\psi}}(\boldsymbol{v}_k)$ is the sum of the polynomial of $\boldsymbol{v}_k$.

For the terms in (6.41), it follows that

$$\begin{aligned}
&\frac{1}{2n}\sum_{i=1}^{2n}(\boldsymbol{\xi}_{i,k} - (\boldsymbol{x}_k)_{\text{CPF}})(\boldsymbol{\delta}_{i,k} - \boldsymbol{f}((\hat{\boldsymbol{x}}_k)_{\text{CPF}}))^{\text{T}} \\
&= \frac{1}{n}\sum_{i=1}^{n}\hat{\boldsymbol{\sigma}}_{i,k}\left(\sum_{j=1}^{\infty}\frac{\boldsymbol{D}_{\hat{\sigma}_{i,k}}^{2j-1}\boldsymbol{f}}{(2j-1)!}\right)^{\text{T}} = \boldsymbol{P}_{\tilde{x},k}\boldsymbol{G}_f^{\text{T}} + \sum_{i=1}^{n}\sum_{j=1}^{\infty}\frac{\hat{\boldsymbol{\sigma}}_{i,k}(\boldsymbol{D}_{\hat{\sigma}_{i,k}}^{2j+1}\boldsymbol{f})^{\text{T}}}{n(2j+1)!}
\end{aligned} \tag{6.42}$$

$$\begin{aligned}
&\frac{1}{2n}\sum_{i=1}^{2n}(\boldsymbol{\delta}_{i,k} - (\boldsymbol{x}_k)_{\text{CPF}})(\boldsymbol{\delta}_{i,k} - \boldsymbol{f}((\hat{\boldsymbol{x}}_k)_{\text{CPF}}))^{\text{T}} \\
&= \frac{1}{2n}\sum_{i=1}^{2n}\boldsymbol{\delta}_{i,k}\boldsymbol{\delta}_{i,k}^{\text{T}} - \boldsymbol{f}((\hat{\boldsymbol{x}}_k)_{\text{CPF}})(\boldsymbol{f}((\hat{\boldsymbol{x}}_k)_{\text{CPF}}))^{\text{T}} \\
&= \boldsymbol{G}_f\boldsymbol{P}_{\tilde{x},k}\boldsymbol{G}_f^{\text{T}} - \mathbb{E}\left(\frac{\boldsymbol{D}_{\Delta x}^2\boldsymbol{f}}{2!}\right)\mathbb{E}\left(\left(\frac{\boldsymbol{D}_{\Delta x}^2\boldsymbol{f}}{2!}\right)^{\text{T}}\right) \\
&\quad + \underbrace{\frac{1}{n}\sum_{l=1}^{n}\sum_{i=1}^{\infty}\sum_{j=1}^{\infty}\frac{1}{i!j!}(\boldsymbol{D}_{\hat{\sigma}_{l,k}}^{i}\boldsymbol{f})(\boldsymbol{D}_{\hat{\sigma}_{l,k}}^{j}\boldsymbol{f})^{\text{T}}}_{\text{condition_1}} \\
&\quad - \underbrace{\sum_{i=1}^{\infty}\sum_{j=1}^{\infty}\frac{1}{(2i)!(2j)!n^2}\sum_{p=1}^{n}\sum_{m=1}^{n}(\boldsymbol{D}_{\hat{\sigma}_{p,k}}^{2i}\boldsymbol{f})(\boldsymbol{D}_{\hat{\sigma}_{m,k}}^{2j}\boldsymbol{f})^{\text{T}}}_{\text{condition_2}}
\end{aligned} \tag{6.43}$$

$$\begin{aligned}
&\frac{1}{2n}\sum_{i=1}^{2n}\left(\boldsymbol{\rho}_{k,i} - \boldsymbol{\eta}(\hat{\boldsymbol{x}}_k)_{\text{CPF}}\right)\left(\boldsymbol{\rho}_{k,i} - \boldsymbol{\eta}(\hat{\boldsymbol{x}}_k)_{\text{CPF}}\right)^{\text{T}} \\
&= \boldsymbol{G}_\eta\boldsymbol{P}_{\tilde{x},k}\boldsymbol{G}_\eta^{\text{T}} - \mathbb{E}\left(\frac{\boldsymbol{D}_{\Delta x}^2\boldsymbol{\eta}}{2!}\right)\mathbb{E}\left(\left(\frac{\boldsymbol{D}_{\Delta x}^2\boldsymbol{\eta}}{2!}\right)^{\text{T}}\right) \\
&\quad + \underbrace{\frac{1}{n}\sum_{l=1}^{n}\sum_{i=1}^{\infty}\sum_{j=1}^{\infty}\frac{1}{i!j!}(\boldsymbol{D}_{\hat{\sigma}_{l,k}}^{i}\boldsymbol{\eta})(\boldsymbol{D}_{\hat{\sigma}_{l,k}}^{j}\boldsymbol{\eta})^{\text{T}}}_{\text{condition_1}} \\
&\quad - \underbrace{\sum_{i=1}^{\infty}\sum_{j=1}^{\infty}\frac{1}{(2i)!(2j)!n^2}\sum_{p=1}^{n}\sum_{m=1}^{n}(\boldsymbol{D}_{\hat{\sigma}_{p,k}}^{2i}\boldsymbol{\eta})(\boldsymbol{D}_{\hat{\sigma}_{m,k}}^{2j}\boldsymbol{\eta})^{\text{T}}}_{\text{condition_2}}
\end{aligned} \tag{6.44}$$

$$\frac{1}{2n}\sum_{i=1}^{2n}(\boldsymbol{\xi}_{i,k}-(\hat{\boldsymbol{x}}_k)_{\mathrm{CPF}})(\boldsymbol{\rho}_{i,k}-\boldsymbol{\eta}(\hat{\boldsymbol{x}}_k)_{\mathrm{CPF}})^{\mathrm{T}}$$
$$=\boldsymbol{P}_{\tilde{x},k}\boldsymbol{G}_{\eta}^{\mathrm{T}}+\sum_{i=1}^{n}\sum_{j=1}^{\infty}\frac{1}{n(2j+1)!}\hat{\boldsymbol{\sigma}}_{i,k}\boldsymbol{D}_{\hat{\sigma}_{i,k}}^{2j+1}\boldsymbol{\eta} \tag{6.45}$$

$$\frac{1}{2n}\sum_{i=1}^{2n}(\boldsymbol{\delta}_{i,k}-\boldsymbol{f}((\hat{\boldsymbol{x}}_k)_{\mathrm{CPF}}))(\boldsymbol{\xi}_{i,k}-(\hat{\boldsymbol{x}}_k)_{\mathrm{CPF}})^{\mathrm{T}}$$
$$=\boldsymbol{G}_f\boldsymbol{P}_{\tilde{x},k}+\sum_{i=1}^{n}\sum_{j=1}^{\infty}\frac{1}{n(2j+1)!}\boldsymbol{D}_{\hat{\sigma}_{i,k}}^{2j+1}\boldsymbol{f}\hat{\boldsymbol{\sigma}}_{i,k} \tag{6.46}$$

$$\frac{1}{2n}\sum_{i=1}^{2n}(\boldsymbol{\rho}_{k,i}-\boldsymbol{\eta}(\hat{\boldsymbol{x}}_k)_{\mathrm{CPF}})(\boldsymbol{\xi}_{k,i}-(\hat{\boldsymbol{x}}_k)_{\mathrm{CPF}})^{\mathrm{T}}$$
$$=\boldsymbol{G}_{\eta}\boldsymbol{P}_{\tilde{x},k}+\sum_{i=1}^{n}\sum_{j=1}^{\infty}\frac{1}{n(2j+1)!}\boldsymbol{D}_{\hat{\sigma}_{i,k}}^{2j+1}\boldsymbol{\eta}\hat{\boldsymbol{\sigma}}_{i,k} \tag{6.47}$$

$$\frac{1}{2n}\sum_{i=1}^{2n}(\boldsymbol{\delta}_{k,i}-\boldsymbol{f}((\hat{\boldsymbol{x}}_k)_{\mathrm{CPF}}))(\boldsymbol{\rho}_{k,i}-\boldsymbol{\eta}(\hat{\boldsymbol{x}}_k)_{\mathrm{CPF}})^{\mathrm{T}}$$
$$=\boldsymbol{G}_f\boldsymbol{P}_{\tilde{x},k}\boldsymbol{G}_{\eta}^{\mathrm{T}}-\mathbb{E}\left(\frac{\boldsymbol{D}_{\Delta x}^2\boldsymbol{f}}{2!}\right)\mathbb{E}\left(\left(\frac{\boldsymbol{D}_{\Delta x}^2\boldsymbol{\eta}}{2!}\right)^{\mathrm{T}}\right)$$
$$+\underbrace{\frac{1}{n}\sum_{l=1}^{n}\sum_{i=1}^{\infty}\sum_{j=1}^{\infty}\frac{1}{i!j!}(\boldsymbol{D}_{\hat{\sigma}_{l,k}}^{i}\boldsymbol{f})(\boldsymbol{D}_{\hat{\sigma}_{l,k}}^{j}\boldsymbol{\eta})^{\mathrm{T}}}_{\text{condition\_1}} \tag{6.48}$$
$$-\underbrace{\sum_{i=1}^{\infty}\sum_{j=1}^{\infty}\frac{1}{(2i)!(2j)!n^2}\sum_{p=1}^{n}\sum_{m=1}^{n}(\boldsymbol{D}_{\hat{\sigma}_{p,k}}^{2i}\boldsymbol{f})(\boldsymbol{D}_{\hat{\sigma}_{m,k}}^{2j}\boldsymbol{\eta})^{\mathrm{T}}}_{\text{condition\_2}}$$

$$\frac{1}{2n}\sum_{i=1}^{2n}\left(\boldsymbol{\rho}_{k,i}-\boldsymbol{\eta}(\hat{\boldsymbol{x}}_k)_{\mathrm{CPF}}\right)\left(\boldsymbol{\delta}_{k,i}-\boldsymbol{f}((\hat{\boldsymbol{x}}_k)_{\mathrm{CPF}})\right)^{\mathrm{T}}$$
$$=\boldsymbol{G}_{\eta}\boldsymbol{P}_{\tilde{x},k}\boldsymbol{G}_f^{\mathrm{T}}-\mathbb{E}\left(\frac{\boldsymbol{D}_{\Delta x}^2\boldsymbol{\eta}}{2!}\right)\mathbb{E}\left(\left(\frac{\boldsymbol{D}_{\Delta x}^2\boldsymbol{f}}{2!}\right)^{\mathrm{T}}\right)$$

$$+\underbrace{\frac{1}{n}\sum_{l=1}^{n}\sum_{i=1}^{\infty}\sum_{j=1}^{\infty}\frac{1}{i!j!}(\boldsymbol{D}^{i}_{\hat{\boldsymbol{\sigma}}_{l,k}}\boldsymbol{\eta})(\boldsymbol{D}^{j}_{\hat{\boldsymbol{\sigma}}_{l,k}}\boldsymbol{f})^{\mathrm{T}}}_{\text{condition\_1}} \tag{6.49}$$

$$-\underbrace{\sum_{i=1}^{\infty}\sum_{j=1}^{\infty}\frac{1}{(2i)!(2j)!n^{2}}\sum_{p=1}^{n}\sum_{m=1}^{n}(\boldsymbol{D}^{2i}_{\hat{\boldsymbol{\sigma}}_{p,k}}\boldsymbol{\eta})(\boldsymbol{D}^{2j}_{\hat{\boldsymbol{\sigma}}_{m,k}}\boldsymbol{f})^{\mathrm{T}}}_{\text{condition\_2}}$$

Applying (6.42)–(6.49), one can simplify (6.41) as

$$\begin{aligned}(\boldsymbol{P}_{\tilde{\boldsymbol{x}},k+1})_{\mathrm{CPF}} &= \boldsymbol{A}_k(\boldsymbol{P}_{\tilde{\boldsymbol{x}},k})_{\mathrm{CPF}}\boldsymbol{A}_k^{\mathrm{T}} - (\Delta t)^2(\mathbb{E}(\frac{\boldsymbol{D}^2_{\Delta x}\boldsymbol{f}}{2!}) + \mathbb{E}(\frac{\boldsymbol{D}^2_{\Delta x}\boldsymbol{\eta}}{2!}))^2 \\ &+ \tilde{\boldsymbol{\Sigma}}_1 + \tilde{\boldsymbol{\Sigma}}_2 + \bar{\boldsymbol{o}}_{\boldsymbol{k}} + \bar{\boldsymbol{\psi}}_{\boldsymbol{k}}\end{aligned} \tag{6.50}$$

where

$$\tilde{\boldsymbol{\Sigma}}_1 = \sum_{i=1}^{n}\sum_{j=1}^{\infty}\frac{\Delta t}{n(2j+1)!}(\hat{\boldsymbol{\sigma}}_{i,k}\boldsymbol{D}^{2j+1}_{\hat{\boldsymbol{\sigma}}_{i,k}}(\boldsymbol{f}+\boldsymbol{\eta}) + \boldsymbol{D}^{2j+1}_{\hat{\boldsymbol{\sigma}}_{i,k}}(\boldsymbol{f}+\boldsymbol{\eta})\hat{\boldsymbol{\sigma}}_{i,k}) \tag{6.51}$$

$$\begin{aligned}\tilde{\boldsymbol{\Sigma}}_2 &= \frac{(\Delta t)^2}{n}\underbrace{\sum_{l=1}^{n}\sum_{i=1}^{\infty}\sum_{j=1}^{\infty}\frac{1}{i!j!}\boldsymbol{D}^{i}_{\hat{\boldsymbol{\sigma}}_{l,k}}(\boldsymbol{f}+\boldsymbol{\eta})(\boldsymbol{D}^{j}_{\hat{\boldsymbol{\sigma}}_{l,k}}(\boldsymbol{f}+\boldsymbol{\eta}))^{\mathrm{T}}}_{\text{condition\_1}} \\ &-\underbrace{\sum_{i=1}^{\infty}\sum_{j=1}^{\infty}\frac{(\Delta t)^2}{(2i)!(2j)!n^{2}}\sum_{p=1}^{n}\sum_{m=1}^{n}\boldsymbol{D}^{2i}_{\hat{\boldsymbol{\sigma}}_{p,k}}(\boldsymbol{f}+\boldsymbol{\eta})(\boldsymbol{D}^{2j}_{\hat{\boldsymbol{\sigma}}_{m,k}}(\boldsymbol{f}+\boldsymbol{\eta}))^{\mathrm{T}}}_{\text{condition\_2}}\end{aligned} \tag{6.52}$$

Moreover, (6.50) can be rewritten as

$$(\boldsymbol{P}_{\tilde{\boldsymbol{x}},k+1})_{\mathrm{CPF}} = \boldsymbol{A}_k(\boldsymbol{P}_{\tilde{\boldsymbol{x}},k})_{\mathrm{CPF}}\boldsymbol{A}_k^{\mathrm{T}} - \boldsymbol{K}_k\boldsymbol{K}_k^{\mathrm{T}} + \boldsymbol{Q}_k \tag{6.53}$$

with

$$\boldsymbol{Q}_k = \tilde{\boldsymbol{\Sigma}}_1 + \tilde{\boldsymbol{\Sigma}}_2 + \bar{\boldsymbol{o}}_k + \bar{\boldsymbol{\psi}}_k \tag{6.54}$$

$$\begin{aligned}\boldsymbol{K}_k &= \frac{(\nabla^{\mathrm{T}}\boldsymbol{P}_k\nabla)\boldsymbol{f}(\boldsymbol{x})\big|_{\boldsymbol{x}=\hat{\boldsymbol{x}}_k}}{2!}\Delta t + \frac{(\nabla^{\mathrm{T}}\boldsymbol{P}_k\nabla)\boldsymbol{\eta}(\boldsymbol{x})\big|_{\boldsymbol{x}=\hat{\boldsymbol{x}}_k}}{2!}\Delta t \\ &= \Delta t\left(\mathbb{E}\left(\frac{\boldsymbol{D}^2_{\Delta x}\boldsymbol{f}}{2!}\right) + \mathbb{E}\left(\frac{\boldsymbol{D}^2_{\Delta x}\boldsymbol{\eta}}{2!}\right)\right)\end{aligned} \tag{6.55}$$

According to (6.53), it is seen that if the initial value satisfies $(\boldsymbol{P}_{\tilde{x},0})_{\text{CPF}} \approx \mathbb{E}(\tilde{\boldsymbol{x}}_0\tilde{\boldsymbol{x}}_0^{\mathrm{T}})$, then $(\boldsymbol{P}_{\tilde{x},k})_{\text{CPF}}$ calculated from (6.53) will approximate the actual value, *i.e.*, $(\boldsymbol{P}_{\tilde{x},k})_{\text{CPF}} \approx \mathbb{E}(\tilde{\boldsymbol{x}}_k\tilde{\boldsymbol{x}}_k^{\mathrm{T}})$. Hence, (6.53) can be used to estimate the covariance of the state estimation error.

Theorem 6.3. *If the CPF is applied to estimate the states of the nonlinear discrete system* (2.32)–(2.33) *with the initial estimation satisfying* $(\boldsymbol{P}_{\tilde{x},0})_{CPF} = (\boldsymbol{P}_{\tilde{x},0})$, *then* $(\boldsymbol{P}_{\tilde{x},k})_{CPF} \leq (\boldsymbol{P}_{\tilde{x},k})_{PF}$ *is guaranteed for every* $k \geq 0$. *The state estimation error covariance of the CPF has a faster convergence rate than the classical PF. Better filtering performance is achieved by the CPF than the classical PF.*

Proof. Comparing (2.74) with (6.53), the conclusions in Theorem 6.3 can be directly proved. □

6.5 Stochastic stability of CPF

The boundedness of the state estimation error covariance ensured by the CPF in Algorithm 6.1 is first analyzed in this section. Then, the states' estimation error is proved to be bounded if the initial error, the disturbing noise terms, and the modeling error are bounded.

6.5.1 Boundedness analysis of state estimation covariance

Theorem 6.4. *Consider the nonlinear stochastic system* (2.32)–(2.33), *with application of the filtering approach in Algorithm 6.1, for every* $k \geq 0$, *if the initial* $(\boldsymbol{P}_{\tilde{x},0})_{CPF}$ *satisfies* $\underline{p}\boldsymbol{I}_n \leq (\boldsymbol{P}_{\tilde{x},0})_{CPF} \leq \bar{p}\boldsymbol{I}_n$ *and there exist positive real numbers* $0 < z \leq 1$, $\bar{k} > 0$, $\bar{r} \geq \underline{r} \geq 0$ *as well as* $\bar{q} \geq \underline{q} \geq 0$ *satisfying*

$$\mathbf{0} \leq \boldsymbol{A}_k\boldsymbol{A}_k^{\mathrm{T}} \leq (1-z)\boldsymbol{I}_n \tag{6.56}$$

$$\mathbf{0} \leq \frac{1}{4}(\Delta t)^2\boldsymbol{\Upsilon}\boldsymbol{\Upsilon}^{\mathrm{T}} \leq \bar{k}\boldsymbol{I}_n \tag{6.57}$$

$$\underline{r}\boldsymbol{I}_n \leq \boldsymbol{R}_k \leq \bar{r}\boldsymbol{I}_n \tag{6.58}$$

$$\underline{q}\boldsymbol{I}_n \leq \boldsymbol{Q}_k \leq \bar{q}\boldsymbol{I}_n \tag{6.59}$$

where $\boldsymbol{\Upsilon} = (\nabla^{\mathrm{T}}\nabla)\boldsymbol{f}(\boldsymbol{x})|_{\boldsymbol{x}=\hat{\boldsymbol{x}}_k} + (\nabla^{\mathrm{T}}\nabla)\boldsymbol{\eta}(\boldsymbol{x})|_{\boldsymbol{x}=\hat{\boldsymbol{x}}_k}$, *then the state estimation error covariance* (6.53) *is bounded as*

$$\underline{p}\boldsymbol{I}_n \leq (\boldsymbol{P}_{\tilde{x},k})_{CPF} \leq \bar{p}\boldsymbol{I}_n,\ k \geq 0 \tag{6.60}$$

Proof. The procedure in the proof of Theorem 5.2 can be followed to prove this theorem. □

6.5.2 Boundedness analysis of state estimation error

Based on obtained state estimation error and its covariance matrix in Section 6.3, we are going to prove the stochastic boundedness of the CPF in this subsection.

Assumption 6.1. For every $k \geq 0$, there exist positive real constants $\bar{a} \in \mathbb{R}_+$, $\underline{a} \in \mathbb{R}_+$, $\bar{p} \in \mathbb{R}_+$, $\underline{p} \in \mathbb{R}_+$, $\bar{q} \in \mathbb{R}_+$, $\underline{q} \in \mathbb{R}_+$, $\underline{r} \in \mathbb{R}_+$, $\bar{r} \in \mathbb{R}_+$, $\underline{\lambda} \in \mathbb{R}_+$, $\bar{\lambda} \in \mathbb{R}_+$, $\underline{s} \in \mathbb{R}_+$, $\bar{s} \in \mathbb{R}_+$, $\bar{k} \in \mathbb{R}_+$, $\delta \in \mathbb{R}_+$ satisfying

$$\underline{a} \leq ||\boldsymbol{A}_k|| \leq \bar{a} \tag{6.61}$$

$$\frac{1}{2} \|\Delta t \boldsymbol{\Upsilon}\| \leq \sqrt{\bar{k}} \tag{6.62}$$

$$\underline{q}\boldsymbol{I}_n \leq \boldsymbol{Q}_k \leq \bar{q}\boldsymbol{I}_n \tag{6.63}$$

$$\underline{r}\boldsymbol{I}_n \leq \boldsymbol{R}_k \tag{6.64}$$

$$\boldsymbol{0} \leq \boldsymbol{W}_E \leq \bar{w}\boldsymbol{I}_n \tag{6.65}$$

$$\underline{\lambda} \leq ||\boldsymbol{\Lambda}(\Delta t)|| \leq \bar{\lambda} \tag{6.66}$$

$$\underline{s} \leq \|(\boldsymbol{U}(\hat{\boldsymbol{x}}_k))_{\mathrm{CPF}}\| \leq \bar{s} \tag{6.67}$$

$$\underline{p}\boldsymbol{I}_n \leq (\boldsymbol{P}_{\tilde{x},k})_{\mathrm{CPF}} \leq \bar{p}\boldsymbol{I}_n \tag{6.68}$$

Assumption 6.2. The matrix $\boldsymbol{A}_k$, $k \geq 0$, in (6.37) is nonsingular.

Assumption 6.3. Two positive real numbers $\kappa_{\bar{\mu}} >$ and $\varepsilon' > 0$ exist such that the following nonlinear function is bounded, *i.e.*,

$$\left\| \bar{\boldsymbol{\mu}}_k(\boldsymbol{x}_k, \boldsymbol{d}_k, (\tilde{\boldsymbol{x}}_k)_{\mathrm{CPF}}, (\hat{\boldsymbol{d}}_k)_{\mathrm{CPF}}) \right\| \leq \kappa_{\bar{\mu}} \left\| \boldsymbol{x}_k - (\hat{\boldsymbol{x}}_k)_{\mathrm{CPF}} \right\|^2 \tag{6.69}$$

$$||\boldsymbol{x}_k - (\hat{\boldsymbol{x}}_k)_{\mathrm{CPF}}|| \leq \varepsilon' \tag{6.70}$$

Lemma 6.1. *If the filtering Algorithm 6.1 is applied, under Assumptions 6.1–6.3, then there is a real and positive scalar* $0 < \alpha < 1$ *guaranteeing that the matrix* $\boldsymbol{\Pi}_k = (\boldsymbol{P}_{\tilde{x},k}^{-1})_{CPF}$ *satisfies*

$$(\boldsymbol{A}_k - \boldsymbol{K}_k\boldsymbol{C}_k)^{\mathrm{T}}\boldsymbol{\Pi}_{k+1}(\boldsymbol{A}_k - \boldsymbol{K}_k\boldsymbol{C}_k) \leq (1-\alpha)\boldsymbol{\Pi}_k,\ k \geq 0 \tag{6.71}$$

where $\boldsymbol{C}_k = ((\boldsymbol{P}_{\tilde{x},k}^{-1})_{CPF}\boldsymbol{A}_k^{-1}\boldsymbol{K}_k)^{\mathrm{T}}$ *and*

$$\alpha = 1 - \frac{1}{1 + \frac{\underline{q}}{\bar{p}(\bar{a}+\bar{p}^2\bar{k}/\underline{a}\underline{p})^2}} \tag{6.72}$$

Proof. Applying the same procedures as in Lemma 5.1, the proof can be directly done. □

Lemma 6.2. *Suppose that Assumptions 6.1–6.3 are satisfied, then there are positive constants $\varepsilon' \in \mathbb{R}_+$ and ε' governing that*

$$\begin{aligned}&\left(\boldsymbol{K}_k\boldsymbol{C}_k\left(\boldsymbol{x}_k-\left(\hat{\boldsymbol{x}}_k\right)_{CPF}\right)\right)^{\mathrm{T}}\boldsymbol{\Pi}_k\left(2(\boldsymbol{A}_k-\boldsymbol{K}_k\boldsymbol{C}_k)(\boldsymbol{x}_k-(\hat{\boldsymbol{x}}_k)_{CPF})\right.\\&\left.+\boldsymbol{K}_k\boldsymbol{C}_k(\boldsymbol{x}_k-(\hat{\boldsymbol{x}}_k)_{CPF})+(\bar{\boldsymbol{\mu}}_k)_{CPF}\right)\leq\kappa_1\left\|\boldsymbol{x}_k-\left(\hat{\boldsymbol{x}}_k\right)_{CDPF}\right\|^2\end{aligned} \tag{6.73}$$

is valid for $||\boldsymbol{x}_k-(\hat{\boldsymbol{x}}_k)_{CPF}||\leq\varepsilon'$.

Proof. It is the same as the proof of Lemma 5.2. □

Lemma 6.3. *If Assumptions 6.1–6.3 are satisfied and $\boldsymbol{\Pi}_k=(\boldsymbol{P}_{\tilde{x},k}^{-1})_{CPF}$ exists, then the following inequality always holds*

$$\begin{aligned}&(\bar{\boldsymbol{\mu}}_k)_{CPF}^{\mathrm{T}}\boldsymbol{\Pi}_k(2(\boldsymbol{A}_k-\boldsymbol{K}_k\boldsymbol{C}_k)(\boldsymbol{x}_k-(\hat{\boldsymbol{x}}_k)_{CPF})\\&+\boldsymbol{K}_k\boldsymbol{C}_k(\boldsymbol{x}_k-(\hat{\boldsymbol{x}}_k)_{CPF})+(\bar{\boldsymbol{\mu}}_k)_{CPF})\leq\kappa_2\left\|\boldsymbol{x}_k-\left(\hat{\boldsymbol{x}}_k\right)_{CPF}\right\|^3\end{aligned} \tag{6.74}$$

where $\kappa_2=\frac{\kappa_{\bar{\mu}}}{\underline{p}}((2\bar{a}+k_1)+\kappa_{\bar{\mu}}\varepsilon')$.

Proof. This can be accomplished by using the same analysis in the proof of Lemma 5.3. □

Lemma 6.4. *If Assumptions 6.1–6.3 are valid, then it follows that $\mathbb{E}((\boldsymbol{s}_k)_{CPF}^{\mathrm{T}}\boldsymbol{\Pi}_k(\boldsymbol{s}_k)_{CPF})\leq\kappa_3\delta$, where $\kappa_3=\frac{\bar{\lambda}^2\bar{s}^2}{\underline{p}\underline{r}^2\underline{\lambda}^4\underline{s}^4}n$ and $\delta=\bar{r}^3$.*

Proof. It is the same as the proof of Lemma 5.4. □

Theorem 6.5. *With the application of the CPF developed Algorithm 6.1 to the nonlinear stochastic system (2.32) and (2.33) with modeling error $\boldsymbol{d}_k$, if Assumptions 6.1–6.3 and $||(\tilde{\boldsymbol{x}}_0)_{CPF}||\leq\varepsilon$ are satisfied for a constant $\varepsilon>0$, then the state estimation error $(\tilde{\boldsymbol{x}}_{k+1})_{CPF}$ given in (6.36) is governed to be exponentially bounded in mean square and bounded with probability one.*

Proof. Applying Definition 2.1, Definition 2.2, Lemma 2.1, Lemma 6.1, Lemma 6.2, Lemma 6.3, and Lemma 6.4, this theorem can be proved by using the same analysis as in Theorem 5.2. □

Remark 6.2. The statement clarified in Remark 5.4 is also valid for the CPF presented in Algorithm 6.1.

6.6 Application to microsatellite attitude control system

By doing the same analysis as carried out in Section 4.7, it can be concluded that the CPF developed in Algorithm 6.1 can be applied and implemented to estimate the states of the microsatellite attitude control system. Based on this conclusion, simulations are carried out for with the CPF applied to a microsatellite with two star sensors and its inertia parameter, the initial attitude, and the initial angular velocity chosen as same as in Section 4.6.2. The microsatellite is assumed to be subject to the modeling error given in (4.119). The initial attitude estimation is set as (4.120).

To demonstrate the theoretical performance proved in Sections 6.4–6.5, comparison between the CPF, the standard CKF in Arasaratnam and Haykin (2009), and the classical PF in Algorithm 2.1 is carried out in simulation for the following two cases with their weighting matrix chosen as $\boldsymbol{W}_E = 10^5 \boldsymbol{I}_{3\times3}$. Moreover, the weighting parameters for the sigma-points of the CPF and the CKF are selected as $W_i^c = W_i^m = \frac{1}{2n}$.

1) Case #1: This is a nominal case, in which $\Theta_0 = 5$ degrees, $\Delta\boldsymbol{N}_e = [-0.003\cos(\omega_T t) \quad 0.002\cos(\omega_T t) \quad 0.003\cos(\omega_T t)]^{\mathrm{T}}$ Nm, $\Delta\boldsymbol{N}_c = [0.004 \quad -0.005 \quad -0.003]^{\mathrm{T}}$ Nm, and $\omega_T = 0.66$ deg/sec are assumed. The initial estimate value of the angular velocity is $[0 \quad 0 \quad 0]^{\mathrm{T}}$ deg/sec. The standard deviation of that two star sensors' measurement noise is assumed as 20 arcseconds.

2) Case #2: $\Theta_0 = 5$ degrees, $\Delta\boldsymbol{N}_c = [0.02 \quad -0.025 \quad -0.0075]^{\mathrm{T}}$ Nm, $\Delta\boldsymbol{N}_e = [-0.015\cos(\omega_T t) \quad 0.01\cos(\omega_T t) \quad 0.015\cos(\omega_T t)]^{\mathrm{T}}$ Nm are assumed with $\omega_T = 0.66$ deg/sec. The initial estimate value of the angular velocity is $[0 \quad 0 \quad 0]^{\mathrm{T}}$ deg/sec. The standard deviation of that two star sensors' measurement noise is assumed as 100 arcseconds.

We use $\boldsymbol{\Theta}_e$, $\boldsymbol{\omega}_E$, and $\boldsymbol{d}_e$ given in Remark 4.5 to denote the attitude state estimation error, the angular velocity estimation error, and the estimation error of the modeling error $\boldsymbol{u}_d$, respectively. The state estimation provided by the CPF, the CKF, and the PF in Case #1 was shown in Figs. 6.1–6.2. It was seen that the CPF and the CKF guaranteed almost the same estimation accuracy for the attitude state. This accuracy was higher than the attitude estimation accuracy achieved by the PF. The convergence time of the state estimation error ensured by the CPF and the CKF was shorter than the PF. A faster state estimation was accomplished by the CPF and the CKF. However, the angular velocity state estimation accuracy established by the CKF was inferior to the CPF and the PF, as we can see in Fig. 6.2. This is because the CKF has weak power in handling the modeling error than the CPF and the PF. Evaluating the state estimation by using the convergence time and accuracy, it was found the proposed CPF achieved better state estimation than the CKF and the PF. Figs. 6.3–6.4 show the state estimation results in Case #2. Comparing with Case #1, the estimation performance of those three filters was deteriorated by the relatively larger modeling error and noises in Case #2. The attitude and the angular velocity estimation accuracy guaranteed by the CPF was still higher than the other two filters. That is

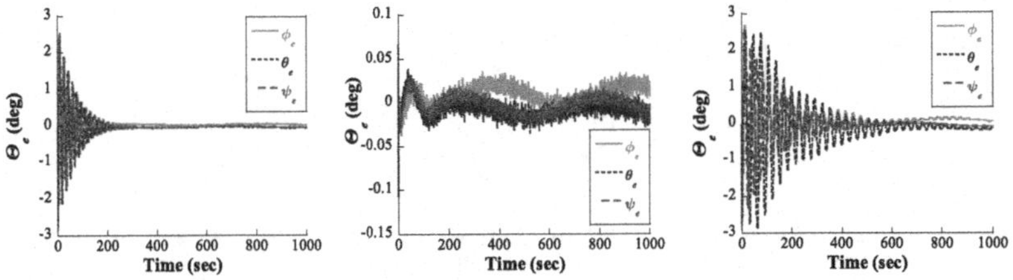

(a) Initial response of CPF (b) Initial response of CKF (c) Initial response of PF

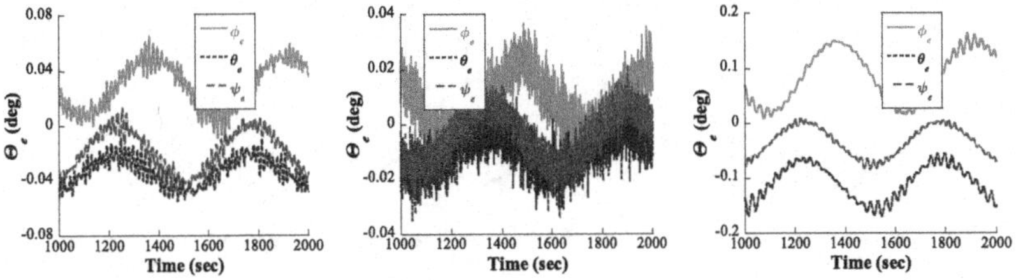

(d) Steady behavior of CPF (e) Steady behavior of UKF (f) Steady behavior of PF

FIGURE 6.1 Attitude estimation error using CDPF, UKF, and PF in Case #1.

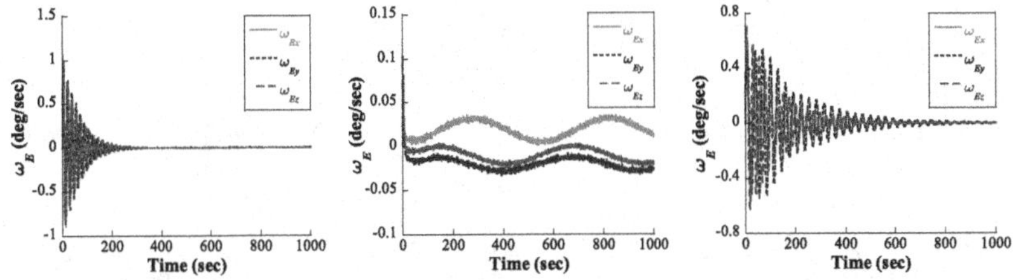

(a) Initial response of CPF (b) Initial response of CKF (c) Initial response of PF

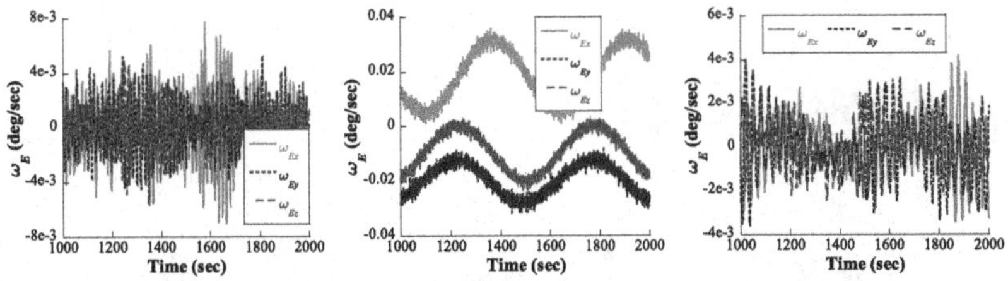

(d) Steady behavior of CPF (e) Steady behavior of CKF (f) Steady behavior of PF

FIGURE 6.2 Angular velocity estimation error using CPF, UKF, and PF in Case #1.

because the CPF has more capability of providing robustness to modeling error and measurement noise than the CKF and the PF.

The initial estimation, measurement noise, and modeling error in Case #1–#2 are relatively small. Hence, stable state estimation can be provided by the CPF. As stated and proved in Theorem 6.5, that stable estimation would not be maintained in the presence of large initial estimation, noise, and modeling error. To

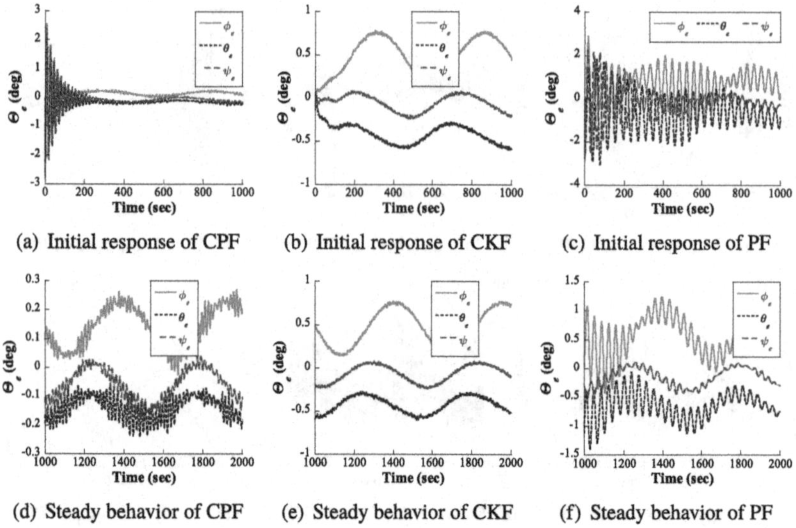

(a) Initial response of CPF (b) Initial response of CKF (c) Initial response of PF

(d) Steady behavior of CPF (e) Steady behavior of CKF (f) Steady behavior of PF

FIGURE 6.3 Attitude estimation error using CPF, UKF, and PF in Case #2.

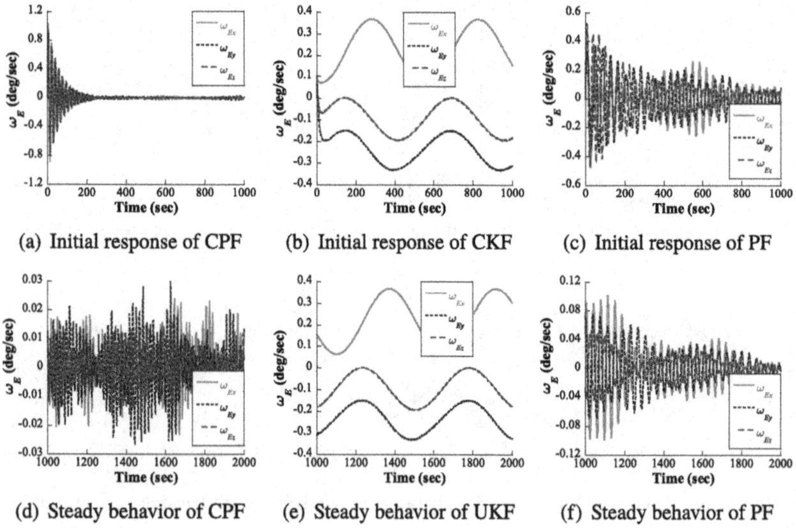

(a) Initial response of CPF (b) Initial response of CKF (c) Initial response of PF

(d) Steady behavior of CPF (e) Steady behavior of UKF (f) Steady behavior of PF

FIGURE 6.4 Angular velocity estimation error using CDPF, UKF, and PF in Case #2.

verify this conclusion, the following three cases were further considered and simulated. It was found in Figs. 6.5–6.6 that the state estimation diverged and was no longer stable in Case #3–#5. This reflects that the proposed CPF has its own weakness that the initial estimation should be carefully chosen for its implementation. Consequently, the conclusions in Theorem 6.5 was validated.

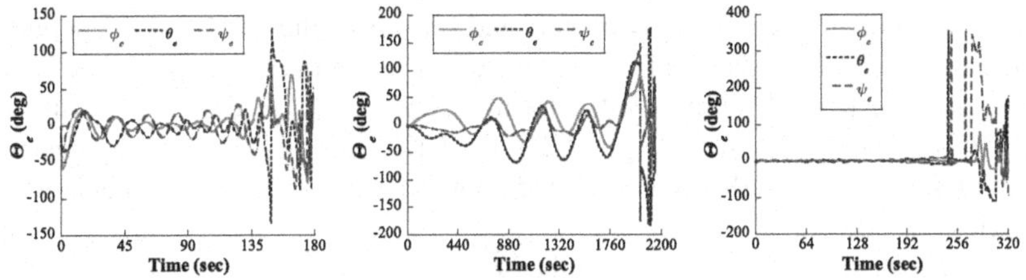

FIGURE 6.5 Attitude estimation error ensured by CPF in Case #3–#5.

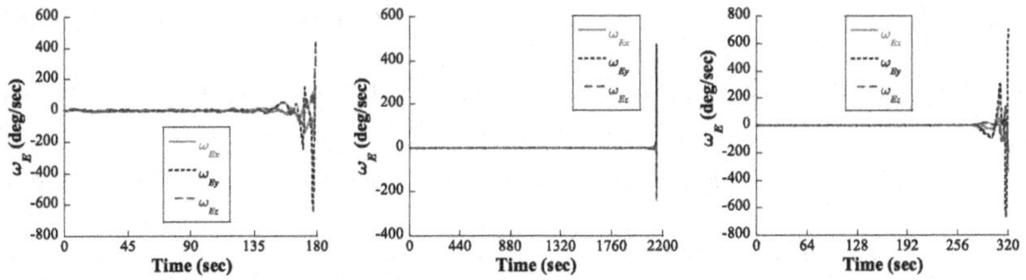

FIGURE 6.6 Angular velocity estimation error ensured by CPF in Case #3–#5.

3) Case #3: Large initial estimation is assumed with $\Theta_0 = 85$ degrees. The initial angular velocity estimation is $[1.7 \quad 1.7 \quad 1.7]^{\mathrm{T}}$ deg/sec. The modeling error and the standard deviation of measurement noise are as same as Case #1.

4) Case #4: Severe measurement noise is considered with the standard deviation of that two star sensors' measurement noise given by 1800 arcseconds. The initial estimation and the modeling error are assumed as same as Case #1.

5) Case #5: Large modeling error is assumed with

$$\Delta N_c = [0.072 \quad -0.115 \quad -0.069]^{\mathrm{T}} \text{ Nm} \tag{6.75}$$

$$\Delta N_e = [-3\bar{b}\cos(\omega_T t) \quad 2\bar{b}\cos(\omega_T t) \quad 3\bar{b}\cos(\omega_T t)]^{\mathrm{T}} \text{ Nm} \tag{6.76}$$

$\bar{b} = 0.023$, and $\omega_T = 0.66$ deg/sec. The initial estimation and the measurement noise are as same as Case #1.

6.7 Application to microsatellite formation flying control system

It is seen in Section 4.7 that the microsatellite formation flying control system can be re-written into the form of the nonlinear system (2.32)–(2.33). Therefore, the CPF developed in Algorithm 6.1 is also applicable to solve the states estimation problem of microsatellite formation flying control system.

Knowing that the CPF can be applied to formation flying control system, numerical simulations are conducted to demonstrate its effectiveness in this section. In simulations, all the parameters and sensors of the chief and the deputy microsatellite, the initial real states, as well as the initial estimation states are set as same as in Section 4.7.4. The modeling error has the same form as (4.167)–(4.168). Furthermore, the CPF, the CKF, and the PF are compared in simulation with the following two modeling error and measurement noise scenarios considered.

1) Scenario #1: This is a nominal case.

$$\Delta \boldsymbol{N}_{c1} = [-6\omega_1 \quad -8\omega_1 \quad 6\omega_1]^{\mathrm{T}} \text{ Nm} \tag{6.77}$$

$$\Delta \boldsymbol{N}_{e1} = [8\omega_1 \cos(\omega_T t) \quad 6\omega_1 \cos(\omega_T t) \quad -4\omega_1 \cos(\omega_T t)]^{\mathrm{T}} \text{ Nm} \tag{6.78}$$

$$\Delta \boldsymbol{N}_{e2} = [5\omega_1 \cos(\omega_T t) \quad -3\omega_1 \cos(\omega_T t) \quad 2\omega_1 \cos(\omega_T t)]^{\mathrm{T}} \text{ N} \tag{6.79}$$

$$\Delta \boldsymbol{N}_{c2} = [3\omega_1 \quad -4\omega_1 \quad 2\omega_1]^{\mathrm{T}} \text{ N} \tag{6.80}$$

are assumed with $\omega_T = 0.66$ deg/sec and $\omega_1 = 0.0005$. The standard deviation of the measurement noise in (4.134) is 0.001 degrees.

2) Scenario #2: This case is with large modeling error.

$$\Delta \boldsymbol{N}_{c1} = [-3\omega_1 \quad -2\omega_1 \quad 3\omega_1]^{\mathrm{T}} \text{ Nm} \tag{6.81}$$

$$\Delta \boldsymbol{N}_{e1} = [3\omega_1 \cos(\omega_T t) \quad \omega_1 \cos(\omega_T t) \quad -2\omega_1 \cos(\omega_T t)]^{\mathrm{T}} \text{ Nm} \tag{6.82}$$

$$\Delta \boldsymbol{N}_{e2} = [-3\omega_1 \cos(\omega_T t) \quad 4\omega_1 \cos(\omega_T t) \quad 2\omega_1 \cos(\omega_T t)]^{\mathrm{T}} \text{ N} \tag{6.83}$$

$$\Delta \boldsymbol{N}_{c2} = [5\omega_1 \quad -3\omega_1 \quad 2\omega_1]^{\mathrm{T}} \text{ N} \tag{6.84}$$

are assumed with $\omega_T = 0.66$ deg/sec and $\omega_1 = 0.005$. The standard deviation of the measurement noise in (4.134) is 0.01 degrees. The terms $\boldsymbol{\Theta}_e$, $\boldsymbol{\omega}_E$, $\boldsymbol{\rho}_e$, and $\dot{\boldsymbol{\rho}}_e$ defined in Remark 4.6 were also used to represent the estimation errors of the relative attitude, the relative angular velocity, the relative position, and the relative velocity, respectively. For Scenario #1, the corresponding state estimation obtained from the CPF, the CKF, and the PF was illustrated in Figs. 6.7–6.10. Although the same relative velocity estimation accuracy was guaranteed by those three filters as shown in Fig. 6.8, the CPF and the PF provided the relative position state with higher estimation accuracy than the CKF, as we can see in Fig. 6.7. The relative position estimation accuracy achieved by the CKF was $x_e = -0.25$ m and $y_e = z_e = -0.05$ m. This inferior relative position estimation performance is owing to the fact that unlike the CPF and the PF, the CKF was not robust to modeling error. Figs. 6.9–6.10 show that the CPF and the CKF achieved better estimation accuracy and estimation rate for the relative attitude and the relative angular velocity states than the PF, because both the CPF and the CKF applied sigma-point sampling strategy to improve the estimation performance in the presence of small measurement noise and modeling errors.

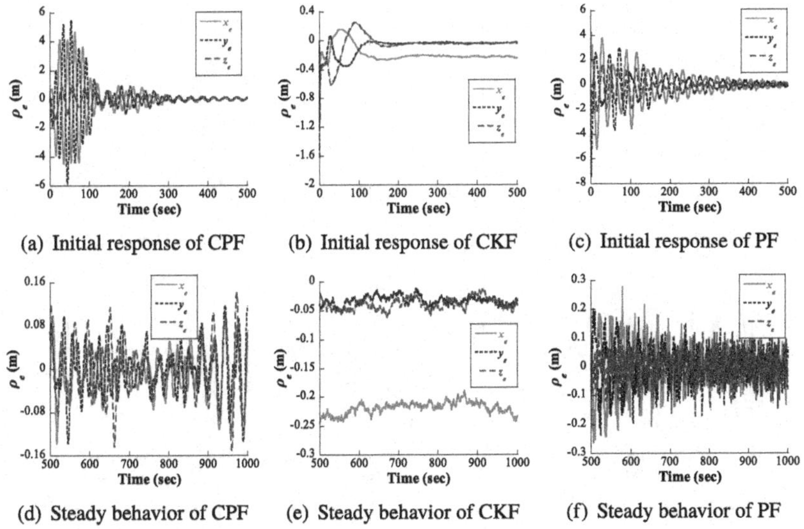

FIGURE 6.7 Relative position state estimation error using CPF, UKF, and PF in Scenario #1.

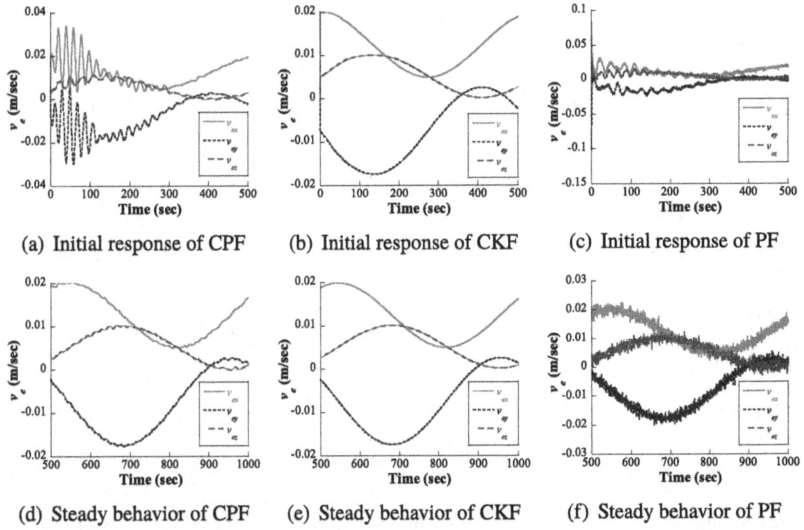

FIGURE 6.8 Relative velocity estimation error using CPF, UKF, and PF in Scenario #1.

When the large modeling error and the severe measurement noise in Scenario #2 were met, the above three filters ensured inferior state estimation performance than Case #1, as we can see in Figs. 6.11–6.14. However, the CPF still led to better estimation performance than the CKF and the PF, especially for the relative position and the relative attitude states. For this case, the estimation performance of the PF was very poor and not appropriate for practical applica-

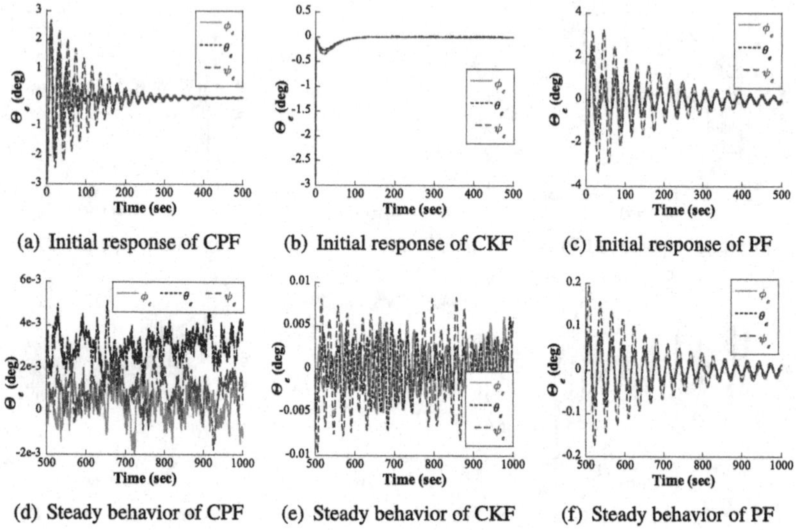

(a) Initial response of CPF (b) Initial response of CKF (c) Initial response of PF

(d) Steady behavior of CPF (e) Steady behavior of CKF (f) Steady behavior of PF

FIGURE 6.9 Relative attitude state estimation error using CPF, CKF, and PF in Scenario #1.

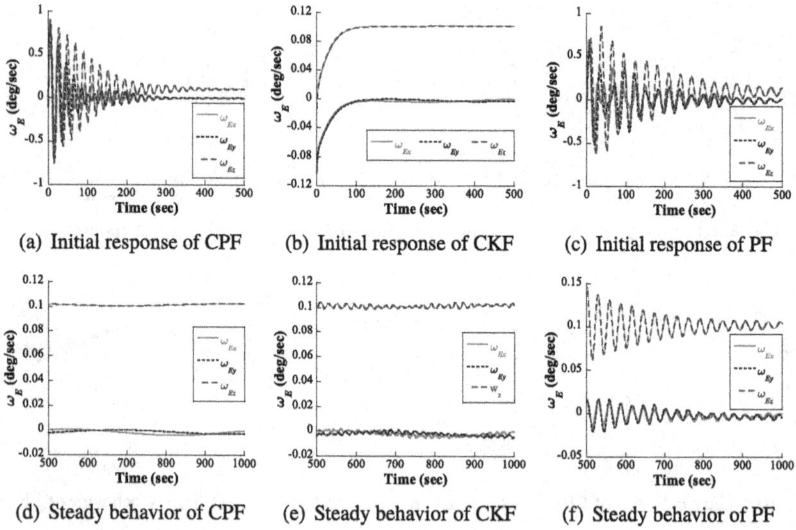

(a) Initial response of CPF (b) Initial response of CKF (c) Initial response of PF

(d) Steady behavior of CPF (e) Steady behavior of CKF (f) Steady behavior of PF

FIGURE 6.10 Relative attitude velocity estimation error using CPF, CKF, and PF in Scenario #1.

tion. Moreover, the CKF resulted in lower estimation accuracy for the relative attitude state; it was shown in Fig. 6.13(b) that the corresponding accuracy was $\phi_e = 0.005$ degrees and $\theta_e = \psi_e = 0.0035$ degrees. This is due to the fact that negative effect of the large modeling error and severe measurement noise can not be appropriately addressed by the CKF. Those results demonstrated that the

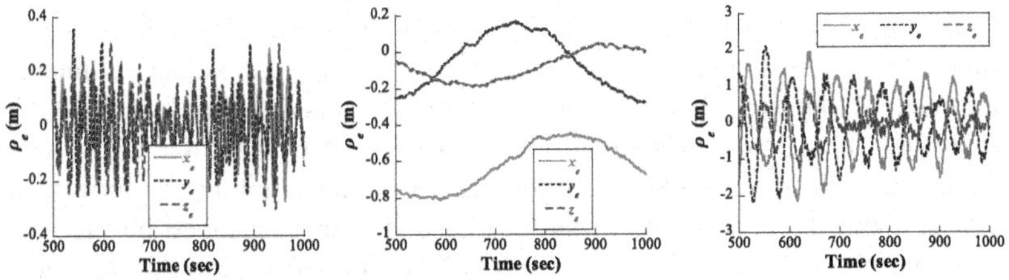

(a) Steady behavior of CPF (b) Steady behavior of CKF (c) Steady behavior of PF

FIGURE 6.11 Relative position state estimation error using CPF, UKF, and PF in Scenario #2.

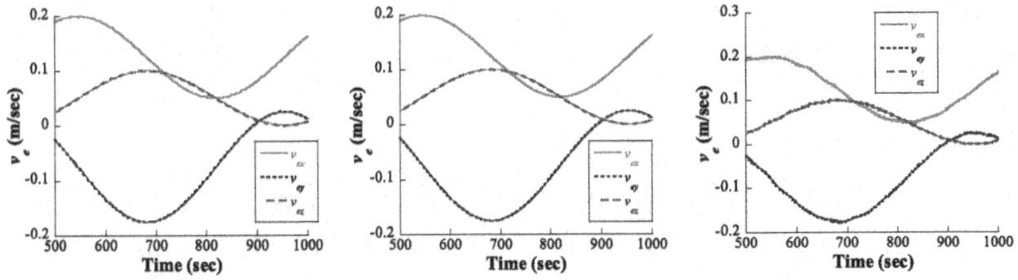

(a) Steady behavior of CPF (b) Steady behavior of CKF (c) Steady behavior of PF

FIGURE 6.12 Relative velocity estimation error using CPF, UKF, and PF in Scenario #2.

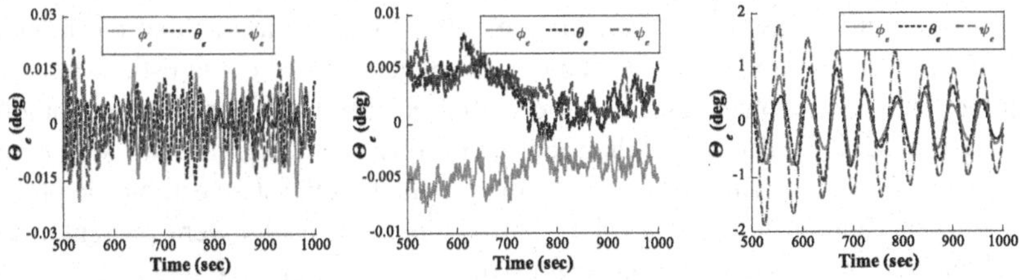

(a) Steady behavior of CPF (b) Steady behavior of CKF (c) Steady behavior of PF

FIGURE 6.13 Relative attitude state estimation error using CPF, CKF, and PF in Scenario #2.

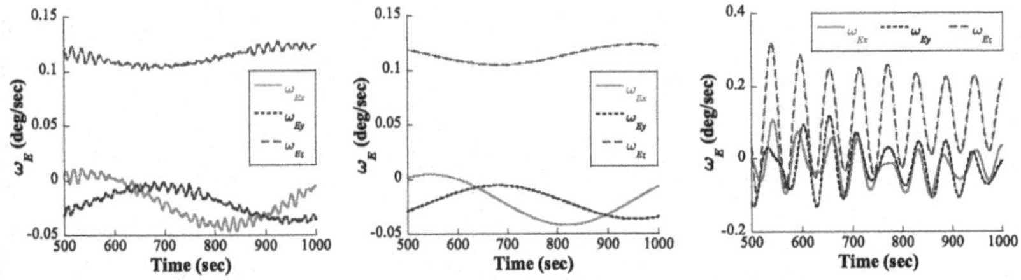

(a) Steady behavior of CPF (b) Steady behavior of CKF (c) Steady behavior of PF

FIGURE 6.14 Relative attitude velocity estimation error using CPF, CKF, and PF in Scenario #2.

CPF has great capability of handling large modeling error and measurement noise as well as ensuring more robustness than the CKF and the PF.

Although the proposed CPF and the CKF both are developed by using the same sigma-point sampling strategy, but the CPF adopts the numerical integration method to propagate the estimation state, which is different from the CKF. Therefore, the calculation burden of CPF is larger than the CKF. In the same simulation environment, for each sampling period, the computation time of the CPF is about 0.023 seconds, while 0.013 seconds are consumed by the CKF. Note that different conditions and input parameters will lead to different computation time. Hence, the preceding comparison for the computation time is just roughly done.

From the above simulation results, it is numerically validated that the proposed CPF is appropriate for estimating the states of the microsatellite formation flying system even in the presence of modeling error and measurement noise.

6.8 Summary

This chapter presented another filtering approach, *i.e.*, CPF, to solve the estimation problem of a class of nonlinear systems and in particular the microsatellite control system subject to modeling error and measurement noise. The stochastic boundedness and the state estimation performance guaranteed by this filter for general nonlinear systems were proved in a stochastic framework. The estimation error was proved to be bounded and the covariance was stable if the system's initial estimation error, the measurement noise terms, and the model error were small enough. It was demonstrated that ensured better estimation accuracy and more robustness than the traditional PF. However, the calculation burden of CPF is also larger than that of PF due to the sigma-point sampling strategy. When applying the proposed filter to address the state estimation problem of the microsatellite control system, simulation results were presented to validate the superior estimation performance of CPF over CKF and PF.

Part III

Predictive variable structure filtering for microsatellite control system

Chapter 7

Predictive variable structure filter

7.1 Introduction

The main feature of the predictive filters is that they predict and correct the modeling error online and in real-time. Moreover, the modeling error or measurement noises handled by the predictive filters are not restricted to the Gaussian distribution. Although the predictive filters presented in Part II achieve better estimation performance for the system states than the classical predictive filters, their performance depends on the choice of the weighting matrix $\boldsymbol{W}_E$ in (2.44). When applying those filters in practical engineering, the weighting matrix is chosen based only on the engineer's experience. This limits their practicability. Motivated by solving this shortcoming, the Minimum Modeling Error (MME) estimation criterion will be extended and three advanced predictive filtering approaches are presented in Part III.

Since the controller developed by using Variable Structure Control (VSC) theory has advantages of rapid response, robustness to parameter variations and uncertainties, and low computational costs (Shtessel et al., 2014), the VSC theory has been adopted to design nonlinear filters with high-accuracy state estimation even in the presence of modeling error. Introducing the VSC to design filtering approaches was firstly seen in Habibi et al. (2002) and Habibi and Burton (2003). The VSC was applied to design the gain matrix of the Kalman filters. By doing this, the robustness of the Kalman filters is greatly improved. Following this methodology, many VSC-based filtering schemes were presented (Habibi, 2006, 2007; Al-Shabi et al., 2013). However, they were all developed in the framework of the Kalman filtering theory framework. Hence, they are unable to solve the inherent drawbacks of the Kalman filters.

Based on the MME estimation criterion and fully taking the advantages of predictive filters and the VSC theory, a predictive variable structure filter is presented in this chapter. This filtering approach is different from the existing filtering theory. It is designed with the assumption of the "Gaussian process" for the modeling errors eliminated, and it does not depend on the weighting matrix in the classical filtering theory. It has a great capability of estimating the modeling errors to update the system model without precise system representation. Moreover, this filter does not require the nonlinear systems and their measurements to be linearized without accuracy loss of high-order items, which ensures the accuracy of the system's model.

Predictive Filtering for Microsatellite Control System. https://doi.org/10.1016/B978-0-12-821865-5.00022-X

7.2 Predictive variable structure filter design

For the nonlinear stochastic discrete-time systems described by (2.32)–(2.33), it is recalled from (2.38)–(2.39) that

$$\boldsymbol{y}_{k+1} = \boldsymbol{y}_k + \boldsymbol{Z}(\boldsymbol{x}_k, \Delta t) + \boldsymbol{\Lambda}(\Delta t)\boldsymbol{U}(\boldsymbol{x}_k)\boldsymbol{d}_k + o(\boldsymbol{e}_{k+1}) \tag{7.1}$$

where $o(\boldsymbol{e}_{k+1}) = \boldsymbol{\varepsilon}(\boldsymbol{x}_k, \boldsymbol{d}_k) + \boldsymbol{v}_{k+1}$. Expanding the measurement $\boldsymbol{y}_{k+1}$ about the state estimation $\hat{\boldsymbol{x}}_k$ in a Taylor series results in

$$\boldsymbol{y}_{k+1} = \hat{\boldsymbol{y}}_k + \boldsymbol{Z}(\hat{\boldsymbol{x}}_k, \Delta t) + \boldsymbol{\Lambda}(\Delta t)\boldsymbol{U}(\hat{\boldsymbol{x}}_k)\hat{\boldsymbol{d}}_k + o(\hat{\boldsymbol{e}}_{k+1}) \tag{7.2}$$

with $o(\hat{\boldsymbol{e}}_{k+1}) = \boldsymbol{\varepsilon}(\hat{\boldsymbol{x}}_k, \hat{\boldsymbol{d}}_k) + \boldsymbol{v}_{k+1}$.

Let $\hat{\boldsymbol{y}}_{k+1}$ denote the estimation of the measurement $\boldsymbol{y}_{k+1}$, *i.e.*,

$$\hat{\boldsymbol{y}}_{k+1} = \hat{\boldsymbol{y}}_k + \boldsymbol{Z}(\hat{\boldsymbol{x}}_k, \Delta t) + \boldsymbol{\Lambda}(\Delta t)\boldsymbol{U}(\hat{\boldsymbol{x}}_k)\hat{\boldsymbol{d}}_k \tag{7.3}$$

Then, it follows from (7.2) and (7.3) that

$$\boldsymbol{y}_{k+1} = \hat{\boldsymbol{y}}_{k+1} + o(\hat{\boldsymbol{e}}_{k+1}) \tag{7.4}$$

Let $\boldsymbol{e}_{\boldsymbol{y}_{k+1/k}}$ and $\boldsymbol{e}_{\boldsymbol{y}_k}$ defined as follows be the prior measurement error and the posterior measurement error, respectively.

$$\boldsymbol{e}_{\boldsymbol{y}_{k+1/k}} = \boldsymbol{y}_{k+1} - \hat{\boldsymbol{y}}_k - \boldsymbol{Z}(\hat{\boldsymbol{x}}_k, \Delta t) \tag{7.5}$$

$$\boldsymbol{e}_{\boldsymbol{y}_k} = \boldsymbol{y}_{k+1} - \hat{\boldsymbol{y}}_k \tag{7.6}$$

Then, the estimation of the modeling error $\boldsymbol{d}_k$ in the nonlinear system (2.32)–(2.33) can be developed as

$$\hat{\boldsymbol{d}}_k = (\boldsymbol{\Lambda}(\Delta t)\boldsymbol{U}(\hat{\boldsymbol{x}}_k))^{-1}\mathbf{diag}(|\boldsymbol{e}_{\boldsymbol{y}_{k+1/k}}| + \boldsymbol{\gamma}|\boldsymbol{e}_{\boldsymbol{y}_k}|)\mathbf{sgn}(\boldsymbol{e}_{\boldsymbol{y}_{k+1/k}}) \tag{7.7}$$

where $\boldsymbol{\gamma} = \mathbf{diag}([\gamma_{11} \quad \gamma_{12} \quad \cdots \quad \gamma_{mm}]^{\mathrm{T}}) \in \mathbb{R}^{m \times m}$ is a constant matrix with $0 < \gamma_{ii} < 1, i = 1, 2, \cdots, m$.

Based on the preceding analysis, we are ready to design a predictive variable structure filter (PVSF) in Algorithm 7.1. The prior measurement error $\boldsymbol{e}_{\boldsymbol{y}_{k+1/k}}$ and the posterior measurement error $\boldsymbol{e}_{\boldsymbol{y}_k}$ are included into the estimation of the modeling error to update the system model and to improve the state estimation accuracy. The constant matrix $\boldsymbol{\gamma}$ is referred as the "memory" of the PVSF. Moreover, the numerical integral method is adopted to update the state and the output estimation in Algorithm 7.1.

Remark 7.1. The key step of the PVSF is the application of (7.7) to update the system model. Then, the state estimation $\hat{\boldsymbol{x}}_k$ can be propagated nonlinearly from time k to time $k+1$ using the updated system model by some numerical integration methods. The PVSF use $\boldsymbol{e}_{\boldsymbol{y}_{k+1/k}}$ and $\boldsymbol{e}_{\boldsymbol{y}_k}$ as two filtering performance

---

Algorithm 7.1: Predictive variable structure filter.

Input Data: Initial estimation $\hat{\boldsymbol{x}}_0$, $\hat{\boldsymbol{y}}_0$, and measurement $\boldsymbol{y}_k$
Result: State estimation $\hat{\boldsymbol{x}}_k$
begin
  for $k = 0, 1, 2 \cdots$ **do**
    Calculating the modeling error's estimation $\hat{\boldsymbol{d}}_k$ by using
      $\boldsymbol{e}_{\boldsymbol{y}_{k+1/k}} = \boldsymbol{y}_{k+1} - \hat{\boldsymbol{y}}_k - \boldsymbol{Z}(\hat{\boldsymbol{x}}_k, \Delta t)$
      $\boldsymbol{e}_{\boldsymbol{y}_k} = \boldsymbol{y}_{k+1} - \hat{\boldsymbol{y}}_k$
      $\hat{\boldsymbol{d}}_k = (\boldsymbol{\Lambda}(\Delta t)\boldsymbol{U}(\hat{\boldsymbol{x}}_k))^{-1}\mathbf{diag}(|\boldsymbol{e}_{\boldsymbol{y}_{k+1/k}}| + \boldsymbol{\gamma}|\boldsymbol{e}_{\boldsymbol{y}_k}|)\mathbf{sgn}(\boldsymbol{e}_{\boldsymbol{y}_{k+1/k}})$
    Updating the state and the output estimation using
      $\hat{\boldsymbol{x}}_{k+1} = \boldsymbol{f}(\hat{\boldsymbol{x}}_k) + \boldsymbol{g}(\hat{\boldsymbol{x}}_k)\hat{\boldsymbol{d}}_k$
      $\hat{\boldsymbol{y}}_k = \boldsymbol{h}\left(\hat{\boldsymbol{x}}_k\right)$
  end
end

---

indices. The term $\boldsymbol{e}_{\boldsymbol{y}_{k+1/k}}$ includes the system modeling error. It can be used to evaluate the estimation performance of the system. $\boldsymbol{e}_{\boldsymbol{y}_k}$ is to evaluate the chattering induced by the discontinuous function $\mathbf{sgn}(\boldsymbol{e}_{\boldsymbol{y}_{k+1/k}})$ in (7.7). Those two terms let the PVSF fully extract the error information to update the system model and increase its accuracy. Consequently, the estimation accuracy of system states can be significantly improved. It should also be stressed that the PVSF is a type of corrector-predictor estimator. The estimation accuracy has great robustness to severe system uncertainties. This is owing to the advantages provided by VSC theory. This will be proved in the subsequent section.

7.3 Theoretical basis of PVSF

It is found that the PVSF in Algorithm 7.1 is similar to the input-output linearization techniques, which view the modeling error as the input for the system. However, the theoretical basis of the PVSF is totally different from the existing filters of nonlinear control theory. In contrast, the PVSF is proposed based on the following performance index function

$$\begin{aligned} J &= \min \mathbb{E}\left(\left(\left|\boldsymbol{y}_k - \hat{\boldsymbol{y}}_k\right|\right)\middle| \boldsymbol{Y}_k, \boldsymbol{d}_k\right) \\ &= \min \mathbb{E}\left(\left(\left|\boldsymbol{y}_k - \boldsymbol{h}(\hat{\boldsymbol{x}}_k)\right|\right)\middle| \boldsymbol{Y}_k, \boldsymbol{d}_k\right) \\ &= \min \mathbb{E}\left(\left(\left|\boldsymbol{h}\left(\boldsymbol{x}_k\right) + \boldsymbol{v}_k - \boldsymbol{h}\left(\hat{\boldsymbol{x}}_k\right)\right|\right) \boldsymbol{Y}_k, \boldsymbol{d}_k\right) \end{aligned} \tag{7.8}$$

where $\boldsymbol{y}_i = [y_{i1} \quad y_{i2} \quad \cdots \quad y_{im}]^{\mathrm{T}} \in \mathbb{R}^m$, $|\boldsymbol{y}_i| = [|y_{i1}| \quad |y_{i2}| \quad \cdots \quad |y_{im}|]^{\mathrm{T}} \in \mathbb{R}^m$, and $\boldsymbol{Y}_k = \left\{\boldsymbol{y}_1, \boldsymbol{y}_2, \cdots, \boldsymbol{y}_k\right\}$ is the measurement vector.

For the PVSF, the error $\boldsymbol{e}_{\boldsymbol{y}_k}$ between the measurement vector $\boldsymbol{y}_k$ and the estimation measurement vector $\hat{\boldsymbol{y}}_k$ is an important performance index parameter. By minimizing $\mathbb{E}((|\boldsymbol{y}_k - \hat{\boldsymbol{y}}_k|)|\boldsymbol{Y}_k, \boldsymbol{d}_k)$, $\boldsymbol{y}_k$ can be obtained. Then, the system model is corrected, and the state estimation is provided using a recursive formula. More specifically, the error's absolute value $|\boldsymbol{e}_{y_k}|$ is decreased by the PVSF to satisfy the index function (7.8). On the other hand, according to the measurement (2.33) in Section 2.5.1, it follows that

$$\mathbb{E}\left((|\boldsymbol{y}_k - \hat{\boldsymbol{y}}_k|)^{\mathrm{T}}(|\boldsymbol{y}_k - \hat{\boldsymbol{y}}_k|)|\boldsymbol{Y}_k, \boldsymbol{d}_k\right) = \hat{\boldsymbol{R}}_k \tag{7.9}$$

Considering the index function (7.8), (7.9) can be updated as

$$E_{\min} = \min \mathbb{E}\left(\left(\left|\boldsymbol{y}_k - \hat{\boldsymbol{y}}_k\right|\right)^{\mathrm{T}}\left(\left|\boldsymbol{y}_k - \hat{\boldsymbol{y}}_k\right|\right)|\boldsymbol{Y}_k, \boldsymbol{d}_k\right) = \mathbb{E}(\boldsymbol{v}_k^{\mathrm{T}}\boldsymbol{v}_k) = \boldsymbol{R}_k \tag{7.10}$$

It is observed that $\hat{\boldsymbol{R}}_k \to \boldsymbol{R}_k$ exist when the index function is satisfied. The covariance constraint in the theory of predictive filter is satisfied. The theoretical premise of this condition (7.10) is to derive the modeling error $\hat{\boldsymbol{d}}_k$ accurately from the index function (7.8). Hence, the index function (7.8) is the theoretical basis for the PVSF.

7.4 Stability of state estimation error

Theorem 7.1. *Applying the PVSF in Algorithm 7.1 to the nonlinear stochastic discrete-time systems described by* (2.32)–(2.33) *with the estimation* $\hat{\boldsymbol{d}}_k$ *being adaptively updated by* (7.7), *if the constant matrix* $\boldsymbol{\gamma}$ *is chosen with* $0 < \gamma_{ii} < 1$, $i = 1, 2, \cdots, m$, *satisfied, then the measurement error* $\boldsymbol{e}_{y_{k+1}} = \boldsymbol{y}_{k+1} - \hat{\boldsymbol{y}}_{k+1}$ *and the state estimation error* $\boldsymbol{e}_{x_{k+1}} = \boldsymbol{x}_{k+1} - \hat{\boldsymbol{x}}_{k+1}$ *will be asymptotically stable despite the modeling error* $\boldsymbol{d}_k$ *and the measurement noise* $\boldsymbol{v}_k$, *i.e.,* $\lim_{k\to\infty} \hat{\boldsymbol{y}}_{k+1} = \boldsymbol{y}_{k+1}$, $\lim_{k\to\infty} \hat{\boldsymbol{x}}_{k+1} = \boldsymbol{x}_{k+1}$.

Proof. Based on (7.3) and (7.6), one has

$$\boldsymbol{e}_{y_{k+1}} = \boldsymbol{y}_{k+1} - \hat{\boldsymbol{y}}_{k+1} = \boldsymbol{y}_{k+1} - \hat{\boldsymbol{y}}_k - \boldsymbol{Z}(\hat{\boldsymbol{x}}_k, \Delta t) - \boldsymbol{\Lambda}(\Delta t)\boldsymbol{U}(\hat{\boldsymbol{x}}_k)\hat{\boldsymbol{d}}_k \tag{7.11}$$

Substituting (7.5) and (7.7) into (7.11) yields

$$\begin{aligned} \boldsymbol{e}_{y_{k+1}} &= \boldsymbol{e}_{y_{k+1/k}} - \mathbf{diag}(|\boldsymbol{e}_{y_{k+1/k}}| + \boldsymbol{\gamma}|\boldsymbol{e}_{y_k}|)\mathbf{sgn}(\boldsymbol{e}_{y_{k+1/k}}) \\ &= -\boldsymbol{\gamma}\,\mathbf{diag}(|\boldsymbol{e}_{y_k}|)\mathbf{sgn}(\boldsymbol{e}_{y_{k+1/k}}) \end{aligned} \tag{7.12}$$

Because the constant matrix $\boldsymbol{\gamma}$ is positive with $0 < \gamma_{ii} < 1$, it can be obtained from (7.12) that

$$|\boldsymbol{e}_{y_{k+1}}| = \boldsymbol{\gamma}|\boldsymbol{e}_{y_k}| < |\boldsymbol{e}_{y_k}| \tag{7.13}$$

which proves that the measurement error $\boldsymbol{e}_{y_{k+1}}$ decays along with time.

Choose a Lyapunov candidate function as

$$V_{y,k+1} = \frac{1}{2}\boldsymbol{e}_{y_{k+1}}^{\mathrm{T}}\boldsymbol{e}_{y_{k+1}} \tag{7.14}$$

It is obtained from (7.13) and (7.14) that

$$V_{y,k+1} = \frac{1}{2}|e_{y_{k+1}}|^{\mathrm{T}}|e_{y_{k+1}}| < \frac{1}{2}|e_{y_k}|^{\mathrm{T}}|e_{y_k}| = V_{y,k} \tag{7.15}$$

Then, it can be proved by using the Lyapunov stability theory in Krstic et al. (1995) that $\boldsymbol{e}_{y_{k+1}}$ is asymptotically stable, *i.e.*, $\lim\limits_{k\to\infty} \boldsymbol{e}_{k+1} = \boldsymbol{0}$, $\lim\limits_{k\to\infty} \hat{y}_{k+1} = y_{k+1}$.

In addition, it can be further got from (7.13) and (7.15) that

$$|\boldsymbol{e}_{y_{k+1}}||\boldsymbol{e}_{y_{k+1}}|^{\mathrm{T}} < |\boldsymbol{e}_{y_{k+1}}||\boldsymbol{e}_{y_k}|^{\mathrm{T}} \tag{7.16}$$

$$\mathbf{diag}(\boldsymbol{e}_{y_{k+1}})\mathbf{diag}(\boldsymbol{e}_{y_{k+1}})^{\mathrm{T}} < \mathbf{diag}(\boldsymbol{e}_{y_k})\mathbf{diag}(\boldsymbol{e}_{y_k})^{\mathrm{T}} \tag{7.17}$$

Because the measurement function $\boldsymbol{h}(\cdot)$ is well defined in (2.32), the measurement error can be calculated as

$$\boldsymbol{e}_{y_{k+1}} = \boldsymbol{h}(\boldsymbol{x}_{k+1}) - \boldsymbol{h}(\hat{\boldsymbol{x}}_{k+1}) + \boldsymbol{v}_{k+1} \approx \boldsymbol{H}\boldsymbol{e}_{x_{k+1}} + \boldsymbol{v}_{k+1} \tag{7.18}$$

where $\boldsymbol{H} \in \mathbb{R}^{m\times m}$ is the linearized measurement matrix of the considered system.

Substituting (7.18) into (7.17) leads to

$$\begin{aligned}
&\mathbf{diag}(\boldsymbol{H}\boldsymbol{e}_{x_{k+1}})\mathbf{diag}(\boldsymbol{H}\boldsymbol{e}_{x_{k+1}}) + \mathbf{diag}(\boldsymbol{v}_{k+1})\mathbf{diag}(\boldsymbol{v}_{k+1}) \\
&\quad + \mathbf{diag}(\boldsymbol{H}\boldsymbol{e}_{x_{k+1}})\mathbf{diag}(\boldsymbol{v}_{k+1}) + \mathbf{diag}(\boldsymbol{v}_{k+1})\mathbf{diag}(\boldsymbol{H}\boldsymbol{e}_{x_{k+1}}) \\
&< \mathbf{diag}(\boldsymbol{H}\boldsymbol{e}_{x_k})\mathbf{diag}(\boldsymbol{H}\boldsymbol{e}_{x_k}) + \mathbf{diag}(\boldsymbol{v}_k)\mathbf{diag}(\boldsymbol{v}_k) \\
&\quad + \mathbf{diag}(\boldsymbol{H}\boldsymbol{e}_{x_{k+1}})\mathbf{diag}(\boldsymbol{v}_{k+1}) + \mathbf{diag}(\boldsymbol{v}_k)\mathbf{diag}(\boldsymbol{H}\boldsymbol{e}_{x_k})
\end{aligned} \tag{7.19}$$

Assuming that the measurement noise $\boldsymbol{v}_{k+1}$ is stationary white, then it leaves (7.19) as

$$\begin{aligned}
&\mathbb{E}\left(\mathbf{diag}(\boldsymbol{H}\boldsymbol{e}_{x_{k+1}})\mathbf{diag}(\boldsymbol{H}\boldsymbol{e}_{x_{k+1}}) + \mathbf{diag}(\boldsymbol{v}_{k+1})\mathbf{diag}(\boldsymbol{v}_{k+1})\right) \\
&< \mathbb{E}\left(\mathbf{diag}(\boldsymbol{H}\boldsymbol{e}_{x_k})\mathbf{diag}(\boldsymbol{H}\boldsymbol{e}_{x_k}) + \mathbf{diag}(\boldsymbol{v}_k)\mathbf{diag}(\boldsymbol{v}_k)\right)
\end{aligned} \tag{7.20}$$

where $\mathbb{E}(\mathbf{diag}(\boldsymbol{H}\boldsymbol{e}_{x_{k+1}})\mathbf{diag}(\boldsymbol{v}_{k+1}))$ and $\mathbb{E}(\mathbf{diag}(\boldsymbol{v}_k)\mathbf{diag}(\boldsymbol{H}\boldsymbol{e}_{x_k}))$ vanish due to the white noise assumption.

For a diagonal, positive and time-invariant measurement matrix, (7.20) can be simplified as

$$\mathbb{E}\left(\mathbf{diag}(\boldsymbol{e}_{x_{k+1}})\mathbf{diag}(\boldsymbol{e}_{x_{k+1}})\right) < \mathbb{E}\left(\mathbf{diag}(\boldsymbol{e}_{x_k})\mathbf{diag}(\boldsymbol{e}_{x_k})\right) \tag{7.21}$$

$$\mathbb{E}(|\boldsymbol{e}_{x_{k+1}}|) < \mathbb{E}(|\boldsymbol{e}_{x_k}|) \tag{7.22}$$

Then, it can be proved from (7.21) and (7.22) that $\boldsymbol{e}_{x_{k+1}}$ is also asymptotically stable, *i.e.*, $\lim\limits_{k\to\infty} \boldsymbol{e}_{k+1} = \boldsymbol{0}$, $\lim\limits_{k\to\infty} \hat{\boldsymbol{x}}_{k+1} = \boldsymbol{x}_{k+1}$. Thereby, the proof is completed here. □

7.5 Practical implementation of PVSF

It is known from (7.7) that the sampling time Δt will affect the estimation performance of the modeling error, and then affect the whole filtering performance of the PVSF. Moreover, due to the use of the discontinuous function $\mathbf{sgn}(\cdot)$ in (7.7), it can be known from the variable structure control theory that the PVSF in Algorithm 7.1 will suffer from chattering. These two issues will deteriorate the performance of PVSF. Hence, they should be solved, when the PVSF is implemented in practice.

7.5.1 Eliminating the effect of sampling time

Because the estimation (7.7) contains the term $\mathbf{\Lambda}(\Delta t)$, if the sampling time satisfies $\Delta t \ll 1$, then the estimated modeling error $\hat{\boldsymbol{d}}_k$ will be quickly increased due to the effect of $(\mathbf{\Lambda}(\Delta t)\boldsymbol{U}(\hat{\boldsymbol{x}}_k))^{-1}$. This may deteriorate the robustness of the PVSF and even leads the state estimation to be unstable. If the sampling time is such that $\Delta t \gg 1$, the sensitivity of the estimated modeling error $\boldsymbol{d}_k$ will be reduced. This will also deteriorate the filtering performance of the PVSF. As a consequence, it is important to eliminate the negative effect of the sampling time Δt on the PVSF.

We can conduct a non-time Taylor expansion to the system measurement $\boldsymbol{y}_k$, then it has

$$\boldsymbol{y}_{k+1} = \boldsymbol{y}_k + \boldsymbol{Z}(\boldsymbol{x}_k) + \boldsymbol{U}(\boldsymbol{x}_k)\boldsymbol{d}_k + o(\boldsymbol{e}_{k+1}^M) \tag{7.23}$$

where $o(\boldsymbol{e}_{k+1}^M) = o(\boldsymbol{e}_{k+1}) + \boldsymbol{Z}(\boldsymbol{x}_k, \Delta t) + \mathbf{\Lambda}(\Delta t)\boldsymbol{U}(\boldsymbol{x}_k)\boldsymbol{d}_k - \boldsymbol{Z}(\boldsymbol{x}_k) + \boldsymbol{U}(\boldsymbol{x}_k)\boldsymbol{d}_k$. By doing the same procedure to (2.39), one can obtain

$$\boldsymbol{y}_{k+1} = \hat{\boldsymbol{y}}_k + \boldsymbol{Z}(\hat{\boldsymbol{x}}_k) + \boldsymbol{U}(\hat{\boldsymbol{x}}_k)\boldsymbol{d}_k + o(\hat{\boldsymbol{e}}_{k+1}^M) \tag{7.24}$$

where $o(\hat{\boldsymbol{e}}_{k+1}^M) = o(\hat{\boldsymbol{e}}_{k+1}) + \boldsymbol{Z}(\hat{\boldsymbol{x}}_k, \Delta t) + \mathbf{\Lambda}(\Delta t)\boldsymbol{U}(\boldsymbol{x}_k)\boldsymbol{d}_k - \boldsymbol{Z}(\boldsymbol{x}_k) + \boldsymbol{U}(\boldsymbol{x}_k)\boldsymbol{d}_k$.

Note that all the approximation errors in (7.23) and (7.24) are included in $o(\boldsymbol{e}_{k+1}^M)$ and $o(\hat{\boldsymbol{e}}_{k+1}^M)$, respectively. The term $\boldsymbol{Z}(\boldsymbol{x}_k)$ of (7.23) is nearly the same as $\boldsymbol{Z}(\boldsymbol{x}_k, \Delta t)$ in (2.38), but does not consider the sampling time Δt. (7.23) and (2.38) are the same for $\Delta t = 1$. As a result, the measurement estimation with a non-time Taylor series expansion can be given by

$$\hat{\boldsymbol{y}}_{k+1} = \hat{\boldsymbol{y}}_k + \boldsymbol{Z}\left(\hat{\boldsymbol{x}}_k\right) + \boldsymbol{U}\left(\hat{\boldsymbol{x}}_k\right)\hat{\boldsymbol{d}}_k \tag{7.25}$$

$$\boldsymbol{y}_{k+1} = \hat{\boldsymbol{y}}_{k+1} + o(\hat{\boldsymbol{e}}_{k+1}^M) \tag{7.26}$$

The prior measurement error $\boldsymbol{e}_{\boldsymbol{y}_{k+1/k}}$ and the posterior measurement error $\boldsymbol{e}_{\boldsymbol{y}_k}$ are determined by

$$\boldsymbol{e}_{\boldsymbol{y}_{k+1/k}} = \boldsymbol{y}_{k+1} - \hat{\boldsymbol{y}}_k - \boldsymbol{Z}\left(\hat{\boldsymbol{x}}_k\right) \tag{7.27}$$

$$\boldsymbol{e}_{\boldsymbol{y}_k} = \boldsymbol{y}_k - \hat{\boldsymbol{y}}_k \tag{7.28}$$

Then, the estimation of the system's modeling error $\boldsymbol{d}_k$ can be re-designed as

$$\hat{\boldsymbol{d}}_k = (\boldsymbol{U}(\hat{\boldsymbol{x}}_k))^{-1}\mathbf{diag}(|\boldsymbol{e}_{\boldsymbol{y}_{k+1/k}}| + \boldsymbol{\gamma}|\boldsymbol{e}_{\boldsymbol{y}_k}|)\mathbf{sgn}(\boldsymbol{e}_{\boldsymbol{y}_{k+1/k}}) \tag{7.29}$$

where $\boldsymbol{\gamma} = \mathbf{diag}([\gamma_{11} \quad \gamma_{12} \quad \cdots \quad \gamma_{mm}]^{\mathrm{T}}) \in \mathbb{R}^{m\times m}$ is a constant matrix with $0 < \gamma_{ii} < 1, i = 1, 2, \cdots, m$.

Based on the preceding analysis, it is ready to present a modified predictive variable structure filtering algorithm (MPVSF) in Algorithm 7.2 with the effect of the sampling time eliminated. Consequently, the estimation performance achieved by the MPVSF is more robust to modeling error. Correspondingly, the estimation is more stable even in the presence of severe modeling error. Moreover, the numerical integral method is adopted to update the state and the output estimation in Algorithm 7.2.

Algorithm 7.2: Modified predictive variable structure filter.

Input Data: Initial estimation $\hat{\boldsymbol{x}}_0$, $\hat{\boldsymbol{y}}_0$, and measurement $\boldsymbol{y}_k$
Result: State estimation $\hat{\boldsymbol{x}}_k$
begin
  for $k = 1, 2 \cdots$ **do**
    Calculating the modeling error's estimation $\hat{\boldsymbol{d}}_k$ by using
      $\boldsymbol{e}_{\boldsymbol{y}_{k+1/k}} = \boldsymbol{y}_{k+1} - \hat{\boldsymbol{y}}_k - \boldsymbol{Z}(\hat{\boldsymbol{x}}_k)$
      $\boldsymbol{e}_{\boldsymbol{y}_k} = \boldsymbol{y}_{k+1} - \hat{\boldsymbol{y}}_k$
      $\hat{\boldsymbol{d}}_k = (\boldsymbol{U}(\hat{\boldsymbol{x}}_k))^{-1}\mathbf{diag}(|\boldsymbol{e}_{\boldsymbol{y}_{k+1/k}}| + \boldsymbol{\gamma}|\boldsymbol{e}_{\boldsymbol{y}_k}|)\mathbf{sgn}(\boldsymbol{e}_{\boldsymbol{y}_{k+1/k}})$
    Updating the state and the output estimation using
      $\hat{\boldsymbol{x}}_{k+1} = \boldsymbol{f}(\hat{\boldsymbol{x}}_k) + \boldsymbol{g}(\hat{\boldsymbol{x}}_k)\hat{\boldsymbol{d}}_k$
      $\hat{\boldsymbol{y}}_k = \boldsymbol{h}(\hat{\boldsymbol{x}}_k)$
  end
end

Theorem 7.2. *Applying the MPVSF developed in Algorithm 7.2 to the nonlinear stochastic discrete-time systems described by* (2.32)–(2.32) *with the estimation* $\hat{\boldsymbol{d}}_k$ *being adaptively updated by* (7.29), *suppose that the constant matrix* $\boldsymbol{\gamma}$ *is chosen with* $0 < \gamma_{ii} < 1$, $i = 1, 2, \cdots, m$, *satisfied, then the measurement error* $\boldsymbol{e}_{\boldsymbol{y}_{k+1}} = \boldsymbol{y}_{k+1} - \hat{\boldsymbol{y}}_{k+1}$ *and the state estimation error* $\boldsymbol{e}_{\boldsymbol{x}_{k+1}} = \boldsymbol{x}_{k+1} - \hat{\boldsymbol{x}}_{k+1}$ *will be asymptotically stable despite the modeling error* $\boldsymbol{d}_k$ *and the measurement noise* $\boldsymbol{v}_k$, *i.e.,* $\lim_{k\to\infty} \hat{\boldsymbol{y}}_{k+1} = \boldsymbol{y}_{k+1}$, $\lim_{k\to\infty} \hat{\boldsymbol{x}}_{k+1} = \boldsymbol{x}_{k+1}$. *Moreover, the filtering performance achieved by the MPVSF is independent of the sampling time* Δt. *Better robustness than the PVSF is guaranteed by the MPVSF.*

Proof. Following the procedure in the proof of Theorem 7.1, this theorem can be directly proved. □

Remark 7.2. When implementing the MPVSF, the generalized inverse theory can be applied to calculate $(\boldsymbol{U}(\boldsymbol{x}_k))^{-1}$. Suppose that the rank of $\boldsymbol{U}(\boldsymbol{x}_k)$ is $\hbar \in \mathbb{R}_+$, $(\boldsymbol{U}(\boldsymbol{x}_k))$ can be decomposed as $\underline{\boldsymbol{U}}(\boldsymbol{x}_k) = \underline{\boldsymbol{U}}_B \underline{\boldsymbol{U}}_C$, where $\underline{\boldsymbol{U}}_B \in \mathbb{R}^{m\times\hbar}$ $\underline{\boldsymbol{U}}_C \in \mathbb{R}^{\hbar\times l}$. Then, $(\boldsymbol{U}(\boldsymbol{x}_k))^{-1}$ can be calculated by

$$(\boldsymbol{U}(\boldsymbol{x}_k))^{-1} = \underline{\boldsymbol{U}}_C^{\mathrm{T}}(\underline{\boldsymbol{U}}_C\underline{\boldsymbol{U}}_C^{\mathrm{T}})^{-1}(\underline{\boldsymbol{U}}_B^{\mathrm{T}}\underline{\boldsymbol{U}}_B)^{-1}\underline{\boldsymbol{U}}_C^{\mathrm{T}} \tag{7.30}$$

This ensures that $(\boldsymbol{U}(\boldsymbol{x}_k))^{-1}$ always exists. Moreover, $\boldsymbol{U}(\boldsymbol{x}_k)$ is always positive and does not affect the filtering stability, as discussed in Li and Zhang (2006).

7.5.2 Reducing chattering

Due to the discontinuous function $\mathbf{sgn}(\cdot)$ in (7.7), the well-known "chattering phenomenon" in the variable structure control theory will be seen in the PVSF. The filtering accuracy will be affected by this phenomenon. More specifically the chattering is induced by the time-varying terms $\mathbf{diag}(|\boldsymbol{e}_{y_{k+1/k}}|)\mathbf{sgn}(\boldsymbol{e}_{y_{k+1/k}})$ and $\mathbf{diag}(|\boldsymbol{e}_{y_k}|)\mathbf{sgn}(\boldsymbol{e}_{y_{k+1/k}})$.

It is got from the proof of Theorem 7.1 and Theorem 7.2 that the posteriori measurement error $\boldsymbol{e}_{y_k}$ decays along with time. The chattering induced by $\mathbf{diag}(|\boldsymbol{e}_{y_k}|)\mathbf{sgn}(\boldsymbol{e}_{y_{k+1/k}})$ will be degraded. This inferred that the chattering is mainly introduced by $\mathbf{diag}(|\boldsymbol{e}_{y_{k+1/k}}|)\mathbf{sgn}(\boldsymbol{e}_{y_{k+1/k}})$. Hence, the effect of $\mathbf{diag}(|\boldsymbol{e}_{y_{k+1/k}}|)\mathbf{sgn}(\boldsymbol{e}_{y_{k+1/k}})$ acting on the chattering should be analyzed. From (7.27), one has

$$\begin{aligned}\boldsymbol{e}_{y_{k+1/k}} &= \boldsymbol{y}_{k+1} - \hat{\boldsymbol{y}}_k - \boldsymbol{Z}(\hat{\boldsymbol{x}}_k)\\ &= \boldsymbol{U}(\boldsymbol{x}_k)\boldsymbol{d}_k + \boldsymbol{o}(\hat{\boldsymbol{e}}_{k+1}^{M})\\ &= \boldsymbol{U}(\boldsymbol{x}_k)\boldsymbol{d}_k + \boldsymbol{\varepsilon}_M(\hat{\boldsymbol{x}}_k, \boldsymbol{d}_k) + \boldsymbol{v}_{k+1}\end{aligned} \tag{7.31}$$

Projecting $\boldsymbol{e}_{y_{k+1/k}}$ of (7.31) into the state estimation error yields

$$\boldsymbol{e}_{y_{k+1/k}} = \boldsymbol{H}^{-1}\boldsymbol{U}(\boldsymbol{x}_k)\boldsymbol{d}_k + \boldsymbol{H}^{-1}\boldsymbol{v}_{k+1} + \boldsymbol{H}^{-1}\boldsymbol{\varepsilon}_M(\hat{\boldsymbol{x}}_k, \boldsymbol{d}_k) \tag{7.32}$$

From (7.32), it can be summarized that the chattering is mainly induced by the modeling errors $\boldsymbol{d}_k$, the measurement noise $\boldsymbol{v}_{k+1}$, and high-order discretization error $\boldsymbol{\varepsilon}_M(\hat{\boldsymbol{x}}_k, \boldsymbol{d}_k)$.

Following the result in the variable structure control theory to eliminate the chattering problem, the sign function in (7.28) can be replaced by a smoothing function (Shtessel et al., 2014). This is called as the boundary lawyer method. Then, (7.29) can be modified as

$$\hat{\boldsymbol{d}}_k = (\boldsymbol{U}(\hat{\boldsymbol{x}}_k))^{-1}\mathbf{diag}(|\boldsymbol{e}_{y_{k+1/k}}| + \gamma|\boldsymbol{e}_{y_k}|)\mathbf{sat}\left(\frac{|\boldsymbol{e}_{y_{k+1/k}}|}{\ell}\right) \tag{7.33}$$

where $\ell \in \mathbb{R}_+$ is a small constant. It represents the error associated with the smooth approximation of the function $\mathbf{sgn}(\cdot)$ by the function $\mathbf{sat}(\cdot)$. Then, it is ready to present a predictive smooth variable structure filtering algorithm

(PSVSF) in Algorithm 7.3 with the effect of sampling time and chattering eliminated simultaneously. Moreover, the numerical integral method is adopted to update the state and the output estimation in Algorithm 7.3.

Remark 7.3. The value 2ℓ denotes the thickness of the boundary layer. In general, the greater the boundary layer thickness, the less the chattering. Although applying a boundary layer alleviates the chattering phenomenon, the sliding condition $\boldsymbol{e}_{\boldsymbol{y}_{k+1/k}} = \mathbf{0}$ is no longer guaranteed. As a sequence, the estimation error converges into a small residual set containing $\boldsymbol{e}_{\boldsymbol{y}_{k+1/k}} = \mathbf{0}$ when a boundary layer is introduced. The size of this residual set is dependent on ℓ, *i.e.*, larger ℓ leads to bigger residual set or inferior estimation accuracy. Therefore, introducing a boundary layer at the sliding surface $\boldsymbol{e}_{\boldsymbol{y}_{k+1/k}} = \mathbf{0}$ leads to a deterioration of estimation accuracy and robustness. A trade-off between chattering and estimation accuracy as well as robustness should be done when choosing ℓ.

Algorithm 7.3: Predictive smooth variable structure filter.

Input Data: Initial estimation $\hat{\boldsymbol{x}}_0$, $\hat{\boldsymbol{y}}_0$, and measurement $\boldsymbol{y}_k$
Result: State estimation $\hat{\boldsymbol{x}}_k$
begin
  for $k = 1, 2 \cdots$ **do**
    Calculating the modeling error's estimation $\hat{\boldsymbol{d}}_k$ by using
      $\boldsymbol{e}_{\boldsymbol{y}_{k+1/k}} = \boldsymbol{y}_{k+1} - \hat{\boldsymbol{y}}_k - \boldsymbol{Z}(\hat{\boldsymbol{x}}_k)$
      $\boldsymbol{e}_{\boldsymbol{y}_k} = \boldsymbol{y}_{k+1} - \hat{\boldsymbol{y}}_k$
      $\hat{\boldsymbol{d}}_k = (\boldsymbol{U}(\hat{\boldsymbol{x}}_k))^{-1}\mathbf{diag}(|\boldsymbol{e}_{\boldsymbol{y}_{k+1/k}}| + \gamma|\boldsymbol{e}_{\boldsymbol{y}_k}|)\mathbf{sat}\left(\frac{|\boldsymbol{e}_{\boldsymbol{y}_{k+1/k}}|}{\ell}\right)$
    Updating the state and the output estimation using
      $\hat{\boldsymbol{x}}_{k+1} = \boldsymbol{f}(\hat{\boldsymbol{x}}_k) + \boldsymbol{g}(\hat{\boldsymbol{x}}_k)\hat{\boldsymbol{d}}_k$
      $\hat{\boldsymbol{y}}_k = \boldsymbol{h}(\hat{\boldsymbol{x}}_k)$
  end
end

7.6 State estimation error covariance of PSVSF

Although the implementation of Algorithms 7.1–7.3 does not need the calculation of the covariance, which is to be computed in this section for evaluating the filtering performance.

Taking Algorithm 7.3 as an example, the states estimation process of the PSVSF can be written into the following Kalman filtering form.

$$\hat{\boldsymbol{x}}_{k+1/k} = \boldsymbol{f}\left(\hat{\boldsymbol{x}}_k\right) \tag{7.34}$$

$$\hat{\boldsymbol{x}}_{k+1} = \hat{\boldsymbol{x}}_{k+1/k} + \boldsymbol{g}(\hat{\boldsymbol{x}}_k)(\boldsymbol{U}(\hat{\boldsymbol{x}}_k))^{-1}\mathbf{diag}(|\boldsymbol{e}_{\boldsymbol{y}_{k+1/k}}| + \gamma|\boldsymbol{e}_{\boldsymbol{y}_k}|)\mathbf{sat}\left(\frac{|\boldsymbol{e}_{\boldsymbol{y}_{k+1/k}}|}{\ell}\right) \tag{7.35}$$

or

$$\hat{\boldsymbol{x}}_{k+1} = \hat{\boldsymbol{x}}_{k+1/k} + \boldsymbol{K}_{k+1}\boldsymbol{e}_{\boldsymbol{y}_{k+1/k}} \tag{7.36}$$

$$\begin{aligned}\boldsymbol{K}_{k+1} = \boldsymbol{g}_U(\hat{\boldsymbol{x}}_k)\mathbf{diag}(|\boldsymbol{e}_{\boldsymbol{y}_{k+1/k}}| + \gamma|\boldsymbol{e}_{\boldsymbol{y}_k}|)\mathbf{diag}\left(\mathbf{sat}\left(\frac{|\boldsymbol{e}_{\boldsymbol{y}_{k+1/k}}|}{\ell}\right)\right) \\ \times (\mathbf{diag}(\boldsymbol{e}_{\boldsymbol{y}_{k+1/k}}))^{-1}\end{aligned} \tag{7.37}$$

where $\boldsymbol{g}_U(\hat{\boldsymbol{x}}_k) = \boldsymbol{g}(\hat{\boldsymbol{x}}_k)(\boldsymbol{U}(\hat{\boldsymbol{x}}_k))^{-1}$. Here, $\boldsymbol{K}_{k+1}$ is defined as the gain matrix of the modeling error.

Based on (7.36)–(7.37), let $\tilde{\boldsymbol{x}}_{\boldsymbol{x}_{k+1/k}} = \boldsymbol{x}_{k+1} - \hat{\boldsymbol{x}}_{k+1/k}$ be the prior state estimation error, then we have

$$\tilde{\boldsymbol{x}}_{\boldsymbol{x}_{k+1/k}} = \boldsymbol{x}_{k+1} - \hat{\boldsymbol{x}}_{k+1/k} = \boldsymbol{f}(\boldsymbol{x}_k) - \boldsymbol{f}\left(\hat{\boldsymbol{x}}_k\right) = \boldsymbol{F}_k\tilde{\boldsymbol{x}}_{k/k} \tag{7.38}$$

where $\boldsymbol{F}_k = \frac{\partial \boldsymbol{f}}{\partial \boldsymbol{x}}|_{\boldsymbol{x}=\boldsymbol{x}_k}$ and $\tilde{\boldsymbol{x}}_{\boldsymbol{x}_{k/k}} = \boldsymbol{x}_k - \hat{\boldsymbol{x}}_{k/k}$.

Using the definition of the prior covariance of state estimation error, it follows that

$$\boldsymbol{P}_{\tilde{x},k+1/k} = \mathbb{E}(\tilde{\boldsymbol{x}}_{k+1/k}\tilde{\boldsymbol{x}}_{k+1/k}^{\mathrm{T}}) = \boldsymbol{F}_k\mathbb{E}(\tilde{\boldsymbol{x}}_{k/k}\tilde{\boldsymbol{x}}_{k/k}^{\mathrm{T}})\boldsymbol{F}_k^{\mathrm{T}} = \boldsymbol{F}_k\boldsymbol{P}_{\tilde{x},k}\boldsymbol{F}_k^{\mathrm{T}} \tag{7.39}$$

$$\boldsymbol{P}_{\tilde{x},k+1} = \mathbb{E}(\tilde{\boldsymbol{x}}_{k+1}\tilde{\boldsymbol{x}}_{k+1}^{\mathrm{T}}) \tag{7.40}$$

with $\tilde{\boldsymbol{x}}_{k+1} = \boldsymbol{x}_{k+1} - \hat{\boldsymbol{x}}_{k+1}$. Then, invoking (7.36)–(7.37) results in

$$\begin{aligned}\tilde{\boldsymbol{x}}_{k+1} &= \boldsymbol{x}_{k+1} - \hat{\boldsymbol{x}}_{k+1/k} - \boldsymbol{K}_{k+1}\boldsymbol{e}_{\boldsymbol{y}_{k+1/k}} \\ &= \tilde{\boldsymbol{x}}_{k+1/k} - \boldsymbol{K}_{k+1}(\boldsymbol{y}_{k+1} - \hat{\boldsymbol{y}}_k - \boldsymbol{Z}(\hat{\boldsymbol{x}}_k)) \\ &= \tilde{\boldsymbol{x}}_{k+1/k} - \boldsymbol{K}_{k+1}(\boldsymbol{h}(\boldsymbol{x}_{k+1}) + \boldsymbol{v}_{k+1} - \boldsymbol{h}(\hat{\boldsymbol{x}}_{k+1/k})) \\ &= \tilde{\boldsymbol{x}}_{k+1/k} - \boldsymbol{K}_{k+1}(\boldsymbol{H}\hat{\boldsymbol{x}}_{k+1/k} + \boldsymbol{v}_{k+1}) \\ &= (\boldsymbol{I}_n - \boldsymbol{K}_{k+1}\boldsymbol{H}_{k+1})\hat{\boldsymbol{x}}_{k+1/k} - \boldsymbol{K}_{k+1}\boldsymbol{v}_{k+1}\end{aligned} \tag{7.41}$$

Combining (7.40) and (7.41), applying the fact that the measurement noise $\boldsymbol{v}_{k+1}$ has zero mean, then the state estimation error covariance $\boldsymbol{P}_{\tilde{x},k+1}$ achieved by the PSVSF can be calculated as

$$\begin{aligned}\boldsymbol{P}_{\tilde{x},k+1} = \mathbb{E}(\boldsymbol{x}_a\boldsymbol{x}_a^{\mathrm{T}}) = (\boldsymbol{I}_n - \boldsymbol{K}_{k+1}\boldsymbol{H}_{k+1})\boldsymbol{P}_{\tilde{x},k+1/k}(\boldsymbol{I}_n - \boldsymbol{K}_{k+1}\boldsymbol{H}_{k+1})^{\mathrm{T}} \\ + \boldsymbol{K}_{k+1}\boldsymbol{R}_{k+1}\boldsymbol{K}_{K+1}^{\mathrm{T}}\end{aligned} \tag{7.42}$$

where $\boldsymbol{x}_a = (\boldsymbol{I}_n - \boldsymbol{K}_{k+1}\boldsymbol{H}_{k+1})\hat{\boldsymbol{x}}_{k+1/k} - \boldsymbol{K}_{k+1}\boldsymbol{v}_{k+1}$.

Based on the above analysis, the covariance estimation included version of the PSVSF can be presented in Algorithm 7.4. Moreover, the numerical integral method is adopted to update the state and the output estimation in Algorithm 7.4.

Algorithm 7.4: Covariance estimation included PSVSF.

Input Data: Initial estimation $\hat{x}_0$, $\hat{y}_0$, and measurement y_k
Result: State estimation $\hat{x}_k$
begin
  for $k = 0, 1, 2 \cdots$ **do**
    Calculating the modeling error's estimation $\hat{d}_k$ by using
$$e_{y_{k+1/k}} = y_{k+1} - \hat{y}_k - Z(\hat{x}_k)$$
$$e_{y_k} = y_{k+1} - \hat{y}_k$$
$$\hat{d}_k = (U(\hat{x}_k))^{-1}\mathbf{diag}(|e_{y_{k+1/k}}| + \gamma |e_{y_k}|)\mathbf{sat}\left(\frac{|e_{y_{k+1/k}}|}{\ell}\right)$$
$$K_{k+1} = g_U(\hat{x}_k)\mathbf{diag}(|e_{y_{k+1/k}}| + \gamma |e_{y_k}|)$$
$$\mathbf{diag}\left(\mathbf{sat}\left(\frac{|e_{y_{k+1/k}}|}{\ell}\right)\right)(\mathbf{diag}(e_{y_{k+1/k}}))^{-1}$$
    Updating the state and the output estimation by using
$$\hat{x}_{k+1/k} = f(\hat{x}_k)$$
$$\hat{x}_{k+1} = \hat{x}_{k+1/k} + K_{k+1}e_{y_{k+1/k}}$$
$$\hat{y}_k = h(\hat{x}_k)$$
    Updating the state estimation error covariance by using
$$F_{k+1} = \frac{\partial f}{\partial x}|_{x=x_{k+1}}$$
$$P_{x,k+1/k} = F_k P_{x,k-1} F_k^{\mathrm{T}}$$
$$P_{x,k+1} = (I_n - K_{k+1}H_{k+1})P_{x,k+1/k}(I_n - K_{k+1}H_{k+1})^{\mathrm{T}} + K_{k+1}R_{k+1}K_{K+1}^{\mathrm{T}}$$
  end
end

Remark 7.4. When applying the PSVSF into practical engineering, updating the estimation of the modeling error and updating the state estimation should be executed only, while updating the error covariance is not necessary. Hence, the implementation of the PSVSF demands fewer computer resources. It has great practical application potential. Moreover, it is very appropriate to solve the state estimation problem of the systems with limited computation power.

Remark 7.5. The linearization technique is applied to estimate $P_{\tilde{x},k+1/k}$. This may affect the estimation accuracy of the covariance $P_{\tilde{x},k+1/k}$. To avoid this shortcoming, the sigma-points schemes in Part II can be applied to improve the covariance's estimation accuracy, those procedures are listed as follows.

- *Method #1: UT-based estimation.* When implementing this method, let us first calculate

$$P_{\tilde{x},k} = S_k S_k^{\mathrm{T}} = \sum_{i=1}^{n} \sigma_{i,k}\sigma_{i,k}^{\mathrm{T}} \tag{7.43}$$

$$\boldsymbol{\xi}_{0,k} = \hat{\boldsymbol{x}}_k,\ W_0 = \frac{\kappa}{n+\kappa} \tag{7.44}$$

$$W_j = \frac{1}{2(n+\kappa)},\ j = 1, 2, 3, \cdots, 2n \tag{7.45}$$

$$\boldsymbol{\xi}_{i,k} = \hat{\boldsymbol{x}}_k + \sqrt{n+\kappa}\boldsymbol{\sigma}_{i,k},\ i = 1, 2, 3, \cdots, n \tag{7.46}$$

$$\boldsymbol{\xi}_{i+n,k} = \hat{\boldsymbol{x}}_k - \sqrt{n+\kappa}\boldsymbol{\sigma}_{i,k} \tag{7.47}$$

$$\hat{\boldsymbol{x}}_{i,k+1/k} = \boldsymbol{f}(\boldsymbol{\xi}_{i,k}),\ i = 1, 2, 3, \cdots, 2n \tag{7.48}$$

then updating the state estimation error covariance by using

$$\boldsymbol{P}_{\tilde{x},k+1/k} = \sum_{i=0}^{2n} W_i(\hat{\boldsymbol{x}}_{i,k+1/k} - \hat{\boldsymbol{x}}_{k+1/k})(\hat{\boldsymbol{x}}_{i,k+1/k} - \hat{\boldsymbol{x}}_{k+1/k})^{\mathrm{T}} \tag{7.49}$$

$$\begin{aligned}\boldsymbol{P}_{\tilde{x},k+1} = {} & (\boldsymbol{I}_n - \boldsymbol{K}_{k+1}\boldsymbol{H}_{k+1})\boldsymbol{P}_{\tilde{x},k+1/k}(\boldsymbol{I}_n - \boldsymbol{K}_{k+1}\boldsymbol{H}_{k+1})^{\mathrm{T}} \\ & + \boldsymbol{K}_{k+1}\boldsymbol{R}_{k+1}\boldsymbol{K}_{k+1}^{\mathrm{T}}\end{aligned} \tag{7.50}$$

- *Method #2: Cubature rule-based estimation.* The following calculations are first conducted to execute this filtering method.

$$\boldsymbol{P}_{\tilde{x},k} = \boldsymbol{S}_k\boldsymbol{S}_k^{\mathrm{T}} = \sum_{i=0}^{n} \boldsymbol{\sigma}_{i,k}\boldsymbol{\sigma}_{i,k}^{\mathrm{T}} \tag{7.51}$$

$$\boldsymbol{\xi}_{i,k} = \hat{\boldsymbol{x}}_k + \sqrt{n}\boldsymbol{\sigma}_{i,k},\ i = 1, 2, 3, \cdots, n \tag{7.52}$$

$$\boldsymbol{\xi}_{i+n,k} = \hat{\boldsymbol{x}}_k - \sqrt{n}\boldsymbol{\sigma}_{i,k},\ i = 1, 2, 3, \cdots, n \tag{7.53}$$

$$\hat{\boldsymbol{x}}_{i,k+1/k} = \boldsymbol{f}(\boldsymbol{\xi}_{i,k}),\ i = 0, 1, 2, \cdots, 2n \tag{7.54}$$

Then, updating the state estimation error covariance by using

$$\boldsymbol{P}_{\tilde{x},k+1/k} = \frac{1}{2n}\sum_{i=0}^{2n}(\hat{\boldsymbol{x}}_{i,k+1/k}\hat{\boldsymbol{x}}_{i,k+1/k}^{\mathrm{T}} - \hat{\boldsymbol{x}}_{k+1/k}\hat{\boldsymbol{x}}_{k+1/k}^{\mathrm{T}}) \tag{7.55}$$

$$\begin{aligned}\boldsymbol{P}_{\tilde{x},k+1} = {} & (\boldsymbol{I}_n - \boldsymbol{K}_{k+1}\boldsymbol{H}_{k+1})\boldsymbol{P}_{\tilde{x},k+1/k}(\boldsymbol{I}_n - \boldsymbol{K}_{k+1}\boldsymbol{H}_{k+1})^{\mathrm{T}} \\ & + \boldsymbol{K}_{k+1}\boldsymbol{R}_{k+1}\boldsymbol{K}_{k+1}^{\mathrm{T}}\end{aligned} \tag{7.56}$$

- *Method #3: Stirling's polynomial interpolation-based estimation.* For this method, the following calculations are carried out first.

$$P_{\tilde{x},k} = S_k S_k^{\mathrm{T}} = \sum_{i=1}^{n} \sigma_{i,k} \sigma_{i,k}^{\mathrm{T}} \tag{7.57}$$

$$\xi_{0,k} = \hat{x}_k,\ \xi_{i,k} = \hat{x}_k + h\sigma_{i,k},\ i = 1, 2, 3, \cdots, n \tag{7.58}$$

$$\xi_{i+n,k} = \hat{x}_k - h\sigma_{i,k},\ i = 1, 2, 3, \cdots, n \tag{7.59}$$

Then, updating the state estimation error covariance by using

$$\begin{aligned} P_{\tilde{x},k+1/k} &= \frac{1}{4h^2} \sum_{i=0}^{2n} (f(\xi_{i,k}) - f(\xi_{i+n,k}))^2 \\ &+ \frac{h^2 - 1}{4h^4} \sum_{i=0}^{2n} (f(\xi_{i,k}) - f(\xi_{i+n,k}) - 2f(\xi_{0,k}))^2 \end{aligned} \tag{7.60}$$

$$\begin{aligned} P_{\tilde{x},k+1} &= (I_n - K_{k+1} H_{k+1}) P_{\tilde{x},k+1/k} (I_n - K_{k+1} H_{k+1})^{\mathrm{T}} \\ &+ K_{k+1} R_{k+1} K_{k+1}^{\mathrm{T}} \end{aligned} \tag{7.61}$$

7.7 Application to microsatellite distributed attitude control system

In this section, the PSVSF developed in Algorithm 7.3 is applied to address the state estimation problem of the distributed microsatellite attitude control system. More specifically, the relative attitude and the relative angular velocity between microsatellites are to be estimated by using the presented PSVSF. Each microsatellite in the distributed attitude control system is equipped with a Charge-Couple Device (CCD) camera-based vision measurement unit/sensor to measure the relative attitude.

7.7.1 Measurement model of CCD camera

Because the measurement technology is not the main focus of this book, it is supposed that the output of the CCD vision measurement unit is the relative attitude information directly, which can be practically achieved. Hence, we can mathematically model the measurement model of the CCD vision measurement unit as

$$\begin{cases} \phi_{i\to j}^{m} = \phi_{i\to j} + \Delta\phi_{i\to j} \\ \theta_{i\to j}^{m} = \theta_{i\to j} + \Delta\theta_{i\to j} \\ \psi_{i\to j}^{m} = \psi_{i\to j} + \Delta\psi_{i\to j} \end{cases} \tag{7.62}$$

where $[\phi_{i\to j} \quad \theta_{i\to j} \quad \psi_{i\to j}]^{\mathrm{T}} \in \mathbb{R}^3$ is the real relative attitude Euler angels between the ith and the jth microsatellite, $[\phi^m_{i\to j} \quad \theta^m_{i\to j} \quad \psi^m_{i\to j}]^{\mathrm{T}} \in \mathbb{R}^3$ is the measured relative attitude Euler angels from the CCD camera between the ith and the jth microsatellite, and $\Delta\Theta_{i\to j} = [\Delta\phi_{i\to j} \quad \Delta\theta_{i\to j} \quad \Delta\psi_{i\to j}]^{\mathrm{T}} \in \mathbb{R}^3$ is the measurement noise or error, $i, j = 0, 1, 2, \cdots, N_{\mathrm{O}}$.

By transforming the relative attitude Euler angles into the relative quaternion, the measurement model can be re-modeled as

$$\boldsymbol{q}^m_{i\to j} = \boldsymbol{q}_{i\to j} \otimes \Delta\boldsymbol{q}_{i\to j} \tag{7.63}$$

where the unit-quaternion $\boldsymbol{q}^m_{i\to j} \in \mathbb{R}^4$ denotes the measured relative attitude quaternion between the ith and the jth microsatellite, $i, j = 0, 1, 2, \cdots, N_{\mathrm{O}}$. This measurement output is contaminated by noises. $\Delta\boldsymbol{q}_{i\to j} \in \mathbb{R}^4$ is the relative attitude quaternion measurement error induced by noise. Moreover, it follows that

$$\boldsymbol{q}^m_{i\to j} = \begin{bmatrix} \cos\frac{\theta^m_{i\to j}}{2}\cos\frac{\phi^m_{i\to j}}{2}\cos\frac{\psi^m_{i\to j}}{2} + \sin\frac{\theta^m_{i\to j}}{2}\sin\frac{\phi^m_{i\to j}}{2}\sin\frac{\psi^m_{i\to j}}{2} \\ \sin\frac{\theta^m_{i\to j}}{2}\cos\frac{\phi_{i\to j}}{2}\cos\frac{\psi^m_{i\to j}}{2} - \cos\frac{\theta^m_{i\to j}}{2}\sin\frac{\phi^m_{i\to j}}{2}\sin\frac{\psi^m_{i\to j}}{2} \\ \cos\frac{\theta^m_{i\to j}}{2}\sin\frac{\phi^m_{i\to j}}{2}\cos\frac{\psi^m_{i\to j}}{2} + \sin\frac{\theta^m_{i\to j}}{2}\cos\frac{\phi^m_{i\to j}}{2}\sin\frac{\psi^m_{i\to j}}{2} \\ \cos\frac{\theta^m_{i\to j}}{2}\cos\frac{\phi^m_{i\to j}}{2}\sin\frac{\psi^m_{i\to j}}{2} - \sin\frac{\theta^m_{i\to j}}{2}\sin\frac{\phi^m_{i\to j}}{2}\cos\frac{\psi^m_{i\to j}}{2} \end{bmatrix} \tag{7.64}$$

$$\boldsymbol{q}_{i\to j} = \begin{bmatrix} \cos\frac{\theta_{i\to j}}{2}\cos\frac{\phi_{i\to j}}{2}\cos\frac{\psi_{i\to j}}{2} + \sin\frac{\theta_{i\to j}}{2}\sin\frac{\phi_{i\to j}}{2}\sin\frac{\psi_{i\to j}}{2} \\ \sin\frac{\theta_{i\to j}}{2}\cos\frac{\phi_{i\to j}}{2}\cos\frac{\psi_{i\to j}}{2} - \cos\frac{\theta_{i\to j}}{2}\sin\frac{\phi_{i\to j}}{2}\sin\frac{\psi_{i\to j}}{2} \\ \cos\frac{\theta_{i\to j}}{2}\sin\frac{\phi_{i\to j}}{2}\cos\frac{\psi_{i\to j}}{2} + \sin\frac{\theta_{i\to j}}{2}\cos\frac{\phi_{i\to j}}{2}\sin\frac{\psi_{i\to j}}{2} \\ \cos\frac{\theta_{i\to j}}{2}\cos\frac{\phi_{i\to j}}{2}\sin\frac{\psi_{i\to j}}{2} - \sin\frac{\theta_{i\to j}}{2}\sin\frac{\phi_{i\to j}}{2}\cos\frac{\psi_{i\to j}}{2} \end{bmatrix} \tag{7.65}$$

and $\Delta\boldsymbol{q}_{i\to j}$ can be denoted by a nonlinear function with $\Delta\phi_{i\to j}$, $\Delta\theta_{i\to j}$, and $\Delta\psi_{i\to j}$ as its variables.

7.7.2 Implementing PSVSF for distributed attitude control system

The distributed microsatellite attitude control system (3.190)–(3.191) in combination with the measurement model (7.62) can be rewritten into the following nonlinear equation.

$$\begin{cases} \dot{\boldsymbol{x}} = \boldsymbol{f}(\boldsymbol{x}) + \boldsymbol{g}(\boldsymbol{x})\boldsymbol{d} \\ \boldsymbol{y} = \boldsymbol{h}(\boldsymbol{x}) + \boldsymbol{v} \end{cases} \tag{7.66}$$

where $x = [q_{i\to j}^{\mathrm{T}} \quad \omega_{i\to j}^{\mathrm{T}}]^{\mathrm{T}}$, $d = [0 \quad u_{dfi}^{\mathrm{T}}]^{\mathrm{T}}$, $v = q_{i\to j} \otimes \Delta q_{i\to j} - q_{i\to j}$, $g(x) = [0 \quad J_i^{-1}]^{\mathrm{T}}$, $h(x) = [I_4 \quad 0]$, and

$$f(x) = \begin{bmatrix} E(q_{i\to j})\omega_{i\to j} \\ f_1(x) \end{bmatrix} \tag{7.67}$$

with $f_1(x) = J_i^{-1}u_{fi} - J_i^{-1}(\omega_{i\to j} + \mathbb{C}(q_{i\to j})\omega_0)^{\times} J_i(\omega_{i\to j} + \mathbb{C}(q_{i\to j})\omega_0) + (\omega_{i\to j}^{\times}\mathbb{C}(q_{i\to j})\omega_0 - \mathbb{C}(q_{i\to j})\dot{\omega}_0)$.

To this end, the distributed microsatellite attitude control system (7.64) can be written into the form of (2.32)–(2.33) after discretization. Then, we can apply the proposed PSVSF in Algorithm 7.3 to estimate the relative attitude and the relative angular velocity states of the distributed attitude control system.

7.7.3 Simulation results

In this subsection, numerical simulations are conducted with the PSVSF applied to a distributed attitude control system of three microsatellites, *i.e.*, $N_0 = 3$. The modeling error given in (4.119) is taken into account. Although the modeling error d_k can be non-Gaussian, the measurement noises of the CCD vision measurement unit are typically Gaussian. Hence, $\Delta\phi_{i\to j}$, $\Delta\theta_{i\to j}$, and $\Delta\psi_{i\to j}$ are assumed to be Gaussian with zero mean and covariance 0.5 degrees, $i, j = 0, 1, 2, \cdots, 3$. The initial value of the relative attitude is set as

$$q_{i\to j}(0) = \frac{\sqrt{3}}{3}\left[\sqrt{3}\cos\frac{\Theta_{i\to j}^0}{2} \quad \sin\frac{\Theta_{i\to j}^0}{2} \quad \sin\frac{\Theta_{i\to j}^0}{2} \quad \sin\frac{\Theta_{i\to j}^0}{2}\right]^{\mathrm{T}} \tag{7.68}$$

where $\Theta_{i\to j}^0 \in \mathbb{R}$ is a constant. Moreover, the other parameters of each microsatellite are listed in Tables 7.1–7.3, respectively.

TABLE 7.1 Simulation parameters of the microsatellite #1.

Parameters	Value	Unit
Inertia matrix	$J_1 = \mathbf{diag}([24.2 \quad 25.2 \quad 22.3]^{\mathrm{T}})$	kg·m^2
Initial relative attitude	Set as (7.66) with $\Theta_{1\to 0}^0 = 5$ degrees	-
Initial relative angular velocity	$\omega_{1\to 0} = [0.1 \quad 0.1 \quad 0.1]^{\mathrm{T}}$	deg/sec
Initial estimated relative attitude	Set as (7.66) with $\Theta_{1\to 0}^0 = 2$ degrees	-
Initial estimated relative angular velocity	$[0 \quad 0 \quad 0]^{\mathrm{T}}$	deg/sec
Constant modeling error ΔN_c	$[-0.003 \quad -0.004 \quad 0.003]^{\mathrm{T}}$	Nm
Periodic modeling error ΔN_e	$0.002\cos(0.0115t)[3 \quad 1 \quad -2]^{\mathrm{T}}$	Nm

TABLE 7.2 Simulation parameters of the microsatellite #2.

Parameters	Value	Unit
Inertia matrix	$\boldsymbol{J}_1 = \textbf{diag}([24.2 \quad 25.2 \quad 22.3]^{\mathrm{T}})$	kg·m^2
Initial relative attitude	Set as (7.66) with $\Theta^0_{2\to1} = 4$ degrees	-
Initial relative angular velocity	$\omega_{2\to1} = [0.08 \quad 0.08 \quad 0.08]^{\mathrm{T}}$	deg/sec
Initial estimated relative attitude	Set as (7.66) with $\Theta^0_{2\to1} = 5$ degrees	-
Initial estimated relative angular velocity	$[0 \quad 0 \quad 0]^{\mathrm{T}}$	deg/sec
Constant modeling error $\Delta\boldsymbol{N}_c$	$[0.003 \quad -0.002 \quad 0.001]^{\mathrm{T}}$	Nm
Periodic modeling error $\Delta\boldsymbol{N}_e$	$0.002\cos(0.0115t)[1 \quad -4 \quad 3]^{\mathrm{T}}$	Nm

TABLE 7.3 Simulation parameters of the microsatellite #3.

Parameters	Value	Unit
Inertia matrix	$\boldsymbol{J}_1 = \textbf{diag}([24.2 \quad 25.2 \quad 22.3]^{\mathrm{T}})$	kg·m^2
Initial relative attitude	Set as (7.66) with $\Theta^0_{3\to2} = 3$ degrees	-
Initial relative angular velocity	$\omega_{3\to2} = [0.05 \quad 0.05 \quad 0.05]^{\mathrm{T}}$	deg/sec
Initial estimated relative attitude	Set as (7.66) with $\Theta^0_{3\to2} = 1$ degree	-
Initial estimated relative angular velocity	$[0 \quad 0 \quad 0]^{\mathrm{T}}$	deg/sec
Constant modeling error $\Delta\boldsymbol{N}_c$	$[0.001 \quad 0.004 \quad 0.003]^{\mathrm{T}}$	Nm
Periodic modeling error $\Delta\boldsymbol{N}_e$	$0.002\cos(0.0115t)[4 \quad 1 \quad -3]^{\mathrm{T}}$	Nm

In simulation, the proposed PSVSF is compared with the standard UKF and the SVSF in Habibi and Burton (2003). The estimation gains of the PSVSF are chosen as $\ell = 0.0005$ and $\gamma_{ii} = 0.1$, $i = 1, 2, \cdots, m$. The estimation gain of UKF is selected as $\kappa = 1$.

Remark 7.6. When applying state estimation approaches to the microsatellite distributed attitude control system, the corresponding state estimation error defined in (2.69) is specified as $\tilde{\boldsymbol{x}} = [\tilde{\boldsymbol{q}}^{\mathrm{T}}_{i\to j} \quad \tilde{\boldsymbol{\omega}}^{\mathrm{T}}_{i\to j}]^{\mathrm{T}}$. As pointed out in Remark 4.5, the state estimation error described by the unit quaternion $\tilde{\boldsymbol{q}}_{i\to j}$ can be transformed into the attitude Euler angles estimation error, which is also denoted as $\boldsymbol{\Theta}_e = [\phi_{eij} \quad \theta_{eij} \quad \psi_{eij}]^{\mathrm{T}}$. The estimation error $\tilde{\boldsymbol{\omega}}_{i\to j}$ of the relative angular velocity state is also denoted as $\boldsymbol{\omega}_E = [\omega_{Exij} \quad \omega_{Eyij} \quad \omega_{Ezij}]^{\mathrm{T}} = \tilde{\boldsymbol{\omega}}_{i\to j}$.

Applying the PSVSF, the SVSF, and the UKF to the considered distributed attitude control system, Figs. 7.1–7.3 show that those three filters almost achieved the same estimation accuracy for the relative attitude states. However,

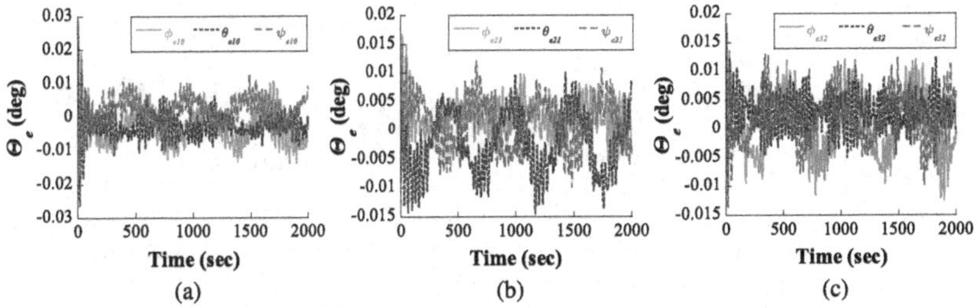

FIGURE 7.1 Relative attitude estimation error ensured by PSVSF.

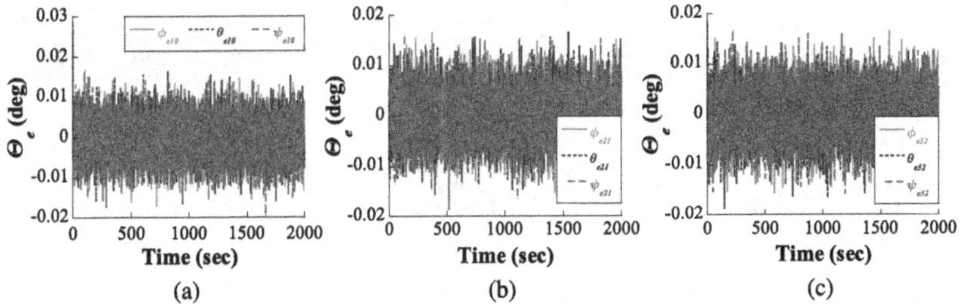

FIGURE 7.2 Relative attitude estimation error ensured by SVSF.

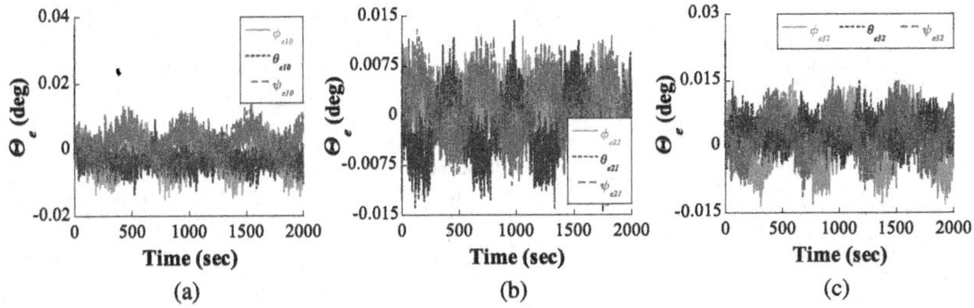

FIGURE 7.3 Relative attitude estimation error ensured by UKF.

the PSVSF achieve higher estimation accuracy for the relative angular velocity state than SVSF and UKF. It was seen in Figs. 7.4–7.6 that the estimated relative angular velocity of SVSF and UKF deviated from the true states. The relative angular velocity estimation of PSVSF approximated the true state well.

The Root-Square-Mean-Error (RSME) of the estimation error was used to quantitatively evaluate state estimation performance further. The comparison results listed in Tables 7.4–7.5 show that the PSVSF achieved better estimation accuracy for the relative attitude and angular velocity states. That is because the PSVSF has the ability to update the modeling error and then to improve the precision of system model.

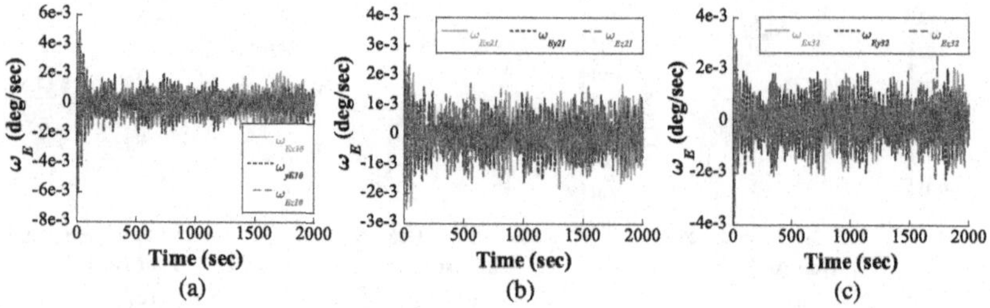

FIGURE 7.4 Relative angular velocity estimation error ensured by PSVSF.

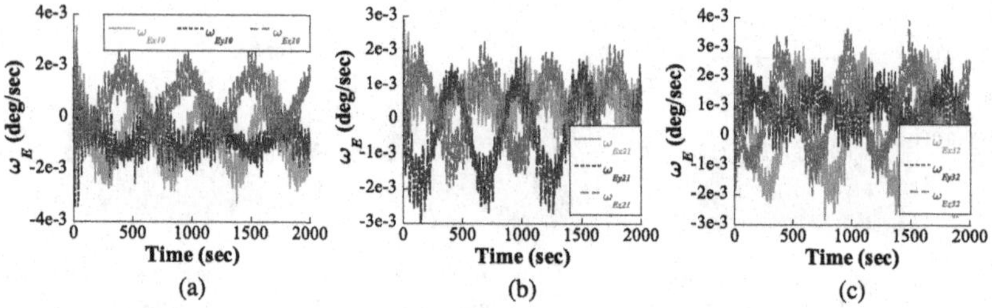

FIGURE 7.5 Relative angular velocity estimation error ensured by SVSF.

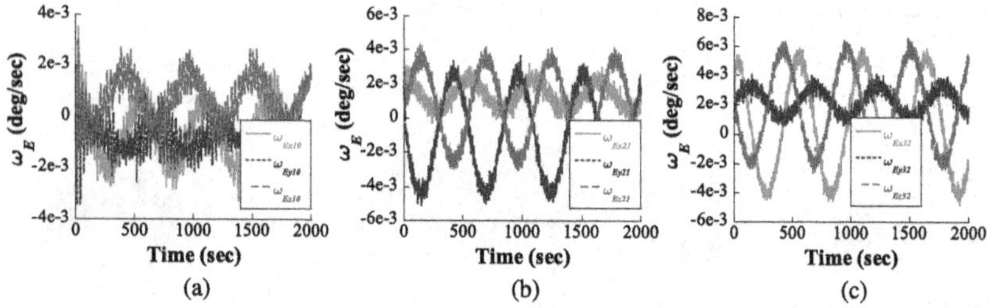

FIGURE 7.6 Relative angular velocity estimation error ensured by UKF.

7.8 Application to microsatellite formation flying control system

It is obtained from Section 4.7 that the microsatellite formation flying control system can be written as the nonlinear system (2.32)–(2.33). The PSVSF is thus applicable to the microsatellite formation flying control system. To validate its effectiveness, numerical simulations are carried out. All the parameters and sensors of that two microsatellites, the initial states, the initial estimation are set as same as in Section 4.7.4. The standard deviation of the measurement noise in (4.134) is assumed as 0.005 degrees. The modeling error is of the form as (4.167)–(4.168) with

$$\Delta N_{c1} = [-3\omega_1 \quad -4\omega_3 \quad 3\omega_1]^{\mathrm{T}}\ \mathrm{Nm} \tag{7.69}$$

$$\Delta N_{c2} = [-3\omega_1 \quad 4\omega_1 \quad 2\omega_1]^{\mathrm{T}}\ \mathrm{N} \tag{7.70}$$

TABLE 7.4 RSME of the relative attitude estimation error using PSVSF, SVSF, and UKF.

RSME (degrees)	State estimation approaches		
	PSVSF	UKF	SVSF
ϕ_{e10}	0.28075	0.25210	0.29794
θ_{e10}	0.18908	0.19481	0.28648
ψ_{e10}	0.18906	0.21199	0.29794
ϕ_{e21}	0.21772	0.16616	0.29221
θ_{e21}	0.29794	0.24637	0.29794
ψ_{e21}	0.21199	0.20626	0.29794
ϕ_{e32}	0.28648	0.28705	0.30367
θ_{e32}	0.21199	0.21772	0.28648
ψ_{e32}	0.22345	0.25210	0.30367

TABLE 7.5 RSME of relative velocity estimation error using PSVSF, SVSF, and UKF.

RSME (deg/sec)	State estimation approaches		
	PSVSF	UKF	SVSF
ω_{Ex10}	0.044691	0.126050	0.068755
ω_{Ey10}	0.032086	0.091673	0.049274
ω_{Ez10}	0.035523	0.097403	0.053285
ω_{Ex21}	0.383880	0.068755	0.040107
ω_{Ey21}	0.034377	0.114590	0.063025
ω_{Ez21}	0.029794	0.097403	0.050420
ω_{Ex32}	0.036669	0.154700	0.080214
ω_{Ey32}	0.042972	0.108860	0.057296
ω_{Ez32}	0.041253	0.137510	0.068755

$$\Delta \boldsymbol{N}_{e1} = [3\omega_1 \cos(\omega_T t) \quad \omega_1 \cos(\omega_T t) \quad -2\omega_1 \cos(\omega_T t)]^{\mathrm{T}} \text{ Nm} \tag{7.71}$$

$$\Delta \boldsymbol{N}_{e2} = [5\omega_1 \cos(\omega_T t) \quad -3\omega_1 \cos(\omega_T t) \quad 2\omega_1 \cos(\omega_T t)]^{\mathrm{T}} \text{ N} \tag{7.72}$$

where $\omega_T = 0.66$ deg/sec and $\omega_1 = 0.0005$.

In simulation, the PSVSF was compared with the UKF. The estimation gains of the PSVSF were chosen as $\ell = 0.0005$ and $\gamma_{ii} = 0.1, i = 1, 2, \cdots, m$. The estimation gains of the UKF were selected as $n = 12$ and $\kappa = 2$. With the PSVSF and the UKF applied to the considered microsatellite formation flying system, the obtained state estimation were shown in Figs. 7.7–7.10. Large overshoot is seen in the initial response of the estimation error achieved by the PSVSF. The reason is that the PSVSF compensates for the modeling error directly. Then,

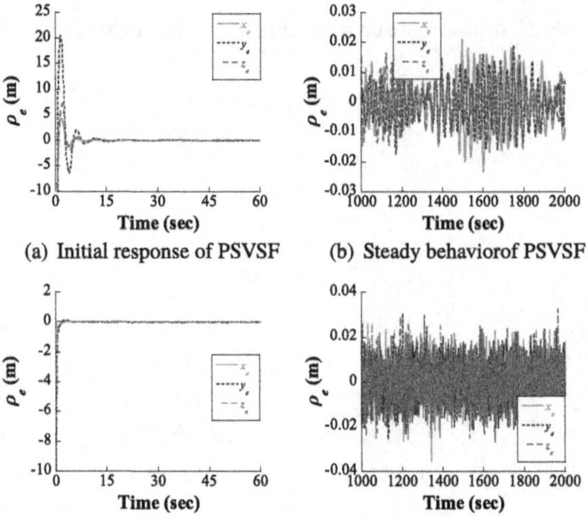

(a) Initial response of PSVSF (b) Steady behaviorof PSVSF

(c) Initial response of UKF (d) Steady behavior of UKF

FIGURE 7.7 Relative position state estimation error using PSVSF and UKF.

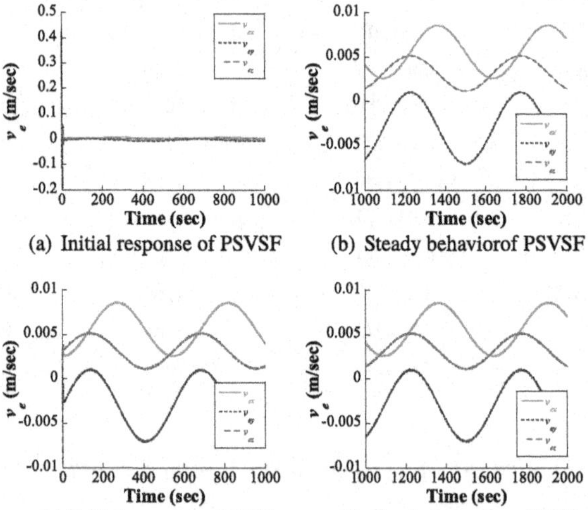

(a) Initial response of PSVSF (b) Steady behaviorof PSVSF

(c) Initial response of UKF (d) Steady behavior of UKF

FIGURE 7.8 Relative velocity estimation error using PSVSF and UKF.

the estimation error existing in the initial stage leads to that large overshoot However, this overshoot is attenuated and stable estimation is established by the PSVSF after a short period of time, *i.e.*, 10 seconds. Although the PSVSF and the UKF achieve almost the same estimation accuracy, most of time the estimation error of the PSVSF is less than the UKF. Compared with the simulation result of Scenario #1 in Section 4.7.4, it is seen that the PSVSF also ensures

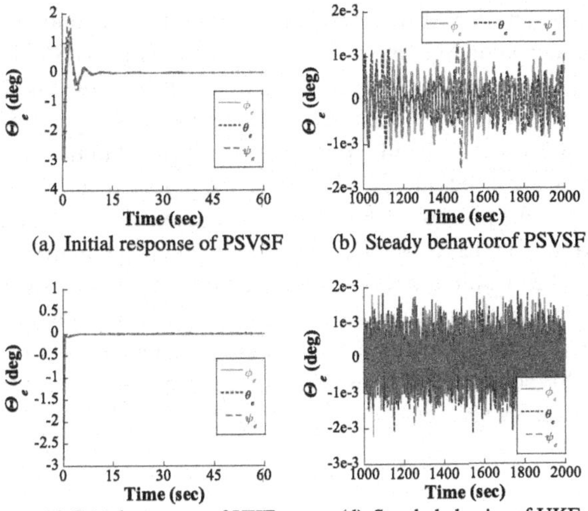

(a) Initial response of PSVSF (b) Steady behaviorof PSVSF

(c) Initial response of UKF (d) Steady behavior of UKF

FIGURE 7.9 Relative attitude state estimation error using PSVSF and UKF.

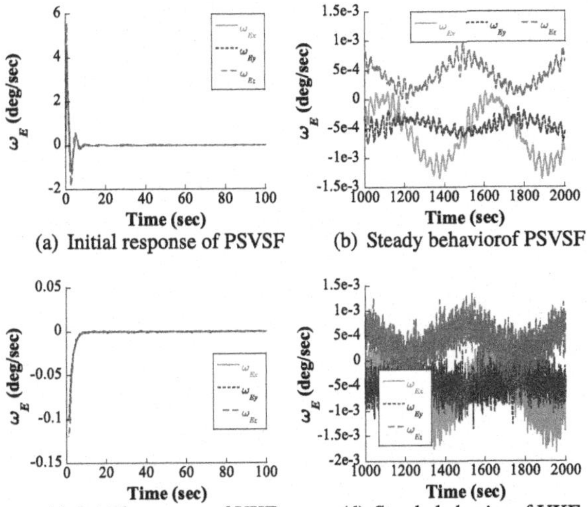

(a) Initial response of PSVSF (b) Steady behaviorof PSVSF

(c) Initial response of UKF (d) Steady behavior of UKF

FIGURE 7.10 Relative angular velocity state estimation error using PSVSF and UKF.

better state estimation performance than the UPF. The advantages of the PSVSF are verified.

7.9 Summary

In this chapter, a predictive variable structure filter (PVSF) was presented for estimating states of the nonlinear systems with non-Gaussian modeling error.

The stability of the PSVSF was proved by using the Lyapunov stability theorem. Moreover, a predictive smooth variable structure filter (PSVSF) was further designed with the negative effect induced by sampling time and chattering in the VSC theory eliminated. Because the modeling error in the PSVSF are not restricted to Gaussian noise only, various kinds of uncertainties, parameter variations, or noises can be addressed. Unlike the sigma points-based predictive filters in Part II, the proposed PSVSF does not demand expensive computation resources. Moreover, its implementation does not depend on the weighting matrix. Based on this result, the state estimation problem of the distributed attitude control system and the relative position control system of microsatellites was solved by the PSVSF. Simulation results were presented further to demonstrate that the PSVSF achieves better estimation accuracy and more robustness than the UKF and the SVSF.

Chapter 8

Predictive adaptive variable structure filter

8.1 Introduction

The state estimation performance achieved by the PSVSF in Chapter 7 strongly depends on the thickness of the boundary layer. This layer should be carefully chosen. If it is chosen to be thicker than the upper bound of modeling error, although less chattering can be resulted, estimation accuracy and robustness will be deteriorated. If a thinner boundary layer is selected, then severe chattering will be induced and estimation accuracy may be lost. This is the main shortcoming of the PSVSF and the VSC theory. Hence, the problem of choosing an appropriate boundary layer should be solved to achieve high-performance state estimation.

To solve the above problem, four different variable-thickness boundary layers were developed in Fuh (2008). Through switching between those four layers, chattering was reduced with desired control performance guaranteed. In Giap and Huang (2020), the thickness of the boundary layer was chosen by using fuzzy control logic. Note that the adaptive control theory is characterized by its capability of handling uncertain parameters adaptively (Krstic et al., 1995), this theory can be introduced into the VSC theory to adaptively determine a dynamic boundary layer (Plestan et al., 2010; Gonzalez et al., 2017). Taking this advantage in mind, four predictive adaptive variable structure filters are developed in this chapter to avoid this problem. Those filters are developed by integrating the PSVSF with the adaptive control theory. Based on the mean-square error principle, the adaptive control theory is applied to adaptively adjust the thickness of the boundary layer. As such, better estimation accuracy and robustness but less chattering is theoretically achieved. As a specific application of this filtering method, its application to the microsatellite control system's state estimation is finally analyzed with simulation results presented.

8.2 Predictive adaptive variable structure filter

The smoothing boundary layer in Chapter 7 has a single term for the corresponding measurement error without considering the coupling effect between them. Thus, the error between the measurement and the state estimation is feedback to calculate its corresponding gain only. Their coupling effects are not explicitly

Predictive Filtering for Microsatellite Control System. https://doi.org/10.1016/B978-0-12-821865-5.00023-1

considered. This may let the optimal estimation of the modeling error be impossible. To obtain a smoothing boundary layer that achieves the optimal state estimation, a full smoothing boundary layer matrix is therefore proposed in this chapter, which is defined as

$$\boldsymbol{\ell} = \begin{bmatrix} \ell_{11} & \ell_{12} & \cdots & \ell_{1m} \\ \ell_{21} & \ell_{22} & \cdots & \ell_{2m} \\ \vdots & \vdots & \ddots & \vdots \\ \ell_{m1} & \ell_{m2} & \cdots & \ell_{mm} \end{bmatrix} \tag{8.1}$$

This smoothing boundary layer matrix includes terms that relate with each other.

Based on the mean-square principle, (7.39), and (7.42), the optimal smoothing boundary layer can be obtained by solving the following equation.

$$\frac{\partial(\mathrm{tr}(\boldsymbol{P}_{\tilde{\boldsymbol{x}},k+1}))}{\partial \boldsymbol{\ell}} = \boldsymbol{0} \tag{8.2}$$

Aiming of solving (8.2), we first rewrite the gain matrix (7.37) of the PSVSF in Chapter 7 as

$$\boldsymbol{K}_{k+1} = \boldsymbol{g}_U(\hat{\boldsymbol{x}}_k)\mathbf{diag}(\boldsymbol{A}_e)\,\mathbf{sat}(\boldsymbol{\ell}^{-1}\mathbf{diag}(\boldsymbol{e}_{y_{k+1/k}}))(\mathbf{diag}(\boldsymbol{e}_{y_{k+1/k}}))^{-1} \tag{8.3}$$

where

$$\boldsymbol{A}_e = |\boldsymbol{e}_{y_{k+1/k}}| + \gamma |\boldsymbol{e}_{y_k}| \tag{8.4}$$

To reduce chattering phenomenon, the modeling error is assumed to be within the boundary layer. Based on this assumption, it is obtained that

$$\mathbf{sat}(\boldsymbol{\ell}^{-1}\mathbf{diag}(\boldsymbol{e}_{y_{k+1/k}})) = \boldsymbol{\ell}^{-1}\mathbf{diag}(\boldsymbol{e}_{y_{k+1/k}}) \tag{8.5}$$

Then, substituting (8.5) into (8.3) and yields

$$\boldsymbol{K}_{k+1} = \boldsymbol{g}_U(\hat{\boldsymbol{x}}_k)\mathbf{diag}(\boldsymbol{A}_e)\boldsymbol{\ell}^{-1} \tag{8.6}$$

In which it is seen that $\boldsymbol{e}_{y_{k+1/k}}$ does not appear in $\boldsymbol{K}_{k+1}$. This implies that $\boldsymbol{e}_{y_{k+1/k}}$ eventually does not affect the gain matrix $\boldsymbol{K}_{k+1}$ of the modeling error.

Because the covariance $\boldsymbol{R}_{k+1}$ of the measurement noise and the state estimation error covariance $\boldsymbol{P}_{\tilde{\boldsymbol{x}},k+1/k}$ are symmetric, $\aleph_{k+1} = \boldsymbol{H}_{k+1}\boldsymbol{P}_{\tilde{\boldsymbol{x}},k+1/k}\boldsymbol{H}_{k+1}^{\mathrm{T}} + \boldsymbol{R}_{k+1}$ is also symmetric. Using (8.6), the covariance $\boldsymbol{P}_{\tilde{\boldsymbol{x}},k+1}$ given in (7.42

achieved by the PSVSF can be rewritten as

$$\begin{aligned}\boldsymbol{P}_{\tilde{\boldsymbol{x}},k+1} &= \boldsymbol{P}_{\tilde{\boldsymbol{x}},k+1/k} - \boldsymbol{K}_{k+1}\boldsymbol{H}_{k+1}\boldsymbol{P}_{\tilde{\boldsymbol{x}},k+1/k} \\ &\quad - \boldsymbol{P}_{\tilde{\boldsymbol{x}},k+1/k}\boldsymbol{H}_{k+1}^{\mathrm{T}}\boldsymbol{K}_{k+1}^{\mathrm{T}} + \boldsymbol{K}_{k+1}\aleph_{k+1}\boldsymbol{K}_{k+1}^{\mathrm{T}} \\ &= \boldsymbol{P}_{\tilde{\boldsymbol{x}},k+1/k} - g_U(\hat{\boldsymbol{x}}_k)\mathbf{diag}(\boldsymbol{A}_e)\boldsymbol{\ell}^{-1}\boldsymbol{H}_{k+1}\boldsymbol{P}_{\tilde{\boldsymbol{x}},k+1/k} \\ &\quad - \boldsymbol{P}_{\tilde{\boldsymbol{x}},k+1/k}\boldsymbol{H}_{k+1}^{\mathrm{T}}(g_U(\hat{\boldsymbol{x}}_k)\mathbf{diag}(\boldsymbol{A}_e)\boldsymbol{\ell}^{-1})^{\mathrm{T}} \\ &\quad + g_U(\hat{\boldsymbol{x}}_k)\mathbf{diag}(\boldsymbol{A}_e)\boldsymbol{\ell}^{-1}\aleph_{k+1}(g_U(\hat{\boldsymbol{x}}_k)\mathbf{diag}(\boldsymbol{A}_e)\boldsymbol{\ell}^{-1})^{\mathrm{T}}\end{aligned} \tag{8.7}$$

Then, the followings are calculated.

$$\frac{\partial(\mathrm{tr}(\boldsymbol{P}_{\tilde{\boldsymbol{x}},k+1/k}))}{\partial\boldsymbol{\ell}} = \mathbf{0} \tag{8.8}$$

$$\begin{aligned}&\frac{\partial(\mathrm{tr}(-g_U(\hat{\boldsymbol{x}}_k)\mathbf{diag}(\boldsymbol{A}_e)\boldsymbol{\ell}^{-1}\boldsymbol{H}_{k+1}\boldsymbol{P}_{\tilde{\boldsymbol{x}},k+1/k}))}{\partial\boldsymbol{\ell}} \\ &= (g_U(\hat{\boldsymbol{x}}_k)\mathbf{diag}(\boldsymbol{A}_e)\boldsymbol{\ell}^{-1})^{\mathrm{T}}\boldsymbol{P}_{\tilde{\boldsymbol{x}},k+1/k}\boldsymbol{H}_{k+1}^{\mathrm{T}}(\boldsymbol{\ell}^{-1})^{\mathrm{T}}\end{aligned} \tag{8.9}$$

$$\begin{aligned}&\frac{\partial(\mathrm{tr}(-\boldsymbol{P}_{\tilde{\boldsymbol{x}},k+1/k}\boldsymbol{H}_{k+1}^{\mathrm{T}}(g_U(\hat{\boldsymbol{x}}_k)\mathbf{diag}(\boldsymbol{A}_e)\boldsymbol{\ell}^{-1})^{\mathrm{T}}))}{\partial\boldsymbol{\ell}} \\ &= (g_U(\hat{\boldsymbol{x}}_k)\mathbf{diag}(\boldsymbol{A}_e)\boldsymbol{\ell}^{-1})^{\mathrm{T}}\boldsymbol{P}_{\tilde{\boldsymbol{x}},k+1/k}\boldsymbol{H}_{k+1}^{\mathrm{T}}(\boldsymbol{\ell}^{-1})^{\mathrm{T}}\end{aligned} \tag{8.10}$$

$$\begin{aligned}&\frac{\partial(\mathrm{tr}(g_U(\hat{\boldsymbol{x}}_k)\mathbf{diag}(\boldsymbol{A}_e)\boldsymbol{\ell}^{-1}\aleph_{k+1}(g_U(\hat{\boldsymbol{x}}_k)\mathbf{diag}(\boldsymbol{A}_e)\boldsymbol{\ell}^{-1})^{\mathrm{T}}))}{\partial\boldsymbol{\ell}} \\ &= -2(g_U(\hat{\boldsymbol{x}}_k)\mathbf{diag}(\boldsymbol{A}_e)\boldsymbol{\ell}^{-1})^{\mathrm{T}}(g_U(\hat{\boldsymbol{x}}_k)\mathbf{diag}(\boldsymbol{A}_e)\boldsymbol{\ell}^{-1})\aleph_{k+1}(\boldsymbol{\ell}^{-1})^{\mathrm{T}}\end{aligned} \tag{8.11}$$

Applying (8.8)–(8.11), (8.2) can be simplified as

$$\begin{aligned}\frac{\partial(\mathrm{tr}(\boldsymbol{P}_{\tilde{\boldsymbol{x}},k+1}))}{\partial\boldsymbol{\ell}} &= 2(g_U(\hat{\boldsymbol{x}}_k)\mathbf{diag}(\boldsymbol{A}_e)\boldsymbol{\ell}^{-1})^{\mathrm{T}}\Big(\boldsymbol{P}_{\tilde{\boldsymbol{x}},k+1/k}\boldsymbol{H}_{k+1}^{\mathrm{T}}(\boldsymbol{\ell}^{-1})^{\mathrm{T}} \\ &\quad - (g_U(\hat{\boldsymbol{x}}_k)\mathbf{diag}(\boldsymbol{A}_e)\boldsymbol{\ell}^{-1})\aleph_{k+1}(\boldsymbol{\ell}^{-1})^{\mathrm{T}}\Big) \\ &= \mathbf{0}\end{aligned} \tag{8.12}$$

Solving (8.12) obtains that the optimal smoothing boundary layer satisfies

$$\boldsymbol{\ell}^{-1} = (\mathbf{diag}(\boldsymbol{A}_e))^{-1}(g_U(\hat{\boldsymbol{x}}_k))^{-1}\boldsymbol{P}_{\tilde{\boldsymbol{x}},k+1/k}\boldsymbol{H}_{k+1}^{\mathrm{T}}\aleph_{k+1}^{-1} \tag{8.13}$$

Let an adaptive smooth boundary layer be developed as

$$\begin{aligned}\boldsymbol{\ell}_{k+1} &= \left((\mathbf{diag}(\boldsymbol{A}_e))^{-1}(g_U(\hat{\boldsymbol{x}}_k))^{-1}\boldsymbol{P}_{\tilde{\boldsymbol{x}},k+1/k}\boldsymbol{H}_{k+1}^{\mathrm{T}}\aleph_{k+1}^{-1}\right)^{-1} \\ &= \aleph_{k+1}(\boldsymbol{H}_{k+1}^{\mathrm{T}})^{-1}\boldsymbol{P}_{\tilde{\boldsymbol{x}},k+1/k}^{-1}g_U(\hat{\boldsymbol{x}}_k)\mathbf{diag}(\boldsymbol{A}_e)\end{aligned} \tag{8.14}$$

Then, it is ready to present the predictive adaptive variable structure filter (PAVSF) in Algorithm 8.1. Note that $\boldsymbol{\ell}_{k+1}$ is a function of the a priori state error covariance $\boldsymbol{P}_{\tilde{x},k+1/k}$, $\aleph_{k+1}$, the measurement matrix $\boldsymbol{H}_{k+1}$, the priori measurement error $\boldsymbol{e}_{y_{k+1/k}}$, the posteriori measurement error $\boldsymbol{e}_{y_k}$, and the "memory" $\boldsymbol{\gamma}$. The thickness of the boundary layer is directly related to the modeling errors, the estimated system, and measurement noises.

Remark 8.1. When implementing the PAVSF in Algorithm 8.1, Algorithm 8.2, Algorithm 8.3, Algorithm 8.4, and Algorithm 8.5, the numerical integral method is adopted to update the state estimation, as stated in Section 6.1.

Algorithm 8.1: Predictive adaptive variable structure filter.

Input Data: Initial estimation $\hat{\boldsymbol{x}}_0$, $\hat{\boldsymbol{y}}_0$, and measurement $\boldsymbol{y}_k$
Result: State estimation $\hat{\boldsymbol{x}}_k$
begin
  for $k = 0, 1, 2 \cdots$ **do**
    Updating the estimation of the modeling error

$$\boldsymbol{e}_{y_{k+1/k}} = \boldsymbol{y}_{k+1} - \hat{\boldsymbol{y}}_k - \boldsymbol{Z}(\hat{\boldsymbol{x}}_k)$$
$$\boldsymbol{P}_{\tilde{x},k+1/k} = \boldsymbol{F}_k \boldsymbol{P}_{\tilde{x},k} \boldsymbol{F}_k^{\mathrm{T}}$$
$$\aleph_{k+1} = \boldsymbol{H}_{k+1} \boldsymbol{P}_{\tilde{x},k+1/k} \boldsymbol{H}_{k+1}^{\mathrm{T}} + \boldsymbol{R}_{k+1}$$
$$\boldsymbol{\ell}_{k+1}^{-1} = (\mathbf{diag}(\boldsymbol{A}_e))^{-1} (\boldsymbol{g}_U(\hat{\boldsymbol{x}}_k))^{-1} \boldsymbol{P}_{\tilde{x},k+1/k} \boldsymbol{H}_{k+1}^{\mathrm{T}} \aleph_{k+1}^{-1}$$
$$\begin{aligned} \boldsymbol{K}_{k+1} &= \boldsymbol{g}_U(\hat{\boldsymbol{x}}_k) \mathbf{diag}(\boldsymbol{A}_e) \boldsymbol{\ell}_{k+1}^{-1} \\ &\approx \boldsymbol{g}_U(\hat{\boldsymbol{x}}_k) \mathbf{diag}(\boldsymbol{A}_e) \mathbf{diag}(\overrightarrow{\boldsymbol{\ell}_{k+1}^{-1}}) \end{aligned}$$

    Updating the state estimation by using

$$\hat{\boldsymbol{x}}_{k+1/k} = \boldsymbol{f}(\hat{\boldsymbol{x}}_k)$$
$$\hat{\boldsymbol{x}}_{k+1} = \hat{\boldsymbol{x}}_{k+1/k} + \boldsymbol{K}_{k+1} \boldsymbol{e}_{y_{k+1/k}}$$
$$\hat{\boldsymbol{y}}_k = \boldsymbol{h}(\hat{\boldsymbol{x}}_k)$$

    Updating the state estimation covariance by using

$$\boldsymbol{F}_{k+1} = \frac{\partial \boldsymbol{f}}{\partial \boldsymbol{x}} |_{\boldsymbol{x}=\boldsymbol{x}_{k+1}}$$
$$\begin{aligned} \boldsymbol{P}_{\tilde{x},k+1} = {} & (\boldsymbol{I}_n - \boldsymbol{K}_{k+1} \boldsymbol{H}_{k+1}) \boldsymbol{P}_{\tilde{x},k+1/k} (\boldsymbol{I}_n - \boldsymbol{K}_{k+1} \boldsymbol{H}_{k+1})^{\mathrm{T}} \\ & + \boldsymbol{K}_{k+1} \boldsymbol{R}_{k+1} \boldsymbol{K}_{K+1}^{\mathrm{T}} \end{aligned}$$

  end
end

Once the adaptive filtering approach is designed in Algorithm 8.1, the low and the upper bound of the $\mathbf{sat}(\cdot)$ in $\boldsymbol{K}_{k+1}$ will be analyzed. Applying (8.13), it is obtained from (8.3) that

$$\begin{aligned} & \mathbf{sat}(\boldsymbol{\ell}^{-1} \mathbf{diag}(\boldsymbol{e}_{y_{k+1/k}})) \\ & = \mathbf{sat}((\mathbf{diag}(\boldsymbol{A}_e))^{-1} (\boldsymbol{g}_U(\hat{\boldsymbol{x}}_k))^{-1} \boldsymbol{P}_{\tilde{x},k+1/k} \boldsymbol{H}_{k+1}^{\mathrm{T}} \aleph_{k+1}^{-1} \mathbf{diag}(\boldsymbol{e}_{y_{k+1/k}})) \end{aligned} \tag{8.15}$$

which can be divided into two terms:

$$\textbf{Term \#1} = \boldsymbol{H}_{k+1}\boldsymbol{P}_{\tilde{x},k+1/k}\boldsymbol{H}_{k+1}^{\mathrm{T}}\aleph_{k+1}^{-1} \tag{8.16}$$

$$\textbf{Term \#2} = (\boldsymbol{H}_{k+1}\boldsymbol{g}_U(\hat{\boldsymbol{x}}_k)\mathbf{diag}(\boldsymbol{A}_e))^{-1}\mathbf{diag}(\boldsymbol{e}_{y_{k+1/k}}) \tag{8.17}$$

Owing to the definition of $\aleph_{k+1}$, we have $\aleph_{k+1} > \boldsymbol{R}_{K+1}$ and we can simplify 'Term #1' as

$$\textbf{Term \#1} = \boldsymbol{H}_{k+1}\boldsymbol{P}_{\tilde{x},k+1/k}\boldsymbol{H}_{k+1}^{\mathrm{T}}\aleph_{k+1}^{-1} = (\aleph_{k+1} - \boldsymbol{R}_{K+1})\aleph_{k+1}^{-1} \tag{8.18}$$

which means that all the elements of 'Term #1' belong to the set $[0, 1]$.

Suppose that the nonlinear system is an second-order system, then it is got from (2.43) and (8.4) that

$$\textbf{Term \#2} = \mathbf{diag}(\boldsymbol{A}_e)^{-1}(\boldsymbol{H}_{k+1}\boldsymbol{f}(\boldsymbol{x}_k)\boldsymbol{H}_{k+1}^{\mathrm{T}})\mathbf{diag}(\boldsymbol{e}_{y_{k+1/k}}) \tag{8.19}$$

Defining $\delta_0 = \max\{\boldsymbol{H}_{k+1}\boldsymbol{f}(\boldsymbol{x}_k)\boldsymbol{H}_{k+1}^{\mathrm{T}}\}$ and applying $\boldsymbol{e}_{y_{k+1}} < |\boldsymbol{e}_{y_{k+1}}| + \gamma|\boldsymbol{e}_{y_k}|$, $\mathbf{diag}(\boldsymbol{A}_e)^{-1}\mathbf{diag}(\boldsymbol{e}_{y_{k+1/k}})$ will be always in the set $[-1, 1]$.

On the basis of the above analysis, when the adaptive boundary layer is updated by using (8.14), the value of the function $\mathbf{sat}(\cdot)$ will be always within the set $[-\delta_0, \delta_0]$.

Remark 8.2. Inserting (8.13) into (8.6) yields

$$\boldsymbol{K}_{k+1} = \boldsymbol{P}_{\tilde{x},k+1/k}\boldsymbol{H}_{k+1}^{\mathrm{T}}\aleph_{k+1}^{-1} \tag{8.20}$$

Comparing (8.20) with the Kalman filter, we find that the matrix $\boldsymbol{K}_{k+1}$ achieved by the PAVSF has the same formula as the Kalman filter. It is therefore inferred that the optimal estimation can be achieved by the PAVSF for linear systems. On the other hand, due to the redundancy of the function $\mathbf{sat}(\cdot)$, the superior performance of robustness to modeling error achieved by the PVSF in Chapter 7 may be lost for PAVSF.

8.3 Derivation of sigma point PAVSF

According to the Kalman filtering theory, the covariance $\boldsymbol{P}_{\tilde{x},k+1/k}$ and $\aleph_{k+1}$ in the PAVSF are obtained with first-order approximation accuracy only. The adaptive smoothing boundary layer thus has the first-order approximation accuracy and reduces the robustness of the PVSF in Chapter 7. To solve these two problems, sigma point scheme and the PAVSF are integrated to present a sigma point predictive adaptive variable filtering approach. The flow chart of this approach is shown in Fig. 8.1. A nominal or ideal boundary layer is pre-determined, which is given by $\ell_{limit} \in \mathbb{R}_+$. When the estimation error is larger than ℓ_{limit}, this case implies that the modeling error is under-estimated. (8.14) should be applied to

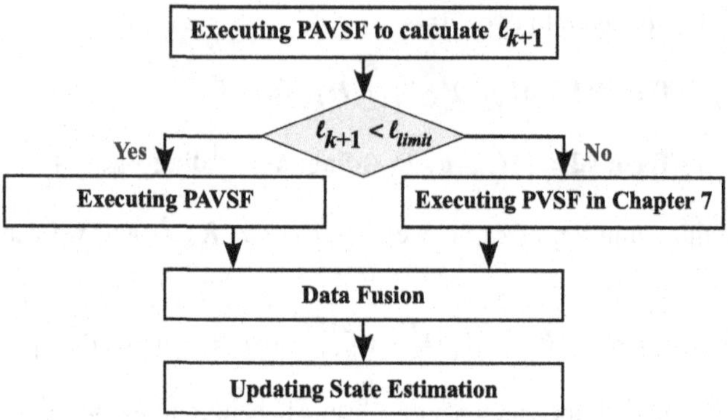

FIGURE 8.1 Flow chart of the sigma point predictive adaptive variable filter.

ensure robustness and stability for the filter. Otherwise, the PVSF is applied to smooth the estimation accuracy.

Because the sigma points can be generated by many methods including the UT sampling, the central difference, and the cubature rule in Part II, the UT sampling and the central difference are applied to specifically design sigma point predictive adaptive variable filtering approaches as follows.

8.3.1 Unscented PAVSF

Applying the UT sampling strategy, it is ready to present the unscented PAVSF (UPAVSF) in Algorithm 8.2, where $W_i^m \in \mathbb{R}$ is the weighting constant, $i = 0, 1, 2, \cdots, 2n$.

8.3.2 Central difference PAVSF

Following the Stirling's polynomial interpolation strategy presented in Chapter 5, we can design the central difference PAVSF in Algorithm 8.3, where $h \in \mathbb{R}_+$ is the weighting scalar.

8.4 Strong tracking PAVSF

Comparing with the standard predictive variable structure filter in Chapter 7, the predictive variable structure filter has less capability of ensuring robustness to system uncertainties. When the filtering is stable, the PAVSF may be unable to estimate abrupt states. That is because when the filtering is stable, the gain matrix $\boldsymbol{K}_k$ is very small. For abrupt states, the small $\boldsymbol{K}_k$ has limited capability of tuning the state estimation. Aiming of increasing state's estimation/tracking capability, in contrast to improve robustness of the PAVSF by using the sigma points strategy, the sequence orthogonal principle is applied to develop a strong tracking PAVSF.

Algorithm 8.2: Unscented predictive adaptive variable structure filter.

Input Data: Initial estimation $\hat{\boldsymbol{x}}_0$, $\hat{\boldsymbol{y}}_0$, and measurement $\boldsymbol{y}_k$
Result: State estimation $\hat{\boldsymbol{x}}_k$
begin
  Initializing the state and the covariance matrix as
    $\hat{\boldsymbol{x}}_0 = \mathbb{E}(\boldsymbol{x}_0)$
    $\boldsymbol{P}_{\tilde{x},0} = \mathrm{Var}(\boldsymbol{x}_0) = \mathbb{E}((\boldsymbol{x}_0 - \hat{\boldsymbol{x}}_0)(\boldsymbol{x}_0 - \hat{\boldsymbol{x}}_0)^{\mathrm{T}})$
  for $k = 0, 1, \cdots$ **do**
    Applying the Cholesky decomposition to the covariance $\boldsymbol{P}_{\tilde{x},k}$

$$\boldsymbol{P}_{\tilde{x},k} = \boldsymbol{S}_k \boldsymbol{S}_k^{\mathrm{T}} = \sum_{i=1}^{n} \boldsymbol{\sigma}_{i,k} \boldsymbol{\sigma}_{i,k}^{\mathrm{T}}$$

    Establishing sigma points

$$\boldsymbol{\xi}_{0,k} = \hat{\boldsymbol{x}}_k$$
$$\boldsymbol{\xi}_{i,k} = \hat{\boldsymbol{x}}_k + \sqrt{n+\kappa}\,\boldsymbol{\sigma}_{i,k} = \hat{\boldsymbol{x}}_k + \hat{\boldsymbol{\sigma}}_{i,k},\ i = 1, 2, \cdots, n$$
$$\boldsymbol{\xi}_{i+n,k} = \hat{\boldsymbol{x}}_k - \sqrt{n+\kappa}\,\boldsymbol{\sigma}_{i,k} = \hat{\boldsymbol{x}}_k - \hat{\boldsymbol{\sigma}}_{i,k},\ i = 1, 2, \cdots, n$$

    Updating the priori state covariance

$$\hat{\boldsymbol{x}}_{i,k+1/k} = \boldsymbol{f}(\boldsymbol{\xi}_{i,k}),\ i = 0, 1, \cdots, 2n$$
$$\hat{\boldsymbol{x}}_{k+1/k} = \sum_{i=0}^{2n} W_i^m \hat{\boldsymbol{x}}_{i,k+1/k}$$
$$\boldsymbol{P}_{\tilde{x},k+1/k} = \sum_{i=0}^{2n} W_i^m (\hat{\boldsymbol{x}}_{i,k+1/k} - \hat{\boldsymbol{x}}_{k+1/k})(\hat{\boldsymbol{x}}_{i,k+1/k} - \hat{\boldsymbol{x}}_{k+1/k})^{\mathrm{T}}$$

    Calculating the adaptive boundary layer

$$\aleph_{i,k+1} = \boldsymbol{H}_{i,k+1} \boldsymbol{P}_{\tilde{x},k+1/k} \boldsymbol{H}_{i,k+1}^{\mathrm{T}} + \boldsymbol{R}_{k+1}$$
$$\boldsymbol{\ell}_{i,k+1} = ((\mathbf{diag}(\boldsymbol{A}_e))^{-1} (\boldsymbol{g}_U(\hat{\boldsymbol{x}}_k))^{-1} \boldsymbol{P}_{\tilde{x},k+1/k} \boldsymbol{H}_{i,k+1}^{\mathrm{T}} \aleph_{i,k+1}^{-1})^{-1}$$

    Updating state estimation
    if $\boldsymbol{\ell}_{i,k+1} < \boldsymbol{\ell}_{limit}$ **then**

$$\boldsymbol{K}_{i,k+1} = \boldsymbol{P}_{\tilde{x},k+1/k} \boldsymbol{H}_{i,k+1}^{\mathrm{T}} \aleph_{k+1}^{-1}$$
$$\boldsymbol{e}_{y_{i,k+1/k}} = \boldsymbol{y}_{k+1} - \hat{\boldsymbol{y}}_{k_i} - \boldsymbol{Z}(\boldsymbol{\xi}_{i,k})$$
$$\hat{\boldsymbol{x}}_{i,k+1/k+1} = \hat{\boldsymbol{x}}_{i,k+1/k} + \boldsymbol{K}_{i,k+1} \boldsymbol{e}_{y_{i,k+1/k}}$$

    end
    if $\boldsymbol{\ell}_{i,k+1} \geq \boldsymbol{\ell}_{limit}$ **then**

$$\boldsymbol{K}_{i,k+1} =$$
$$\boldsymbol{g}_U(\hat{\boldsymbol{x}}_k) \mathbf{diag}(\boldsymbol{A}_e)\, \mathbf{sat}(\overrightarrow{\boldsymbol{\ell}_{i,k+1}^{-1} \mathbf{diag}(\boldsymbol{e}_{y_{k+1/k}})}) (\mathbf{diag}(\boldsymbol{e}_{y_{k+1/k}}))^{-1}$$
$$\boldsymbol{e}_{y_{i,k+1/k}} = \boldsymbol{y}_{k+1} - \hat{\boldsymbol{y}}_{k_i} - \boldsymbol{Z}(\boldsymbol{\xi}_{i,k})$$
$$\hat{\boldsymbol{x}}_{i,k+1/k+1} = \hat{\boldsymbol{x}}_{i,k+1/k} + \boldsymbol{K}_{i,k+1} \boldsymbol{e}_{y_{i,k+1/k}}$$

    end
    Updating the state estimation error covariance

$$\hat{\boldsymbol{x}}_{k+1} = \sum_{i=0}^{2n} W_i^m \hat{\boldsymbol{x}}_{i,k+1/k+1}$$
$$\boldsymbol{P}_{\tilde{x},k+1/k+1} = \sum_{i=0}^{2n} W_i^m (\hat{\boldsymbol{x}}_{i,k+1/k+1} - \hat{\boldsymbol{x}}_{k+1})(\hat{\boldsymbol{x}}_{i,k+1/k+1} - \hat{\boldsymbol{x}}_{k+1})^{\mathrm{T}}$$

  end
end

Algorithm 8.3: Central difference PAVSF.

Input Data: Initial estimation $\hat{\boldsymbol{x}}_0$, $\hat{\boldsymbol{y}}_0$, and measurement $\boldsymbol{y}_k$
Result: State estimation $\hat{\boldsymbol{x}}_k$
begin
  Initializing the state and the covariance matrix as
    $\hat{\boldsymbol{x}}_0 = \mathbb{E}(\boldsymbol{x}_0)$
    $\boldsymbol{P}_{\tilde{x},0} = \mathrm{Var}(\boldsymbol{x}_0) = \mathbb{E}((\boldsymbol{x}_0 - \hat{\boldsymbol{x}}_0)(\boldsymbol{x}_0 - \hat{\boldsymbol{x}}_0)^{\mathrm{T}})$
  for $k = 0, 1, \cdots$ **do**
    Applying the Cholesky decomposition to the covariance $\boldsymbol{P}_{\tilde{x},k}$
      $\boldsymbol{P}_{\tilde{x},k} = \boldsymbol{S}_k \boldsymbol{S}_k^{\mathrm{T}} = \sum_{i=1}^{n} \boldsymbol{\sigma}_{i,k} \boldsymbol{\sigma}_{i,k}^{\mathrm{T}}$
    Getting sigma points via the Stirling's polynomial interpolation
      $\boldsymbol{\xi}_{0,k} = \hat{\boldsymbol{x}}_k$
      $\boldsymbol{\xi}_{i,k} = \hat{\boldsymbol{x}}_k + h\boldsymbol{\sigma}_{i,k},\ i = 1, 2, \cdots, n$
      $\boldsymbol{\xi}_{i+n,k} = \hat{\boldsymbol{x}}_k - h\boldsymbol{\sigma}_{i,k},\ i = 1, 2, \cdots, n$
    Updating the priori state covariance
      $\hat{\boldsymbol{x}}_{i,k+1/k} = \boldsymbol{f}(\boldsymbol{\xi}_{i,k}), i = 0, 1, \cdots, 2n$
      $\hat{\boldsymbol{x}}_{k+1/k} = \frac{h^2-n}{h^2}\hat{\boldsymbol{x}}_{0,k+1/k} + \frac{1}{2h^2}\sum_{i=1}^{n}(\hat{\boldsymbol{x}}_{i,k+1/k} + \hat{\boldsymbol{x}}_{i+n,k+1/k})$
      $\boldsymbol{P}_{\tilde{x},k+1/k} = \frac{1}{4h^2}\sum_{i=1}^{n}(\hat{\boldsymbol{x}}_{i,k+1/k} - \hat{\boldsymbol{x}}_{i+n,k+1/k})^2 + \frac{h^2-1}{4h^2}\sum_{i=1}^{n}(\hat{\boldsymbol{x}}_{i,k+1/k} - \hat{\boldsymbol{x}}_{i+n,k+1/k} - 2\hat{\boldsymbol{x}}_{0,k+1/k})^2$
    Calculating the adaptive boundary layer
      $\aleph_{i,k+1} = \boldsymbol{H}_{i,k+1}\boldsymbol{P}_{\tilde{x},k+1/k}\boldsymbol{H}_{i,k+1}^{\mathrm{T}} + \boldsymbol{R}_{k+1}$
      $\boldsymbol{\ell}_{i,k+1} = ((\mathbf{diag}(\boldsymbol{A}_e))^{-1}(\boldsymbol{g}_U(\hat{\boldsymbol{x}}_k))^{-1}\boldsymbol{P}_{\tilde{x},k+1/k}\boldsymbol{H}_{i,k+1}^{\mathrm{T}}\aleph_{i,k+1}^{-1})^{-1}$
    Updating state estimation
    if $\boldsymbol{\ell}_{i,k+1} < \ell_{limit}$ **then**
      $\boldsymbol{K}_{i,k+1} = \boldsymbol{P}_{\tilde{x},k+1/k}\boldsymbol{H}_{i,k+1}^{\mathrm{T}}\aleph_{k+1}^{-1}$
      $\boldsymbol{e}_{y_{i,k+1/k}} = \boldsymbol{y}_{k+1} - \hat{\boldsymbol{y}}_{k_i} - \boldsymbol{Z}(\boldsymbol{\xi}_{i,k})$
      $\hat{\boldsymbol{x}}_{i,k+1/k+1} = \hat{\boldsymbol{x}}_{i,k+1/k} + \boldsymbol{K}_{i,k+1}\boldsymbol{e}_{y_{i,k+1/k}}$
    end
    if $\boldsymbol{\ell}_{i,k+1} \geq \ell_{limit}$ **then**
      $\boldsymbol{K}_{i,k+1} =$
      $\boldsymbol{g}_U(\hat{\boldsymbol{x}}_k)\mathbf{diag}(\boldsymbol{A}_e)\,\mathbf{sat}(\overrightarrow{\boldsymbol{\ell}_{i,k+1}^{-1}\mathbf{diag}(\boldsymbol{e}_{y_{k+1/k}})})(\mathbf{diag}(\boldsymbol{e}_{y_{k+1/k}}))^{-1}$
      $\boldsymbol{e}_{y_{i,k+1/k}} = \boldsymbol{y}_{k+1} - \hat{\boldsymbol{y}}_{k_i} - \boldsymbol{Z}(\boldsymbol{\xi}_{i,k})$
      $\hat{\boldsymbol{x}}_{i,k+1/k+1} = \hat{\boldsymbol{x}}_{i,k+1/k} + \boldsymbol{K}_{i,k+1}\boldsymbol{e}_{y_{i,k+1/k}}$
    end
    Updating the state estimation error covariance
      $\hat{\boldsymbol{x}}_{k+1} = \frac{h^2-n}{h^2}\hat{\boldsymbol{x}}_{0,k+1/k+1} + \frac{1}{2h^2}\sum_{i=1}^{n}(\hat{\boldsymbol{x}}_{i,k+1/k+1} + \hat{\boldsymbol{x}}_{i+n,k+1/k+1})$
      $\boldsymbol{P}_{\tilde{x},k+1} = \frac{1}{4h^2}\sum_{i=1}^{n}(\hat{\boldsymbol{x}}_{i,k+1/k+1} - \hat{\boldsymbol{x}}_{i+n,k+1/k+1})^2 + \frac{h^2-1}{4h^2}\sum_{i=1}^{n}(\hat{\boldsymbol{x}}_{i,k+1/k+1} - \hat{\boldsymbol{x}}_{i+n,k+1/k+1} - 2\hat{\boldsymbol{x}}_{0,k+1/k+1})^2$
  end
end

8.4.1 Sequence orthogonal principle

The basic theory of the sequence orthogonal principle is that the measurement errors at any time are orthogonal to each other. This can extract valuable information from the measurements. Hence, the current and the previous measurement residuals can be applied to improve the filtering performance. More specifically, the sequence orthogonal principle is formulated as

$$\mathbb{E}(\boldsymbol{\varepsilon}_{k+j}\boldsymbol{\varepsilon}_k^{\mathrm{T}}) = \mathbf{0}, k \in \mathbb{N}_0,\ j \in \mathbb{N}_0 \tag{8.21}$$

where

$$\boldsymbol{\varepsilon}_k = \boldsymbol{y}_k - \hat{\boldsymbol{y}}_{k/k+1} = \boldsymbol{y}_k - \hat{\boldsymbol{y}}_k - \boldsymbol{Z}\left(\hat{\boldsymbol{x}}_k\right) \tag{8.22}$$

If this principle is satisfied, then the output residual is ensured to be orthogonal with all the output residuals, while better estimation accuracy is guaranteed.

Lemma 8.1. *(Maybeck, 1982; Zhou and Park, 1996) For the nonlinear system described by* (2.32)–(2.33), *if* $\left\|\boldsymbol{x}_k - \hat{\boldsymbol{x}}_k\right\| \ll \|\boldsymbol{x}_k\|$, *then the following condition is satisfied for the output residual series:*

$$\mathbb{E}(\boldsymbol{\varepsilon}_{k+j}\boldsymbol{\varepsilon}_k^{\mathrm{T}}) = \boldsymbol{H}_{k+j}\left(\prod_{m=1}^{j}\boldsymbol{F}_{k+j+1-m,k+j-m}\boldsymbol{A}_{KH}\right)\mathbb{E}((\boldsymbol{x}_k - \hat{\boldsymbol{x}}_k)\boldsymbol{\varepsilon}_{y_k}^{\mathrm{T}}) \tag{8.23}$$

where $\boldsymbol{A}_{KH} = (I - \boldsymbol{K}_{k+j-m}\boldsymbol{H}_{k+j-m})\boldsymbol{F}_{k+j+1-m,k+j-m}$. $\boldsymbol{F}$ *and* $\boldsymbol{H}$ *are the Jacobian matrices of the functions* $\boldsymbol{f}(\cdot)$ *and* $\boldsymbol{h}(\cdot)$, *respectively.*

8.4.2 Derivation of strong tracking PAVSF

To guarantee that the filter has great capability of handling modeling error, one common approach is to introduce a time-varying fading factor to alleviate the old measurement's effect on the estimation and to enhance the use efficiency of new measurements. Let $\boldsymbol{\lambda}_{k+1} = \mathbf{diag}([\lambda_{k+1}^1 \quad \lambda_{k+1}^2 \quad \cdots \quad \lambda_{k+1}^n]^{\mathrm{T}})$, $\lambda_{k+1}^i \geq 1$, $i = 1, 2, \cdots, n$, be the fading factor and introduce it into the priori state error covariance (7.39). Then, the state estimation error covariance after correction is tuned as

$$\boldsymbol{P}_{\tilde{x},k+1/k} = \boldsymbol{\lambda}_{k+1}\boldsymbol{F}_k\boldsymbol{P}_{\tilde{x},k}\boldsymbol{F}_k^{\mathrm{T}} + \boldsymbol{Q}_k^P \tag{8.24}$$

where $\boldsymbol{Q}_k^P$ is the positive-definite symmetric matrix of the process noise, which is introduced to tune the filtering performance.

From (7.41) and (8.22), the estimation error of the system states can be calculated as

$$\begin{aligned}\boldsymbol{x}_{k+1} - \hat{\boldsymbol{x}}_{k+1} &= \boldsymbol{x}_{k+1} - \hat{\boldsymbol{x}}_{k+1/k} - \boldsymbol{K}_{k+1}\boldsymbol{\varepsilon}_{k+1} \\ &= \tilde{\boldsymbol{x}}_{k+1/k} - \boldsymbol{K}_{k+1}\boldsymbol{\varepsilon}_{k+1}\end{aligned} \tag{8.25}$$

where

$$\boldsymbol{\varepsilon}_{k+1} = \boldsymbol{y}_k - \hat{\boldsymbol{y}}_{k/k+1} = \boldsymbol{H}_{k+1}\hat{\boldsymbol{x}}_{k+1/k} + \boldsymbol{v}_{k+1} \tag{8.26}$$

Introducing a new variable $\boldsymbol{V}_{k+1} = \mathbb{E}(\boldsymbol{\varepsilon}_{k+1}\boldsymbol{\varepsilon}_{k+1}^{\mathrm{T}})$ and assuming that $\tilde{\boldsymbol{x}}_{k+1/k}$ is uncorrelated with $\boldsymbol{v}_k$, it is got from (8.25) that

$$\begin{aligned}\mathbb{E}((\boldsymbol{x}_{k+1} - \hat{\boldsymbol{x}}_{k+1})\boldsymbol{\varepsilon}_{k+1}^{\mathrm{T}}) &= \mathbb{E}((\tilde{\boldsymbol{x}}_{k+1/k} - \boldsymbol{K}_{k+1}\boldsymbol{\varepsilon}_{k+1})\boldsymbol{\varepsilon}_{k+1}^{\mathrm{T}}) \\ &= \mathbb{E}(\tilde{\boldsymbol{x}}_{k+1/k}(\boldsymbol{H}_{k+1}\tilde{\boldsymbol{x}}_{k+1/k} + \boldsymbol{v}_{k+1}^{\mathrm{T}})) \\ &\quad - \boldsymbol{K}_{k+1}\boldsymbol{x}_{k+1}\mathbb{E}(\boldsymbol{\varepsilon}_{k+1}\boldsymbol{\varepsilon}_{k+1}^{\mathrm{T}}) \\ &= \boldsymbol{P}_{\tilde{x},k+1/k}\boldsymbol{H}_{k+1}^{\mathrm{T}} - \boldsymbol{K}_{k+1}\boldsymbol{V}_{k+1}\end{aligned} \tag{8.27}$$

Based on (8.27), we can establish a sufficient condition for the sequence orthogonal principle in (8.21) as

$$\mathbb{E}\left(\left(\boldsymbol{x}_{k+1} - \hat{\boldsymbol{x}}_{k+1}\right)\boldsymbol{\varepsilon}_{k+1}^{\mathrm{T}}\right) = \boldsymbol{P}_{\tilde{x},k+1/k}\boldsymbol{H}_{k+1}^{\mathrm{T}} - \boldsymbol{K}_{k+1}\boldsymbol{V}_{k+1} = \boldsymbol{0} \tag{8.28}$$

If there is modeling error or estimation error, their effects will be projected into the residuals $\boldsymbol{\varepsilon}_{k+1}$. It leads $\boldsymbol{V}_{k+1}$ to vary. The sufficient condition (8.28) will not be satisfied, and $\mathbb{E}(\boldsymbol{\varepsilon}_{k+j}\boldsymbol{\varepsilon}_k^{\mathrm{T}}) \neq \boldsymbol{0}$, *i.e.*, the output residual sequence is not orthogonal. At this time, the fading factor matrix $\boldsymbol{\lambda}_{k+1}$ should be introduced to adjust the gain matrix $\boldsymbol{K}_{k+1}$ to satisfy the sufficient condition. As such, good estimation can be provided by the filter.

Substituting (8.20) into (8.28) leads to

$$\begin{aligned}&\boldsymbol{P}_{\tilde{x},k+1/k}\boldsymbol{H}_{k+1}^{\mathrm{T}} - \boldsymbol{P}_{\tilde{x},k+1/k}\boldsymbol{H}_{k+1}^{\mathrm{T}}(\boldsymbol{H}_{k+1}\boldsymbol{P}_{\tilde{x},k+1/k}\boldsymbol{H}_{k+1}^{\mathrm{T}} + \boldsymbol{R}_{k+1})^{-1}\boldsymbol{V}_{k+1} \\ &= \boldsymbol{P}_{\tilde{x},k+1/k}\boldsymbol{H}_{k+1}^{\mathrm{T}}(\boldsymbol{I}_n - (\boldsymbol{H}_{k+1}\boldsymbol{P}_{\tilde{x},k+1/k}\boldsymbol{H}_{k+1}^{\mathrm{T}} + \boldsymbol{R}_{k+1})^{-1}V_{k+1}) = \boldsymbol{0}\end{aligned} \tag{8.29}$$

Hence, the following equation should be satisfied to ensure that (8.28) is always valid.

$$\boldsymbol{I}_n - (\boldsymbol{H}_{k+1}\boldsymbol{P}_{\tilde{x},k+1/k}\boldsymbol{H}_{k+1}^{\mathrm{T}} + \boldsymbol{R}_{k+1})^{-1}\boldsymbol{V}_{k+1} = \boldsymbol{0} \tag{8.30}$$

or

$$\boldsymbol{H}_{k+1}\boldsymbol{\lambda}_{k+1}\boldsymbol{F}_k\boldsymbol{P}_{\tilde{x},k}\boldsymbol{F}_k^{\mathrm{T}}\boldsymbol{H}_{k+1}^{\mathrm{T}} = \boldsymbol{V}_{k+1} - \boldsymbol{R}_{k+1} - \boldsymbol{H}_{k+1}\boldsymbol{Q}_k^P\boldsymbol{H}_{k+1}^{\mathrm{T}} \tag{8.31}$$

where (8.24) is applied.

To avoid serious regulating effect of the fading factor, a weakening factor $\mu \in \mathbb{R}_+$, $\mu > 1$ is introduced into (8.31) to smooth the state estimation. Then (8.31) is rewritten as

$$\boldsymbol{H}_{k+1}\boldsymbol{\lambda}_{k+1}\boldsymbol{F}_k\boldsymbol{P}_{\tilde{x},k}\boldsymbol{F}_k^{\mathrm{T}}\boldsymbol{H}_{k+1}^{\mathrm{T}} = \boldsymbol{V}_{k+1} - \mu\boldsymbol{R}_{k+1} - \boldsymbol{H}_{k+1}\boldsymbol{Q}_k^P\boldsymbol{H}_{k+1}^{\mathrm{T}} \tag{8.32}$$

The weakening factor μ can be chosen by the experiences or calculated as

$$\mu = \min \sum_{k=1}^{Z} \sum_{i=1}^{n} |\boldsymbol{x}_{i,k} - \hat{\boldsymbol{x}}_{i,k}| \tag{8.33}$$

where $Z \in \mathbb{N}$ is the simulation steps. It reflects the cumulative errors in the filtering process.

Applying the property that

$$\mathrm{tr}(\boldsymbol{H}_{k+1}\boldsymbol{F}_k\boldsymbol{P}_{\tilde{x},k}\boldsymbol{F}_k^{\mathrm{T}}\boldsymbol{H}_{k+1}^{\mathrm{T}}) = \mathrm{tr}(\boldsymbol{F}_k\boldsymbol{P}_{\tilde{x},k}\boldsymbol{F}_k^{\mathrm{T}}\boldsymbol{H}_{k+1}^{\mathrm{T}}\boldsymbol{H}_{k+1}) \tag{8.34}$$

The trace of both sides of (8.32) is calculated as

$$\mathrm{tr}(\boldsymbol{\lambda}_{k+1}\boldsymbol{H}_{k+1}\boldsymbol{F}_k\boldsymbol{P}_{\tilde{x},k}\boldsymbol{F}_k^{\mathrm{T}}\boldsymbol{H}_{k+1}^{\mathrm{T}}) = \mathrm{tr}(\boldsymbol{V}_{k+1} - \mu\boldsymbol{R}_{k+1} - \boldsymbol{H}_{k+1}\boldsymbol{Q}_k^P\boldsymbol{H}_{k+1}^{\mathrm{T}}) \tag{8.35}$$

or

$$\mathrm{tr}(\boldsymbol{\lambda}_{k+1}\underline{\boldsymbol{M}}_{k+1}) = \mathrm{tr}(\underline{\boldsymbol{N}}_{k+1}) \tag{8.36}$$

where $\underline{\boldsymbol{M}}_{k+1} = \boldsymbol{H}_{k+1}\boldsymbol{F}_k\boldsymbol{P}_{\tilde{x},k}\boldsymbol{F}_k^{\mathrm{T}}\boldsymbol{H}_{k+1}^{\mathrm{T}}$ and $\underline{\boldsymbol{N}}_{k+1} = \boldsymbol{V}_{k+1} - \mu\boldsymbol{R}_{k+1} - \boldsymbol{H}_{k+1}\boldsymbol{Q}_k^P\boldsymbol{H}_{k+1}^{\mathrm{T}}$.

According to the prior knowledge, the designer can roughly determine the following relationship for the multiple fading factors.

$$\lambda_{k+1}^i = \alpha_i c_{k+1}, i = 1, 2, \cdots, n \tag{8.37}$$

where $\alpha_i \geq 1, i = 1, 2, \cdots, n$, is a constant preliminarily chosen by the designer. c_{k+1} is an factor to be determined.

Inserting (8.37) into (8.36) results in

$$\begin{aligned}
\mathrm{tr}(\boldsymbol{\lambda}_{k+1}\underline{\boldsymbol{M}}_{k+1}) &= \mathrm{tr}(\mathbf{diag}([\lambda_{k+1}^1 \quad \lambda_{k+1}^2 \quad \cdots \quad \lambda_{k+1}^n]^{\mathrm{T}})\underline{\boldsymbol{M}}_{k+1}) \\
&= \mathrm{tr}(\mathbf{diag}([\alpha_1 c_{k+1} \quad \alpha_2 c_{k+1} \quad \cdots \quad \alpha_n c_{k+1}]^{\mathrm{T}})\underline{\boldsymbol{M}}_{k+1}) \\
&= \mathrm{tr}(\underline{\boldsymbol{N}}_{k+1})
\end{aligned} \tag{8.38}$$

Solving (8.38) obtains that c_{k+1} is given by

$$c_{k+1} = \frac{\mathrm{tr}(\underline{\boldsymbol{N}}_{k+1})}{\sum_{i=1}^{n} \alpha_i \underline{\boldsymbol{M}}_{k+1}^{ii}} \tag{8.39}$$

where $\underline{\boldsymbol{M}}_{k+1}^{ii}$ is the main diagonal element of the matrix $\underline{\boldsymbol{M}}_{k+1}, i = 1, 2, \cdots, n$.

Applying (8.40), the general algorithm of the fading factors can be designed as

$$\lambda_{k+1}^i = \begin{cases} \alpha_i c_{k+1}, & \text{if} \quad \alpha_i c_{k+1} > 1 \\ 1, & \text{if} \quad \alpha_i c_{k+1} \leq 1 \end{cases}, \quad i = 1, 2, \cdots, n \tag{8.40}$$

It should be pointed out the residual matrix V_{k+1} in (8.32) cannot be obtained directly. Hence, the following equation is applied to approximate its real value.

$$V_{k+1} = \begin{cases} \varepsilon_1 \varepsilon_1^{\mathrm{T}}, & \text{if } k = 1 \\ \dfrac{\rho V_k + \varepsilon_k \varepsilon_k^{\mathrm{T}}}{1+\rho}, & \text{if } k \geq 2 \end{cases} \tag{8.41}$$

where $\rho \in \mathbb{R}_+$ is the forgetting factor satisfying $0 < \rho \leq 1$, and it is usually chosen as 0.95.

Summarizing the preceding analysis, the sequence orthogonal principle can be incorporated into the PAVSF. This adjusts the gain matrix in order to satisfy (8.23). Then, the strong tracking PAVSF (ST-PAVSF) can be presented in Algorithm 8.4.

Algorithm 8.4: Strong tracking PAVSF.

Input Data: Initial estimation $\hat{x}_0$, $\hat{y}_0$, and measurement y_k
Result: State estimation $\hat{x}_k$
begin
  for $k = 0, 1, 2 \cdots$ **do**
    Updating the estimation of the modeling error

$$e_{y_{k+1/k}} = y_{k+1} - \hat{y}_k - Z(\hat{x}_k)$$
$$P_{\tilde{x},k+1/k} = \lambda_{k+1} F_k P_{\tilde{x},k} F_k^{\mathrm{T}}$$
$$\aleph_{k+1} = H_{k+1} P_{\tilde{x},k+1/k} H_{k+1}^{\mathrm{T}} + R_{k+1}$$
$$\ell_{k+1}^{-1} = (\mathbf{diag}(A_e))^{-1} (g_U(\hat{x}_k))^{-1} P_{\tilde{x},k+1/k} H_{k+1}^{\mathrm{T}} \aleph_{k+1}^{-1}$$
$$K_{k+1} = g_U(\hat{x}_k) \mathbf{diag}(A_e) \ell_{k+1}^{-1} \approx g_U(\hat{x}_k) \mathbf{diag}(A_e) \mathbf{diag}(\overrightarrow{\ell_{k+1}^{-1}})$$

    Updating the state estimation by using

$$\hat{x}_{k+1/k} = f(\hat{x}_k)$$
$$\hat{x}_{k+1} = \hat{x}_{k+1/k} + K_{k+1} e_{y_{k+1/k}}$$

    Updating the covariance by using

$$F_{k+1} = \frac{\partial f}{\partial x}\Big|_{x = x_{k+1}}$$
$$P_{\tilde{x},k+1} = (I_n - K_{k+1} H_{k+1}) P_{\tilde{x},k+1/k} (I_n - K_{k+1} H_{k+1})^{\mathrm{T}} + K_{k+1} R_{k+1} K_{K+1}^{\mathrm{T}}$$

    Updating the fading factor

$$\underline{M}_{k+1} = H_{k+1} F_k P_{\tilde{x},k} F_k^{\mathrm{T}} H_{k+1}^{\mathrm{T}}$$
$$\underline{N}_{k+1} = V_{k+1} - \mu R_{k+1} - H_{k+1} Q_k^P H_{k+1}^{\mathrm{T}}$$
$$c_{k+1} = \frac{\mathrm{tr}(\underline{N}_{k+1})}{\sum_{i=1}^{n} \alpha_i \underline{M}_{k+1}^{ii}}$$

    Calculating V_{k+1} using (8.41)
    Calculating λ_{k+1}^i using (8.40), $i = 1, 2, \cdots, n$
  end
end

8.4.3 Derivation of strong tracking sigma point PAVSF

Although good estimation performance is guaranteed by the ST-PAVSF with the fading factor included, its implementation requests the continuity of the function $\boldsymbol{f}(\cdot)$ and the calculation of the Jacobian matrix of $\boldsymbol{f}(\cdot)$. This limits its application. Following the method in Section 8.3, the UT sampling, the Cubature rule, or the Stirling polynomial interpolation is available to design a sigma point integrated ST-PAVSF. In this chapter, as an example, the UT sampling approach is incorporated into the ST-PAVSF to develop a sigma point ST-PAVSF (ST-SP-PAVSF). Its development does not need the Jacobian matrix of $\boldsymbol{f}(\cdot)$. Hence, this filter has great application potential.

Let the priori state error covariance, the priori measurement error covariance, and the cross covariance before introducing the fading factor be denoted as $\underline{\boldsymbol{P}}_{x,k+1/k}$, $\underline{\boldsymbol{P}}_{yy,k+1}$, and $\underline{\boldsymbol{P}}_{xy,k+1}$, respectively. Then, it follows that

$$\boldsymbol{P}_{\tilde{x},k+1/k} = \mathbb{E}((\boldsymbol{x}_{k+1} - \hat{\boldsymbol{x}}_{k+1/k})(\boldsymbol{x}_{k+1} - \hat{\boldsymbol{x}}_{k+1/k})^{\mathrm{T}}) \tag{8.42}$$

$$\underline{\boldsymbol{P}}_{yy,k+1} = \mathbb{E}((\boldsymbol{y}_{k+1} - \hat{\boldsymbol{y}}_{k+1/k})(\boldsymbol{y}_{k+1} - \hat{\boldsymbol{y}}_{k+1/k})^{\mathrm{T}}) \tag{8.43}$$

$$\underline{\boldsymbol{P}}_{xy,k+1} = \mathbb{E}((\boldsymbol{x}_{k+1} - \hat{\boldsymbol{x}}_{k+1/k})(\boldsymbol{y}_{k+1} - \hat{\boldsymbol{y}}_{k+1/k})^{\mathrm{T}}) \tag{8.44}$$

$$\boldsymbol{P}_{\tilde{x},k+1/k} = \boldsymbol{F}_k \boldsymbol{P}_{\tilde{x},k} \boldsymbol{F}_k^{\mathrm{T}} + \boldsymbol{Q}_k^P \tag{8.45}$$

Invoking the assumption that $\hat{\boldsymbol{x}}_{k+1/k}$ defined in (7.38) is uncorrelated with $\boldsymbol{v}_{k+1}$, (8.43) and (8.44) can be further calculated as

$$\begin{aligned}
\underline{\boldsymbol{P}}_{yy,k+1} &= \mathbb{E}((\boldsymbol{y}_{k+1} - \hat{\boldsymbol{y}}_{k+1/k})(\boldsymbol{y}_{k+1} - \hat{\boldsymbol{y}}_{k+1/k})^{\mathrm{T}}) \\
&= \mathbb{E}((\boldsymbol{H}_{k+1}(\boldsymbol{x}_{k+1} - \hat{\boldsymbol{x}}_{k+1/k}) + \boldsymbol{v}_{k+1})^2) \\
&= \boldsymbol{H}_{k+1}\mathbb{E}((\boldsymbol{x}_{k+1} - \hat{\boldsymbol{x}}_{k+1/k})^2)\boldsymbol{H}_{k+1}^{\mathrm{T}} + \mathbb{E}(\boldsymbol{v}_{k+1}\boldsymbol{v}_{k+1}^{\mathrm{T}}) \\
&= \boldsymbol{H}_{k+1}\underline{\boldsymbol{P}}_{x,k+1/k}\boldsymbol{H}_{k+1}^{\mathrm{T}} + \boldsymbol{R}_{k+1}
\end{aligned} \tag{8.46}$$

$$\begin{aligned}
\underline{\boldsymbol{P}}_{xy,k+1} &= \mathbb{E}((\boldsymbol{x}_{k+1} - \hat{\boldsymbol{x}}_{k+1/k})(\boldsymbol{y}_{k+1} - \hat{\boldsymbol{y}}_{k+1/k})^{\mathrm{T}}) \\
&= \mathbb{E}((\boldsymbol{x}_{k+1} - \hat{\boldsymbol{x}}_{k+1/k})(\boldsymbol{H}_{k+1}(\boldsymbol{x}_{k+1} - \hat{\boldsymbol{x}}_{k+1/k}) + \boldsymbol{v}_{k+1})^{\mathrm{T}}) \\
&= \underline{\boldsymbol{P}}_{x,k+1/k}\boldsymbol{H}_{k+1}^{\mathrm{T}}
\end{aligned} \tag{8.47}$$

Let $\boldsymbol{P}_{\tilde{x},k+1/k}$, $\boldsymbol{P}_{yy,k+1}$, and $\boldsymbol{P}_{xy,k+1}$ be the priori state error covariance, the priori measurement error covariance, and the cross covariance after introducing the fading factor, respectively. Based on (8.46) and (8.47), it gives

$$\boldsymbol{P}_{yy,k+1} = \boldsymbol{H}_{k+1}\boldsymbol{P}_{\tilde{x},k+1/k}\boldsymbol{H}_{k+1}^{\mathrm{T}} + \boldsymbol{R}_{k+1} \tag{8.48}$$

$$\boldsymbol{P}_{xy,k+1} = \boldsymbol{P}_{\tilde{x},k+1/k}\boldsymbol{H}_{k+1}^{\mathrm{T}} \tag{8.49}$$

Since $\boldsymbol{R}_{k+1}$ is the positive-definite and symmetric, $\underline{\boldsymbol{P}}_{x,k+1/k}^{-1}$ and $\boldsymbol{P}_{\tilde{x},k+1/k}^{-1}$ exist. We can solve the Jacobi matrix of measurement equation $\boldsymbol{H}_{k+1}$ from (8.47) and (8.49) that

$$\boldsymbol{H}_{k+1} = \underline{\boldsymbol{P}}_{xy,k+1}^{\mathrm{T}} \underline{\boldsymbol{P}}_{x,k+1/k}^{-1} \tag{8.50}$$

$$\boldsymbol{H}_{k+1} = \boldsymbol{P}_{xy,k+1}^{\mathrm{T}} \boldsymbol{P}_{x,k+1/k}^{-1} \tag{8.51}$$

Inserting (8.48)–(8.51) into (7.42) and (8.20) gives

$$\boldsymbol{K}_{k+1} = \boldsymbol{P}_{xy,k+1} \boldsymbol{P}_{yy,k+1}^{-1} \tag{8.52}$$

$$\boldsymbol{P}_{\tilde{x},k+1} = \boldsymbol{P}_{\tilde{x},k+1/k} - \boldsymbol{K}_{k+1} \boldsymbol{P}_{\tilde{x},k+1/k} \boldsymbol{K}_{k+1}^{\mathrm{T}} \tag{8.53}$$

Moreover, it follows that

$$\underline{\boldsymbol{N}}_{k+1} = \boldsymbol{V}_{k+1} - \boldsymbol{R}_{k+1} - \underline{\boldsymbol{P}}_{xy,k+1}^{\mathrm{T}} \underline{\boldsymbol{P}}_{x,k+1/k}^{-1} (\boldsymbol{Q}_k^P)^{\mathrm{T}} \underline{\boldsymbol{P}}_{x,k+1/k}^{-1} \underline{\boldsymbol{P}}_{xy,k+1} \tag{8.54}$$

$$\begin{aligned}
\underline{\boldsymbol{M}}_{k+1} &= \boldsymbol{F}_k \boldsymbol{P}_{\tilde{x},k} \boldsymbol{F}_k^{\mathrm{T}} \boldsymbol{H}_{k+1}^{\mathrm{T}} \boldsymbol{H}_{k+1} \\
&= \boldsymbol{H}_{k+1} (\underline{\boldsymbol{P}}_{x,k+1/k} - \boldsymbol{Q}_k^P) \boldsymbol{H}_{k+1}^{\mathrm{T}} \\
&= \boldsymbol{H}_{k+1} \underline{\boldsymbol{P}}_{x,k+1/k} \boldsymbol{H}_{k+1}^{\mathrm{T}} - \boldsymbol{H}_{k+1} \boldsymbol{Q}_k^P \boldsymbol{H}_{k+1}^{\mathrm{T}} \\
&= \boldsymbol{H}_{k+1} \underline{\boldsymbol{P}}_{x,k+1/k} \boldsymbol{H}_{k+1}^{\mathrm{T}} + \boldsymbol{R}_{k+1} - \boldsymbol{V}_{k+1} + \underline{\boldsymbol{N}}_{k+1} \\
&= \underline{\boldsymbol{P}}_{yy,k+1} - \boldsymbol{V}_{k+1} + \underline{\boldsymbol{N}}_{k+1}
\end{aligned} \tag{8.55}$$

Based on the above analysis, the UT sampling strategy can be invoked to compute the Jacobian matrix of $\boldsymbol{f}(\cdot)$ and the posterior mean and covariance. Then, the ST-SP-PAVSF can be presented in Algorithm 8.5, where $W_i^m \in \mathbb{R}$ is the weighting constant, $i = 0, 1, 2, \cdots, 2n$.

8.5 Application to microsatellite attitude control system

8.5.1 Application to attitude control system with low precision sensors

According to transformed system discussed in Section 5.5, it is known that the PAVSF in Algorithm 8.1 and the unscented PAVSF in Algorithm 8.2 can be applied to estimate the states of the microsatellite attitude control system with a low-precision three-axis magnetometer and a low-precision sun sensor as measurement unit. To validate this application, numerical simulation is done with all the simulation parameters and the modeling error chosen as Section 5.5.3. Moreover, the UPAVSF, the UKF, and the PAVSF are compared in simulation with following two modeling error and measurement noise scenarios considered. The estimation gains $\ell_{limit} = 0.0005$ and $\gamma_{ii} = 0.01$ are chosen for the PAVSF and its unscented version.

Algorithm 8.5: Strong tracking sigma point PAVSF.

Input Data: Initial estimation $\hat{\boldsymbol{x}}_0$, $\hat{\boldsymbol{y}}_0$, and measurement $\boldsymbol{y}_k$
Result: State estimation $\hat{\boldsymbol{x}}_k$
begin
  Initializing the state and the covariance matrix as
    $\hat{\boldsymbol{x}}_0 = \mathbb{E}(\boldsymbol{x}_0)$
    $\boldsymbol{P}_{\tilde{x},0} = \mathrm{Var}(\boldsymbol{x}_0) = \mathbb{E}((\boldsymbol{x}_0 - \hat{\boldsymbol{x}}_0)(\boldsymbol{x}_0 - \hat{\boldsymbol{x}}_0)^{\mathrm{T}})$
  for $k = 0, 1, \cdots$ **do**
    Applying the Cholesky factorization to the covariance $\boldsymbol{P}_{\tilde{x},k}$
    $$\boldsymbol{P}_{\tilde{x},k} = \boldsymbol{S}_k \boldsymbol{S}_k^{\mathrm{T}} = \sum_{i=1}^{n} \boldsymbol{\sigma}_{i,k} \boldsymbol{\sigma}_{i,k}^{\mathrm{T}}$$
    Establishing sigma points
    $\boldsymbol{\xi}_{0,k} = \hat{\boldsymbol{x}}_k$
    $\boldsymbol{\xi}_{i,k} = \hat{\boldsymbol{x}}_k + \sqrt{n+\kappa}\,\boldsymbol{\sigma}_{i,k},\ i = 1, 2, \cdots, n$
    $\boldsymbol{\xi}_{i+n,k} = \hat{\boldsymbol{x}}_k - \sqrt{n+\kappa}\,\boldsymbol{\sigma}_{i,k},\ i = 1, 2, \cdots, n$
    Updating the priori state covariance
    $\hat{\boldsymbol{x}}_{i,k+1/k} = \boldsymbol{f}(\boldsymbol{\xi}_{i,k}),\ i = 0, 1, \cdots, 2n$
    $$\hat{\boldsymbol{x}}_{k+1/k} = \sum_{i=0}^{2n} W_i^m \hat{\boldsymbol{x}}_{i,k+1/k}$$
    $$\underline{\boldsymbol{P}}_{x,k+1/k} = \sum_{i=0}^{2n} W_i^m (\hat{\boldsymbol{x}}_{i,k+1/k} - \hat{\boldsymbol{x}}_{k+1/k})(\hat{\boldsymbol{x}}_{i,k+1/k} - \hat{\boldsymbol{x}}_{k+1/k})^{\mathrm{T}}$$
    $\boldsymbol{P}_{\tilde{x},k+1/k} = \boldsymbol{\lambda}_{k+1} \underline{\boldsymbol{P}}_{x,k+1/k}$
    Using $\hat{\boldsymbol{x}}_{k+1/k}$ and $\underline{\boldsymbol{P}}_{x,k+1/k}$ to generate new sigma points:
    $\underline{\hat{\boldsymbol{x}}}_{i,k+1/k},\ i = 0, 1, 2, \cdots, 2n$
    Updating measurement
    $\underline{\hat{\boldsymbol{y}}}_{i,k+1/k} = \boldsymbol{h}(\underline{\hat{\boldsymbol{x}}}_{i,k+1/k}),\ i = 0, 1, \cdots, 2n$
    $$\underline{\hat{\boldsymbol{y}}}_{k+1/k} = \sum_{i=0}^{2n} W_i^m \underline{\hat{\boldsymbol{y}}}_{i,k+1/k}$$
    $$\underline{\boldsymbol{P}}_{yy,k+1/k} = \sum_{i=0}^{2n} W_i^m (\underline{\hat{\boldsymbol{y}}}_{i,k+1/k} - \underline{\hat{\boldsymbol{y}}}_{k+1/k})^2 + \boldsymbol{R}_{k+1}$$
    $$\underline{\boldsymbol{P}}_{xy,k+1/k} = \sum_{i=0}^{2n} W_i^m (\underline{\hat{\boldsymbol{x}}}_{i,k+1/k} - \underline{\hat{\boldsymbol{x}}}_{k+1/k})(\underline{\hat{\boldsymbol{y}}}_{i,k+1/k} - \underline{\hat{\boldsymbol{y}}}_{k+1/k})^{\mathrm{T}}$$
    Calculating the adaptive boundary layer
    $\aleph_{i,k+1} = \boldsymbol{H}_{i,k+1} \boldsymbol{P}_{\tilde{x},k+1/k} \boldsymbol{H}_{i,k+1}^{\mathrm{T}} + \boldsymbol{R}_{k+1}$
    $\boldsymbol{\ell}_{i,k+1} = ((\mathbf{diag}(\boldsymbol{A}_e))^{-1} (\boldsymbol{g}_U(\hat{\boldsymbol{x}}_k))^{-1} \boldsymbol{P}_{\tilde{x},k+1/k} \boldsymbol{H}_{i,k+1}^{\mathrm{T}} \aleph_{i,k+1}^{-1})^{-1}$
    Updating state estimation
    $\boldsymbol{e}_{y_{i,k+1/k}} = \boldsymbol{y}_{k+1} - \hat{\boldsymbol{y}}_{k_i} - \boldsymbol{Z}(\underline{\hat{\boldsymbol{x}}}_{i,k+1/k})$
    $\boldsymbol{K}_{i,k+1} = \boldsymbol{g}_U(\hat{\boldsymbol{x}}_k) \mathbf{diag}(\boldsymbol{A}_e) \mathbf{diag}(\overrightarrow{\boldsymbol{\ell}_{i,k+1}^{-1}})$
    $\hat{\boldsymbol{x}}_{i,k+1/k+1} = \hat{\boldsymbol{x}}_{i,k+1/k} + \boldsymbol{K}_{i,k+1} \boldsymbol{e}_{y_{i,k+1/k}}$
    Updating the state estimation error covariance
    $$\hat{\boldsymbol{x}}_{k+1} = \sum_{i=0}^{2n} W_i^m \hat{\boldsymbol{x}}_{i,k+1/k+1}$$
    $$\boldsymbol{P}_{\tilde{x},k+1} = \sum_{i=0}^{2n} W_i^m (\hat{\boldsymbol{x}}_{i,k+1/k+1} - \hat{\boldsymbol{x}}_{k+1})(\hat{\boldsymbol{x}}_{i,k+1/k+1} - \hat{\boldsymbol{x}}_{k+1})^{\mathrm{T}}$$
  end
end

1) Nominal case: The modeling error is assumed as

$$\Delta \boldsymbol{N}_c = [-2\omega_1 \quad -4\omega_1 \quad 3\omega_1]^{\mathrm{T}}\ \mathrm{Nm} \tag{8.56}$$

$$\Delta \boldsymbol{N}_e = [3\omega_1 \cos(\omega_T t) \quad \omega_1 \cos(\omega_T t) \quad -2\omega_1 \cos(\omega_T t)]^{\mathrm{T}}\ \mathrm{Nm} \tag{8.57}$$

with $\omega_T = 0.66$ deg/sec and $\omega_1 = 0.0003$. The standard deviation of the measurement noises of the sun sensor and the magnetometer in (5.102) is assumed as 0.03 degrees and 50 nT, respectively.

2) Severe case: In this case, severe measurement noise and large modeling error are considered with

$$\Delta \boldsymbol{N}_e = [3\omega_1 \cos(\omega_T t) \quad \omega_1 \cos(\omega_T t) \quad -2\omega_1 \cos(\omega_T t)]^{\mathrm{T}}\ \mathrm{Nm} \tag{8.58}$$

$$\Delta \boldsymbol{N}_c = [-3\omega_1 \quad -4\omega_1 \quad 3\omega_1]^{\mathrm{T}}\ \mathrm{Nm} \tag{8.59}$$

where $\omega_T = 0.66$ deg/sec and $\omega_1 = 0.001$. The standard deviation of the measurement noises of the sun sensor and the magnetometer in (5.102) is 0.1 degrees and 150 nT, respectively.

The terms $\boldsymbol{\Theta}_e$ and $\boldsymbol{\omega}_E$ defined in Remark 4.5 are used in this section to denote the estimation error of the attitude and the angular velocity, respectively. In the nominal case, the state estimation results obtained from the UPAVSF, the PAVSF, and the UKF were shown in Figs. 8.2–8.3. The UPAVSF achieved the highest estimation accuracy for the attitude and the angular velocity states than the PAVSF and the UKF. The UKF led its attitude and angular velocity estimation accuracy to be inferior to the UPAVSF and the PAVSF, although the UKF provided a faster estimation for that two states. That is because the UKF has the weak capability of providing high-accuracy estimation for states, when the system is subject to modeling error. In contrast, both the UPAVSF and the PAVSF are capable of achieving this. It is seen in Fig. 8.3 that the estimated angular velocity provided by the UPAVSF and the PAVSF is bounded within the set containing the actual angular velocity. This is owing the superior capability of the UPAVSF and the PAVSF in estimating the modeling error and compensating for its effect. Consequently, the model precision of the microsatellite attitude control system is improved. Based on this more accurate model, a high-accuracy estimation of the attitude and the angular velocity was thus ensured by the UPAVSF and the PAVSF. However, due to the use of the UT sampling strategy, the UPAVSF achieved a better state estimation in accuracy than the PAVSF.

In the severe case, the magnitude of the measurement noise and the large modeling error was three times larger than the nominal case. Correspondingly the state estimation results from the UPAVSF, the PAVSF, and the UKF were illustrated in Figs. 8.4–8.5. The attitude and the angular velocity estimation accuracy was inferior to the nominal case. The UPAVSF and the PAVSF still achieved a stable estimation for the states, especially for the angular velocity state, as shown in Fig. 8.5. However, due to the periodic modeling error in thi

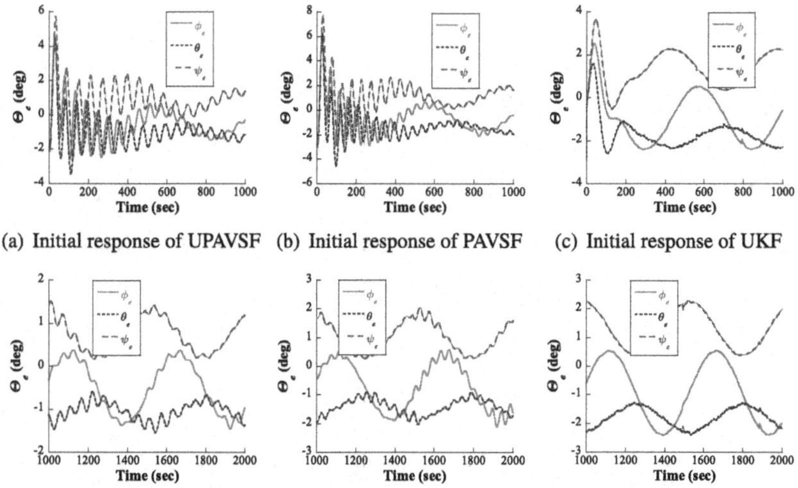

(a) Initial response of UPAVSF (b) Initial response of PAVSF (c) Initial response of UKF

(d) Steady behavior of UPAVSF (e) Steady behavior of PAVSF (f) Steady behavior of UKF

FIGURE 8.2 Attitude estimation error using UPAVSF, PAVSF, and UKF in the nominal case.

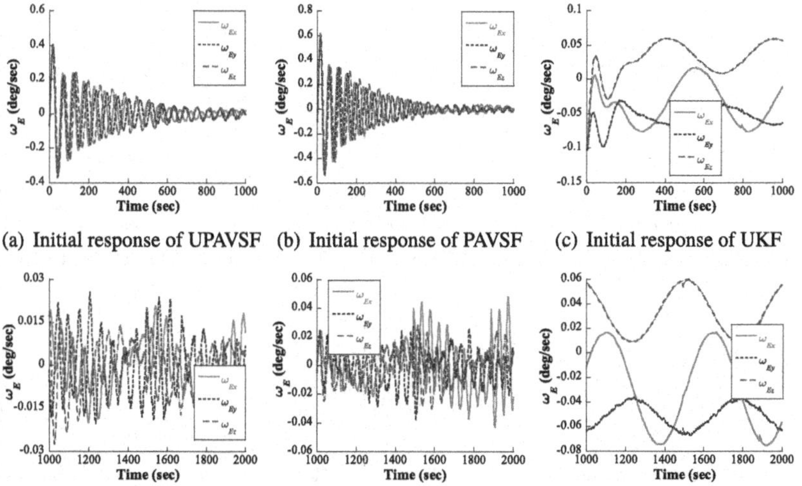

(a) Initial response of UPAVSF (b) Initial response of PAVSF (c) Initial response of UKF

(d) Steady behavior of UPAVSF (e) Steady behavior of PAVSF (f) Steady behavior of UKF

FIGURE 8.3 Angular velocity estimation error using UPAVSF, PAVSF, and UKF in the nominal case.

case, the UKF was not able to guarantee such stable estimation. Periodic fluctuation was seen in the angular velocity estimation for the UKF.

The above simulation results validated the theoretical conclusion that UPAVSF and the PAVSF provides better state estimation performance than the UKF in the presence of modeling error and measurement noise.

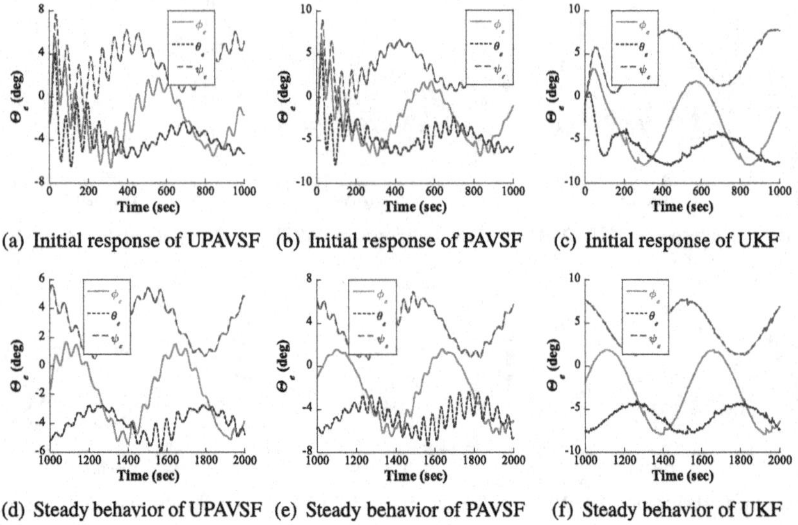

(a) Initial response of UPAVSF (b) Initial response of PAVSF (c) Initial response of UKF

(d) Steady behavior of UPAVSF (e) Steady behavior of PAVSF (f) Steady behavior of UKF

FIGURE 8.4 Attitude estimation error using UPAVSF, PAVSF, and UKF in the case of severe measurement noise and large modeling error.

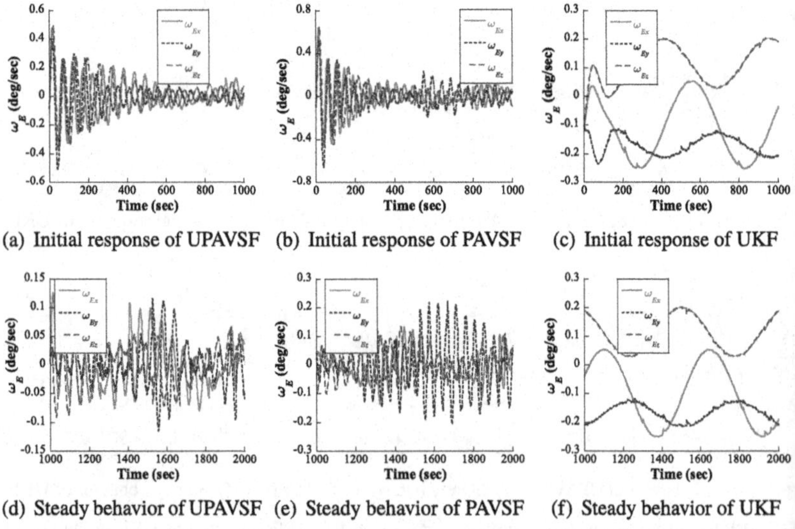

(a) Initial response of UPAVSF (b) Initial response of PAVSF (c) Initial response of UKF

(d) Steady behavior of UPAVSF (e) Steady behavior of PAVSF (f) Steady behavior of UKF

FIGURE 8.5 Angular velocity estimation error using UPAVSF, PAVSF, and UKF in the case o severe measurement noise and large modeling error.

8.5.2 Application to distributed attitude control system

It is analyzed in Section 7.7 that the microsatellite distributed attitude system can be transformed into the nonlinear system (2.32)–(2.33). Hence, it is concluded that the ST-SP-PAVSF in Algorithm 8.5 and the PAVSF in Algorithm 8.

can be used to solve the state estimation problem of the microsatellite distributed attitude system. Based on this conclusion, numerical simulations are conducted to demonstrate their effectiveness in comparison with the standard UKF. In simulations, all the simulation parameters are chosen as same as Section 7.7.3.

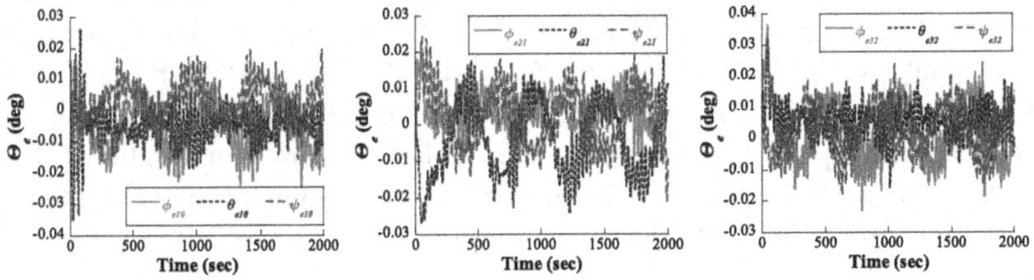

(a) (b) (c)

FIGURE 8.6 Relative attitude estimation error ensured by ST-SP-PAVSF.

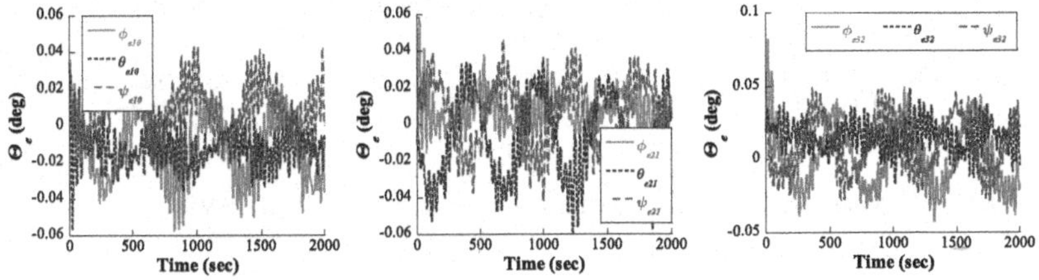

(a) (b) (c)

FIGURE 8.7 Relative attitude estimation error ensured by PAVSF.

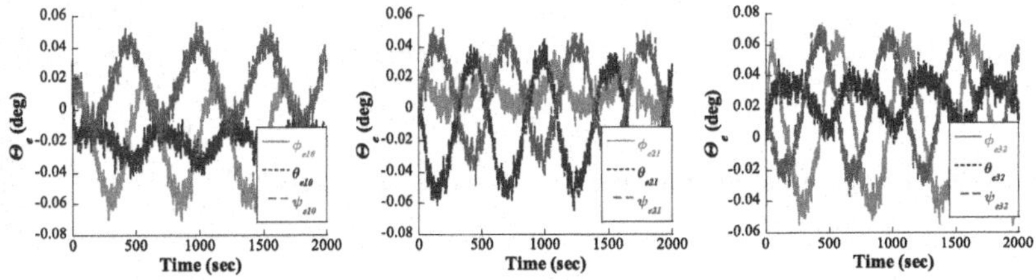

(a) (b) (c)

FIGURE 8.8 Relative attitude estimation error ensured by UKF.

Based on Remark 7.6, $\boldsymbol{\Theta}_e$ and $\boldsymbol{\omega}_E$ were used in simulation to denote the relative attitude the relative angular velocity estimation errors, respectively. Figs. 8.6–8.11 show the state estimation errors govern by the ST-SP-PAVSF, the PAVSF, and the UKF. It is seen Figs. 8.6–8.8 that the ST-SP-PAVSF improved the PAVSF in the relative attitude state estimation accuracy. While the relative attitude state estimation accuracy obtained from the UKF was inferior to the ST-SP-PAVSF and the PAVSF. Moreover, the relative angular velocity state estimation accuracy obtained from the UKF was also inferior to the ST-SP-PAVSF and the PAVSF, as shown in Figs. 8.9–8.11. It is seen the magnitude of the relative angular velocity estimation error achieved by the UKF has a order of 0.01, while the magnitude of the relative angular velocity estimation error achieved by the ST-SP-PAVSF and the PAVSF has a order of 0.001. The relative angular velocity accuracy of the ST-SP-PAVSF and the PAVSF is ten times higher than the UKF. This is due to the UT sampling strategy and the orthogonal sequence principle incorporated in the ST-SP-PAVSF. Based on this, the ST-SP-PAVSF guaranteed higher state estimation accuracy and more robustness to the modeling error than the PAVSF.

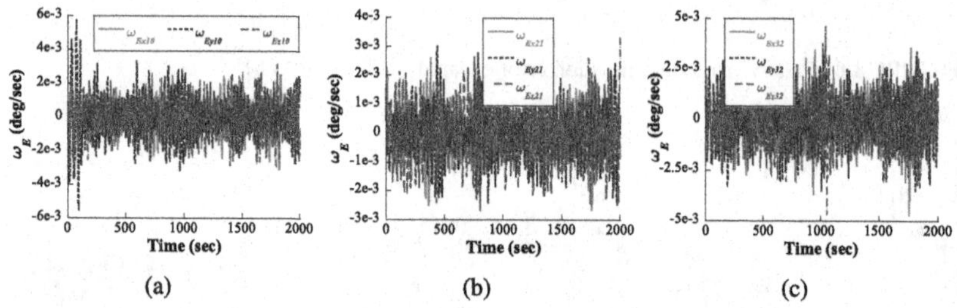

FIGURE 8.9 Relative angular velocity estimation error ensured by ST-SP-PAVSF.

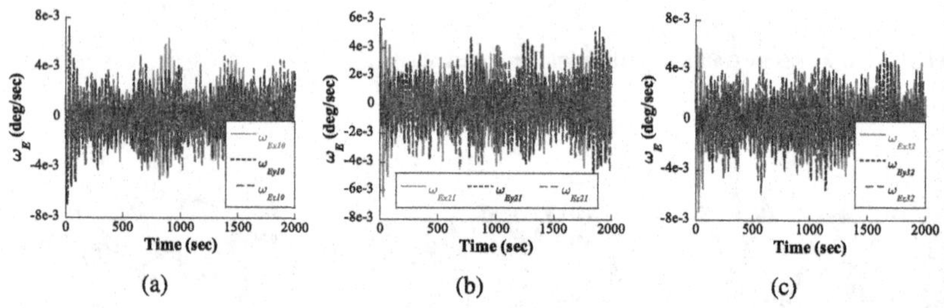

FIGURE 8.10 Relative angular velocity estimation error ensured by PAVSF.

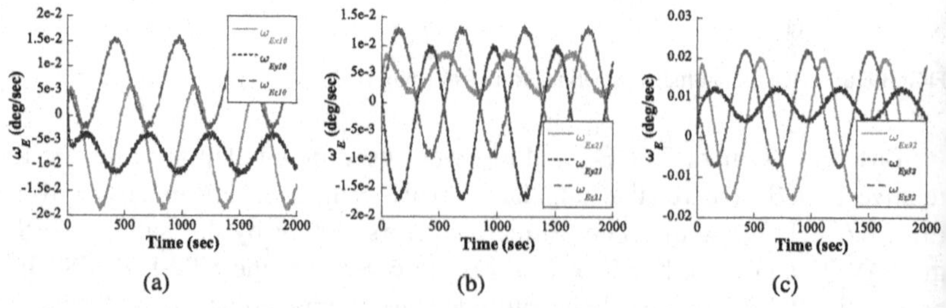

FIGURE 8.11 Relative angular velocity estimation error ensured by UKF.

Numerical simulations were carried out further by enlarging the modeling error. Simulation results show that the ST-SP-PAVSF and the PAVSF still achieved a stable and acceptable estimation for the relative attitude and the relative angular velocity states. However, the UKF leads the state estimation to diverge. That is because the UKF has a weak capability of handing large modeling error. The preceding simulation results validated the theoretical conclusion that ST-SP-PAVSF and the PAVSF provides better state estimation performance than the UKF in the presence of modeling error and measurement noise.

8.6 Summary

In this chapter following the result in Chapter 7 the state estimation problem of nonlinear systems with modeling error was further investigated by using variable structure control theory. A predictive adaptive variable structure filter and its sigma-point version were first developed to eliminate the drawback of the predictive smooth variable structure filter in Chapter 7. This was achieved by adaptively tuning the thickness of the boundary layer. Then, another filtering approach called as the strong tracking predictive adaptive variable structure filter and its sigma-point version were synthesized. This approach guaranteed better robustness and higher estimation accuracy, because it can capture the 2nd central moment's information of the modeling error. The restriction of the limit boundary layer in the predictive adaptive variable structure filter was eliminated by this approach. The main features of the approach were that it was robust to modeling error and insensitive to measurement noise as well as the initial statistical property. This approach has great estimation capability for abrupt states and retain this capability even during the stable filtering process. Moreover, this filtering approach has faster estimation rate and its computation is not complex. All the filters designed in this chapter are applicable to the state estimation problem of the microsatellite attitude control system.

Chapter 9

Predictive high-order variable structure filter

9.1 Introduction

When applying the predictive variable structure filters presented in Chapter 7, the chattering phenomenon would be induced due to the use of the discontinuous switching function. Although the predictive adaptive variable structure filters developed in Chapter 8 are robust to system uncertainties, the estimation accuracy was reduced. Hence, these two issues should be solved to improve the application of predictive variable structure filtering method.

It is known from the variable structure control theory that the boundary layer method in Chapter 7 and its adaptive version in Chapter 8 are not the only solution to reduce chattering. Instead, the high-order variable structure control is another widely utilized approach to avoid chattering phenomenon. In this approach, the discontinuous inputs are applied to high-order derivative of sliding surface. The discontinuity is hidden and a much smoother performance could be achieved in the sliding surface. For example, a second-order variable structure controller (Bartolini et al., 1998) and a super-twisting variable structure control scheme (Shtessel et al., 2012; Levant, 1993; Chalanga et al., 2016) were available to eliminate chattering, while great robustness to system uncertainties was guaranteed. Inspired by the advantages of the high-order variable structure control, a predictive high-order variable structure filtering is presented in this chapter. The estimation error is proved to be asymptotically stable. Moreover, to improve the robustness of the filter to the measurement and modeling errors, a robust version of this filtering approach is finally presented by integrating the sequence orthogonal principle and the Huber's generalized maximum likelihood estimation theory with the sigma-point sampling strategy. When using those filters to the microsatellite control system, simulation results are shown to verify their state estimation performance.

9.2 Predictive high-order variable structure filter

Consider the nonlinear discrete systems described by (2.32)–(2.33), for their to-be-estimated model (2.36)–(2.37), let $\Delta \boldsymbol{e}_{y_k}$ be defined as the error between $\boldsymbol{e}_{y_k}$ and $\boldsymbol{e}_{y_{k-1}}$, where $\boldsymbol{e}_{y_k}$ is the posterior estimation error (7.28), *i.e.*,

$$\Delta \boldsymbol{e}_{y_k} = \boldsymbol{e}_{y_k} - \boldsymbol{e}_{y_{k-1}} \tag{9.1}$$

Predictive Filtering for Microsatellite Control System. https://doi.org/10.1016/B978-0-12-821865-5.00024-3

Then, invoking the prior estimation error given by (7.27), it is ready to design a predictive high-order variable structure filter (PHVSF) in Algorithm 9.1, where $\boldsymbol{\gamma} = [\gamma_{11} \quad \gamma_{22} \quad \cdots \quad \gamma_{mm}]^{\mathrm{T}} \in \mathbb{R}^m$ is a constant vector such that $0 < \sum_{i=1}^{m} \gamma_{ii}^2 < 1$ and $\gamma_{ii} \in \mathbb{R}_+$, $i = 1, 2, \cdots, m$. Moreover, the numerical integral method is adopted to update the state and the output estimation in Algorithm 9.1.

Algorithm 9.1: Predictive high-order variable structure filter.

Input Data: Initial estimation $\hat{\boldsymbol{x}}_0$, $\hat{\boldsymbol{y}}_0$, and measurement $\boldsymbol{y}_k$
Result: State estimation $\hat{\boldsymbol{x}}_k$
begin
  for $k = 1, 2 \cdots$ **do**
    Calculating the modeling error's estimation $\hat{\boldsymbol{d}}_k$ by using

$$\boldsymbol{e}_{\boldsymbol{y}_{k+1/k}} = \boldsymbol{y}_{k+1} - \hat{\boldsymbol{y}}_k - \boldsymbol{Z}(\hat{\boldsymbol{x}}_k)$$

$$\boldsymbol{e}_{\boldsymbol{y}_k} = \boldsymbol{y}_{k+1} - \hat{\boldsymbol{y}}_k$$

$$\Delta \boldsymbol{e}_{\boldsymbol{y}_k} = \boldsymbol{e}_{\boldsymbol{y}_k} - \boldsymbol{e}_{\boldsymbol{y}_{k-1}}$$

$$\hat{\boldsymbol{d}}_k = (\boldsymbol{U}(\hat{\boldsymbol{x}}_k))^{-1} \left(\boldsymbol{e}_{\boldsymbol{y}_{k+1/k}} - \frac{\boldsymbol{e}_{\boldsymbol{y}_k}}{2} - \sqrt{\frac{\boldsymbol{e}_{\boldsymbol{y}_k}^{\mathrm{T}} \boldsymbol{e}_{\boldsymbol{y}_k}}{4} + \frac{(\Delta \boldsymbol{e}_{\boldsymbol{y}_k})^{\mathrm{T}} \Delta \boldsymbol{e}_{\boldsymbol{y}_k}}{2}}\, \boldsymbol{\gamma} \right)$$

    Updating the state and the output estimation using

$$\hat{\boldsymbol{x}}_{k+1} = \boldsymbol{f}(\hat{\boldsymbol{x}}_k) + \boldsymbol{g}(\hat{\boldsymbol{x}}_k)\hat{\boldsymbol{d}}_k$$

$$\hat{\boldsymbol{y}}_k = \boldsymbol{h}\left(\hat{\boldsymbol{x}}_k\right)$$

  end
end

Theorem 9.1. *For the nonlinear discrete systems* (2.32)–(2.33), *with the application of the filtering approach in Algorithm 9.1, if the constant matrix* $\boldsymbol{\gamma}$ *is chosen to satisfy* $0 < \sum_{i=1}^{m} \gamma_{ii}^2 < 1$, *then the estimation error* $\boldsymbol{e}_{\boldsymbol{y}_k}$ *and its difference* $\Delta \boldsymbol{e}_{\boldsymbol{y}_k}$ *are asymptotically stable despite the modeling error* $\boldsymbol{d}_k$ *and the measurement noise* $\boldsymbol{v}_k$. *The state estimation* $\hat{\boldsymbol{x}}_k$ *asymptotically converges to the state* $\boldsymbol{x}_k$, *i.e.,* $\lim_{k \to \infty} \hat{\boldsymbol{x}}_k = \boldsymbol{x}_k$.

Proof. Inserting $\hat{\boldsymbol{d}}_k$ in Algorithm 9.1 into (7.25), it can be obtained from (7.28) that

$$\begin{aligned} \boldsymbol{e}_{\boldsymbol{y}_{k+1}} =& \boldsymbol{y}_{k+1} - \hat{\boldsymbol{y}}_k - \boldsymbol{Z}\left(\hat{\boldsymbol{x}}_k\right) - \boldsymbol{U}\left(\hat{\boldsymbol{x}}_k\right)\hat{\boldsymbol{d}}_k \\ =& \boldsymbol{y}_{k+1} - \hat{\boldsymbol{y}}_k - \boldsymbol{Z}\left(\hat{\boldsymbol{x}}_k\right) - \boldsymbol{e}_{\boldsymbol{y}_{k+1/k}} \\ &+ \frac{\boldsymbol{e}_{\boldsymbol{y}_k}}{2} + \sqrt{\frac{\boldsymbol{e}_{\boldsymbol{y}_k}{}^{\mathrm{T}} \boldsymbol{e}_{\boldsymbol{y}_k}}{4} + \frac{(\Delta \boldsymbol{e}_{\boldsymbol{y}_k})^{\mathrm{T}} \Delta \boldsymbol{e}_{\boldsymbol{y}_k}}{2}}\, \boldsymbol{\gamma} \end{aligned} \quad (9.2$$

Applying (7.27), (9.2) can be simplified as

$$e_{y_{k+1}} = \frac{e_{y_k}}{2} - \sqrt{\frac{{e_{y_k}}^{\mathrm{T}} e_{y_k}}{4} + \frac{(\Delta e_{y_k})^{\mathrm{T}} \Delta e_{y_k}}{2}}\gamma \tag{9.3}$$

Due to $0 < \sum_{i=1}^{m} \gamma_{ii}^2 < 1$, it is got from (9.3) that

$$\left(e_{y_{k+1}} - \frac{e_{y_k}}{2}\right)^{\mathrm{T}} \left(e_{y_{k+1}} - \frac{e_{y_k}}{2}\right) < \frac{e_{y_k}^{\mathrm{T}} e_{y_k}}{4} + \frac{(\Delta e_{y_k})^{\mathrm{T}} \Delta e_{y_k}}{2} \tag{9.4}$$

and

$$e_{y_{k+1}}^{\mathrm{T}} e_{y_{k+1}} - e_{y_{k+1}}^{\mathrm{T}} e_{y_k} < \frac{1}{2}(\Delta e_{y_k})^{\mathrm{T}} \Delta e_{y_k} \tag{9.5}$$

Based on (9.5), it follows that

$$e_{y_{k+1}}^{\mathrm{T}} e_{y_{k1}} + (e_{y_{k+1}} - e_{y_k})^{\mathrm{T}} (e_{y_{k+1}} - e_{y_k}) < e_{y_k}^{\mathrm{T}} e_{y_k} + (\Delta e_{y_k})^{\mathrm{T}} \Delta e_{y_k} \tag{9.6}$$

or

$$e_{y_{k+1}}^{\mathrm{T}} e_{y_{k+1}} + (\Delta e_{y_{k+1}})^{\mathrm{T}} \Delta e_{y_{k+1}} < e_{y_k}^{\mathrm{T}} e_{y_k} + (\Delta e_{y_k})^{\mathrm{T}} \Delta e_{y_k} \tag{9.7}$$

Choose a Lyapunov candidate function as

$$\bar{V}_k = e_{y_k}^{\mathrm{T}} e_{y_k} + (\Delta e_{y_k})^{\mathrm{T}} \Delta e_{y_k} \tag{9.8}$$

which yields $\bar{V}_{k+1} < \bar{V}_k$ from (9.7). Hence, it can be proved that $\lim_{k\to\infty} e_{y_k} = \mathbf{0}$ and $\lim_{k\to\infty} \Delta e_{y_k} = \mathbf{0}$. The estimation error e_{y_k} and its difference Δe_{y_k} are asymptotically stable. The output $\hat{y}_k$ provided by Algorithm 9.1 asymptotically converges to y_k despite the modeling error and the measurement noises. Meanwhile, following the proof of Theorem 7.1, $\lim_{k\to\infty} \hat{x}_k = x_k$ can be proved. Then, the estimation $\hat{x}_k$ is governed to asymptotically converge to the state x_k. Thereby, the proof is completed here. □

It is seen in Algorithm 9.1 that the proposed predictive high-order variable structure filter is dependent of the prior measurement error $e_{y_{k+1/k}}$, the posterior measurement error e_{y_k}, and its difference Δe_{y_k}. The output of this filter is adaptively updated by $e_{y_{k+1/k}}$, e_{y_k}, and Δe_{y_k}. This lets this filter have the form of the corrector-predictor.

Remark 9.1. It can be obtained from (9.1) and (7.28) that

$$e_{y_{k+1}} = y_{k+1} - \hat{y}_{k+1} = o(\hat{e}_{k+1}) \tag{9.9}$$

$$\Delta e_{y_{k+1}} = o(\hat{e}_{k+1}) - o(\hat{e}_k) = \Delta o(\hat{e}_{k+1}) \tag{9.10}$$

Then, based on $\bar{V}_{k+1} < \bar{V}_k$ in the proof of Theorem 9.1, applying (9.8)–(9.10) establishes that

$$\bar{V}_{k+1} = ||o(\hat{\boldsymbol{e}}_{k+1})||^2 + ||o(\hat{\boldsymbol{e}}_{k+1})||^2 < ||o(\hat{\boldsymbol{e}}_k)||^2 + ||o(\hat{\boldsymbol{e}}_k)||^2 = \bar{V}_k \quad (9.11)$$

which implies that the accuracy of the non-time Taylor expansion of (2.38) and (2.39) is improved with the convergence of the state estimation being asymptotically stable. Therefore, the non-time Taylor expansion (2.38) and (2.39) are reasonable and suitable for the PHVSF. Moreover, it is seen in the Algorithm 9.1 that the input $\boldsymbol{d}_k$ is independent of the sampling time Δt. Therefore, the developed PHVSF does not depend on the sampling time Δt. The sampling time will not affect the estimation accuracy. When implementing the proposed approach, there is no constraint or assumption on the measurement noises. The noise handled is not necessary to be Gaussian white type, and it can be the heavy-tailed noise. This method is hence suitable for any linear/nonlinear estimation systems with any type of noise. The modeling error $\boldsymbol{d}_k$ is also considered. Hence, the proposed filter is capable of providing the states with high-accuracy estimation despite any measurement noise and modeling errors.

Remark 9.2. When implementing the PHVSF, the constant vector $\boldsymbol{\gamma}$ should be properly selected such that $0 < \sum_{i=1}^{m} \gamma_{ii}^2 < 1$. It can be viewed as the PHVSF "memory" and represents the convergence rate of the estimation error. The proposed PHVSF is quite different from the traditional Kalman filter. The PHVSF does not need to calculate the state estimation error covariance. Hence, the PHVSF demands less calculation than the traditional Kalman filter. In comparison with the filters presented in Chapters 7–8, the PHVSF does not have chattering phenomenon. The drawback of the filters in Chapters 7–8 is thus eliminated.

9.3 Derivation of orthogonal PHVSF

To improve the robustness and accuracy of the PHVSF to the modeling error and measurement noise, a robust version of the PHVSF, called the orthogonal PHVSF (OPHVSF), is developed in this section by using the sequence orthogonal principle. This is achieved by introducing a fading factor into $\hat{\boldsymbol{d}}_k$. As such the filtering capability can be enhanced despite any modeling error or any heavy-tailed noise.

Following the method in Section 8.4, the condition (8.23) of Lemma 8.1 in the sequence orthogonal principle is equivalent to the following relation.

$$\mathbb{E}((\boldsymbol{x}_k - \hat{\boldsymbol{x}}_k)\boldsymbol{e}_{y_k}^{\mathrm{T}}) = \boldsymbol{0} \quad (9.12)$$

Introducing a fading factor $\boldsymbol{\lambda}_k = \mathbf{diag}([\lambda_k^1 \quad \lambda_k^2 \quad \cdots \quad \lambda_k^p]) \in \mathbb{R}^{p\times p}$, $\lambda_k^i \geq 1$, $i = 1, 2, \cdots, p$, the term $\hat{\boldsymbol{d}}_k$ in Algorithm 9.1 is then modified as

$$\hat{\boldsymbol{d}}_k = \boldsymbol{\lambda}_k(\boldsymbol{U}(\hat{\boldsymbol{x}}_k))^{-1}\left(\boldsymbol{e}_{\boldsymbol{y}_{k+1/k}} - \frac{\boldsymbol{e}_{\boldsymbol{y}_k}}{2} - \sqrt{\frac{\boldsymbol{e}_{\boldsymbol{y}_k}^{\mathrm{T}}\boldsymbol{e}_{\boldsymbol{y}_k}}{4} + \frac{(\Delta\boldsymbol{e}_{\boldsymbol{y}_k})^{\mathrm{T}}\Delta\boldsymbol{e}_{\boldsymbol{y}_k}}{2}\gamma}\right) \tag{9.13}$$

Inserting (9.3) into (9.13) yields

$$\hat{\boldsymbol{d}}_k = \boldsymbol{\lambda}_k(\boldsymbol{U}(\hat{\boldsymbol{x}}_k))^{-1}(\boldsymbol{e}_{\boldsymbol{y}_{k+1/k}} - \boldsymbol{e}_{\boldsymbol{y}_{k+1}}) = \boldsymbol{K}_k(\boldsymbol{e}_{\boldsymbol{y}_{k+1/k}} - \boldsymbol{e}_{\boldsymbol{y}_{k+1}}) \tag{9.14}$$

where $\boldsymbol{K}_k = \boldsymbol{\lambda}_k(\boldsymbol{U}(\hat{\boldsymbol{x}}_k))^{-1}$.

From (9.12) and (9.14), we have

$$\boldsymbol{\lambda}_k\mathbb{E}((\boldsymbol{x}_{k+1} - \hat{\boldsymbol{x}}_{k+1})\boldsymbol{e}_{\boldsymbol{y}_{k+1/k}}^{\mathrm{T}}) = \mathbb{E}((\boldsymbol{x}_{k+1} - \hat{\boldsymbol{x}}_{k+1})(\boldsymbol{U}(\hat{\boldsymbol{x}}_k)\hat{\boldsymbol{d}}_k)^{\mathrm{T}}) \tag{9.15}$$

In practice, it may be difficult to get the expected values in (9.15). To solve this problem, the sigma-point extension technique is considered to expand the two sides of (9.15). More specifically, the following cubature rule is applied:

$$\begin{aligned}\underline{\boldsymbol{M}}_k &= \mathbb{E}((\boldsymbol{x}_{k+1} - \hat{\boldsymbol{x}}_{k+1})\boldsymbol{e}_{\boldsymbol{y}_{k+1/k}}^{\mathrm{T}}) \\ &\approx \sum_{i=1}^{L} W_i(\hat{\boldsymbol{x}}_{i,k+1} - \hat{\boldsymbol{x}}_{k+1})(\boldsymbol{y}_{k+1} - \hat{\boldsymbol{y}}_{i,k} - \boldsymbol{Z}_M(\hat{\boldsymbol{x}}_{i,k}))^{\mathrm{T}}\end{aligned} \tag{9.16}$$

$$\begin{aligned}\underline{\boldsymbol{N}}_k &= \mathbb{E}((\boldsymbol{x}_{k+1} - \hat{\boldsymbol{x}}_{k+1})(\boldsymbol{U}(\hat{\boldsymbol{x}}_k)\hat{\boldsymbol{d}}_k)^{\mathrm{T}}) \\ &\approx \sum_{i=1}^{L} W_i(\hat{\boldsymbol{x}}_{i,k+1} - \hat{\boldsymbol{x}}_{k+1})(\boldsymbol{U}(\hat{\boldsymbol{x}}_k)\hat{\boldsymbol{d}}_{i,k})^{\mathrm{T}}\end{aligned} \tag{9.17}$$

where $\underline{\boldsymbol{M}}_k \in \mathbb{R}^{m\times n}$ and $\underline{\boldsymbol{N}}_k \in \mathbb{R}^{m\times n}$. The sigma-point $\hat{\boldsymbol{x}}_{i,k}$, $i = 1, 2, \cdots, L$ is generated from its mean $\hat{\boldsymbol{x}}_k$ and covariance $\boldsymbol{P}_{\tilde{\boldsymbol{x}},k}$.

To this end, using the same procedures in Section 8.4.2, the fading factor $\boldsymbol{\lambda}_k$ can be calculated as

$$\lambda_k^i = \begin{cases} \alpha_i c_k, & \text{if} \quad \alpha_i c_k > 1 \\ 1, & \text{if} \quad \alpha_i c_k \leq 1 \end{cases}, i = 1, 2, \cdots, n \tag{9.18}$$

where

$$c_k = \frac{\mathrm{tr}(\underline{\boldsymbol{N}}_k)}{\sum_{i=1}^{p} \alpha_i \underline{\boldsymbol{M}}_k^{ii}} \tag{9.19}$$

Based on the proof in Zhou and Park (1996), it is got that the OPHVSF s inherent a strong estimation filter. The OPHVSF has great robustness and provide higher estimation accuracy than PHVSF in the presence of measurement noises and modeling errors. The pseudo code of the OPHVSF can be listed in the Algorithm 9.2. It is noted that the propagation of the state and covariance are similar to CKF without numerical integration. Hence, the OPHVSF requires almost the same computational burden as the CKF or the UKF.

Algorithm 9.2: Orthogonal PHVSF.

Input Data: Initial estimation $\hat{x}_0$, $\hat{y}_0$, and measurement y_k
Result: State estimation $\hat{x}_k$
begin
  Initializing the state and the covariance matrix as
    $\hat{x}_0 = \mathbb{E}(x_0)$
    $P_{\tilde{x},0} = \mathrm{Var}(x_0) = \mathbb{E}((x_0 - \hat{x}_0)(x_0 - \hat{x}_0)^{\mathrm{T}})$
  for $k = 0, 1, 2 \cdots$ **do**
    Applying the Cholesky factorization to the covariance $P_{\tilde{x},k}$
      $P_{\tilde{x},k} = S_k S_k^{\mathrm{T}} = \sum_{i=1}^{n} \sigma_{i,k} \sigma_{i,k}^{\mathrm{T}}$
    Establishing sigma points
      $\hat{x}_{i,k}, i = 1, 2, \cdots, L$
    Calculating the modeling error's estimation $\hat{d}_k$ by using
      $e_{y_{i,k+1/k}} = y_{k+1} - \hat{y}_{i,k} - Z(\hat{x}_{i,k})$
      $e_{y_{i,k}} = y_{k+1} - \hat{y}_{i,k}$
      $\Delta e_{y_{i,k}} = e_{y_{i,k}} - e_{y_{k-1}}$
      $\hat{d}_{i,k} =$

$$\lambda_k (U(\hat{x}_{i,k}))^{-1} \left(e_{y_{i,k+1/k}} - \frac{e_{y_{i,k}}}{2} - \sqrt{\frac{e_{y_{i,k}}^{\mathrm{T}} e_{y_{i,k}}}{4} + \frac{\left(\Delta e_{y_{i,k}}\right)^{\mathrm{T}} \Delta e_{y_{i,k}}}{2}} \gamma \right)$$

    Calculating the fading factor
      $\hat{x}_{i,k+1} = f(\hat{x}_{i,k}) + g(\hat{x}_{i,k}) \hat{d}_{i,k}$
      $\hat{x}_{k+1} = \sum_{i=1}^{L} W_i^m \hat{x}_{i,k+1}$
      $\underline{M}_k = \sum_{i=1}^{L} W_i (\hat{x}_{i,k+1} - \hat{x}_{k+1})(y_{k+1} - \hat{y}_{i,k} - Z_M(\hat{x}_{i,k}))^{\mathrm{T}}$
      $\underline{N}_k = \sum_{i=1}^{L} W_i (\hat{x}_{i,k+1} - \hat{x}_{k+1})(U(\hat{x}_k) \hat{d}_{i,k})^{\mathrm{T}}$
      Calculating λ_k^i using (9.18), $i = 1, 2, \cdots, p$
    Updating the state and the output estimation using
      $\hat{x}_{k+1} = f(\hat{x}_k) + g(\hat{x}_k) \hat{d}_k$
      $\hat{y}_k = h(\hat{x}_k)$
  end
end

9.4 Huber PHVSF design

The generalized maximum likelihood estimation theory invented by Peter J. Huber is a robust estimator (Huber, 1964, 2004). For this theory, the measurements and the state correction attained transforms into a linear regression problem (Gao et al., 2014). It is a robust method to solve the problem of symmetry disturbance near the Gaussian distribution. The corresponding filter is developed by using the hybrid cost function ℓ_1/ℓ_2 norm to solve the problem of interfering Gaussian distribution (Karlgaard and Schaub, 2007). This estimation method is effective to improve robustness and accuracy of the PHVSF. Hence, a roust version of the PHVSF call as the HPHVSF is developed by integrating the PHVSF and the generalized maximum likelihood estimation theory. When developing the HPHVSF, the measurement update should be re-taken as the regression problem between the estimation and the state prediction.

Let the state estimation error at the time $k+1$ be defined as

$$\boldsymbol{\varepsilon}_{x,k+1} = \boldsymbol{x}_{k+1} - \hat{\boldsymbol{x}}_{k+1} \tag{9.20}$$

where $\hat{\boldsymbol{x}}_{k+1}$ is the state estimation after updating the HPHVSF.

The measurement equation is linearized based on the estimated state $\hat{\boldsymbol{x}}_{k+1}$.

$$\boldsymbol{y}_{k+1} = \boldsymbol{h}(\hat{\boldsymbol{x}}_{k+1}) + \boldsymbol{H}_{k+1}\left(\boldsymbol{x}_{k+1} - \hat{\boldsymbol{x}}_{k+1}\right) + \boldsymbol{v}_{k+1} \tag{9.21}$$

where $\boldsymbol{H}_{k+1}$ is the Jacobian matrix of $\boldsymbol{h}(\cdot)$ with respect to the $\boldsymbol{x}_{k+1}$. Moreover, $\boldsymbol{H}_{k+1}$ can be calculated by using (8.50) or (8.51).

Based on (9.20) and (9.21), the nonlinear regression equations can be constructed as

$$\begin{bmatrix} \boldsymbol{y}_{k+1} - \boldsymbol{h}(\hat{\boldsymbol{x}}_{k+1}) + \boldsymbol{H}_{k+1}\hat{\boldsymbol{x}}_{k+1} \\ \hat{\boldsymbol{x}}_{k+1} \end{bmatrix} = \begin{bmatrix} \boldsymbol{H}_{k+1} \\ \boldsymbol{I}_n \end{bmatrix} \boldsymbol{x}_{k+1} + \begin{bmatrix} \boldsymbol{v}_{k+1} \\ -\boldsymbol{\varepsilon}_{x,k+1} \end{bmatrix} \tag{9.22}$$

or

$$\boldsymbol{\Xi}_{k+1} = \boldsymbol{\Gamma}_{k+1}\boldsymbol{x}_{k+1} + \boldsymbol{\varsigma}_{k+1} \tag{9.23}$$

where

$$\boldsymbol{\Xi}_{k+1} = \begin{bmatrix} \boldsymbol{R}_{k+1} & \boldsymbol{0} \\ \boldsymbol{0} & \boldsymbol{P}_{x,k+1} \end{bmatrix}^{-\frac{1}{2}} \begin{bmatrix} \boldsymbol{y}_{k+1} - \boldsymbol{h}(\hat{\boldsymbol{x}}_{k+1}) + \boldsymbol{H}_{k+1}\hat{\boldsymbol{x}}_{k+1} \\ \hat{\boldsymbol{x}}_{k+1} \end{bmatrix} \tag{9.24}$$

$$\boldsymbol{\Gamma}_{k+1} = \begin{bmatrix} \boldsymbol{R}_{k+1} & \boldsymbol{0} \\ \boldsymbol{0} & \boldsymbol{P}_{x,k+1} \end{bmatrix}^{-\frac{1}{2}} \begin{bmatrix} \boldsymbol{H}_{k+1} \\ \boldsymbol{I}_n \end{bmatrix} \tag{9.25}$$

$$\boldsymbol{\varsigma}_{k+1} = \begin{bmatrix} \boldsymbol{R}_{k+1} & \boldsymbol{0} \\ \boldsymbol{0} & \boldsymbol{P}_{\tilde{x},k+1} \end{bmatrix}^{-\frac{1}{2}} \begin{bmatrix} \boldsymbol{v}_{k+1} \\ -\boldsymbol{\varepsilon}_{x,k+1} \end{bmatrix} \tag{9.26}$$

Hence, the generalized maximum likelihood estimation theory (Huber, 1964, 2004) can be introduced by ensuring the following cost function to be minimum (Karlgaard and Schaub, 2011).

$$J(\boldsymbol{x}_{k+1}) = \sum_{i=1}^{m+n} \rho(\eta_{k+1,i}) \tag{9.27}$$

where $\varsigma_{k+1,i}$ is the ith component of $\boldsymbol{\varsigma}_{k+1}$, and the score function $\rho(\varsigma_{k+1,i})$ is chosen as

$$\varsigma(\eta_{k+1,i}) = \begin{cases} \frac{1}{2}\varsigma_{k+1,i}^2, & \text{if } \ |\varsigma_{k+1,i}| \le \lambda \\ -\lambda^2\left(1 + \frac{|\varsigma_{k+1,i}|}{\lambda}\right)\exp\left(1 - \frac{|\varsigma_{k+1,i}|}{\lambda}\right), & \text{if } \ |\varsigma_{k+1,i}| > \lambda \end{cases} \tag{9.28}$$

with an error threshold $\lambda \in \mathbb{R}_+$. When this method is applied to contaminated Gaussian or outlier measurements, the estimation minimizes the maximum asymptotic estimation variance. Robustness is guaranteed (Huber, 2004).

To solve the regression problem, one can differentiate (9.27) as

$$\begin{aligned} \frac{\partial J(\boldsymbol{x}_{k+1})}{\partial \boldsymbol{x}_{k+1}} &= \sum_{i=1}^{m+n} \frac{\rho(\varsigma_{k+1,i})}{\partial \varsigma_{k+1,i}} \frac{\partial \varsigma_{k+1,i}}{\partial \boldsymbol{x}_{k+1}} \\ &= \sum_{i=1}^{m+n} \psi(\varsigma_{k+1,i})\varsigma_{k+1,i} \frac{\partial \varsigma_{k+1,i}}{\partial \boldsymbol{x}_{k+1}} = \mathbf{0} \end{aligned} \tag{9.29}$$

where $\psi(\varsigma_{k+1,i})$ is derived with respect to the score function (9.28), respectively.

$$\psi(\varsigma_{k+1,i}) = \begin{cases} 1, & \text{if } \ |\varsigma_{k+1,i}| \le \lambda \\ \exp\left(1 - \frac{|\varsigma_{k+1,i}|}{\lambda}\right), & \text{if } \ |\varsigma_{k+1,i}| > \lambda \end{cases} \tag{9.30}$$

Applying the iteratively re-weighting method in Karlgaard and Schaub (2011), the solution of (9.29) can be obtained. Actually, by introducing a new matrix as

$$\boldsymbol{\psi} = \mathbf{diag}([\psi(\varsigma_{k+1,1}) \quad \psi(\varsigma_{k+1,2}) \quad \cdots \quad \psi(\varsigma_{k+1,m+n})]^{\mathrm{T}}) \tag{9.31}$$

the implicit equation can be rewritten into the following form

$$\boldsymbol{\Gamma}_{k+1}^{\mathrm{T}}\boldsymbol{\psi}(\boldsymbol{\Gamma}_{k+1}\boldsymbol{x}_{k+1} - \boldsymbol{\Xi}_{k+1}) = \mathbf{0} \tag{9.32}$$

Solving (9.32), one has

$$\boldsymbol{x}_{k+1} = (\boldsymbol{\Gamma}_{k+1}^{\mathrm{T}}\boldsymbol{\psi}\boldsymbol{\Gamma}_{k+1})^{-1}\boldsymbol{\Gamma}_{k+1}^{\mathrm{T}}\boldsymbol{\psi}\boldsymbol{\Xi}_{k+1} \tag{9.33}$$

Because the matrix $\boldsymbol{\psi}$ depends on the residual $\boldsymbol{\delta}_{k+1}$, the iterative solution can be derived as

$$\boldsymbol{x}_{k+1}^{(j+1)} = (\boldsymbol{\Gamma}_{k+1}^{\mathrm{T}} \boldsymbol{\psi}^{(j)} \boldsymbol{\Gamma}_{k+1})^{-1} \Gamma_{k+1}^{\mathrm{T}} \boldsymbol{\psi}^{(j)} \boldsymbol{\Xi}_{k+1} \tag{9.34}$$

where $\boldsymbol{\psi}^{(j)}$ is the iteration value at the jth step. The estimation error covariance can be computed by

$$\boldsymbol{P}_{\tilde{x},k+1} = (\boldsymbol{\Gamma}_{k+1}^{\mathrm{T}} \boldsymbol{\psi}^{(final)} \boldsymbol{\Gamma}_{k+1})^{-1} \tag{9.35}$$

with $\boldsymbol{\psi}^{(final)}$ being the final value of $\boldsymbol{\psi}$.

Remark 9.3. It should be stressed that $\lambda \to \infty$ ensures the matrix $\boldsymbol{\psi}$ converging to the identity matrix. The generalized maximum likelihood estimation method becomes the least squares estimator. Then, the HPHVSF becomes the original PHVSF. In contrast, if $\lambda \to \infty$, this method is going to be the absolute value estimator (Karlgaard and Schaub, 2007, 2011). The hybrid estimator provides the HPHVSF with robustness to the Gaussian distributed noise or the heavy-tailed noise by making full use of the residual error. Particularly, the large residuals are down-weighted by taking the inverse value of the residual magnitude (Li et al., 2014).

Since the generalized maximum likelihood estimation approach needs the covariance information, it cannot be provided by the HPHVSF. Therefore, the strategy of sigma-point extension is employed to update the covariance, when implementing the HPHVSF. In this book, the cubature rule in Chapter 6 applied to generate sigma-points and the other theories (such as the UT in Chapter 4 or the Stirling's polynomial interpolation in Chapter 5) are applicable in a similar way. Hence, the practical HPHVSF can be summarized in Algorithm 9.3.

Remark 9.4. The propagation of the state and covariance in are similar to the KF without the numerical integration. Moreover, the number of iterations only takes three to five times, which usually obtains the higher accuracy. Therefore, the calculation burden of proposed HPHVSF is equivalent to the traditional CKF or UKF. In addition, the error threshold λ is very important for filtering performance. Larger λ will lead to a faster rate of convergence. However, a smaller threshold value λ is usually chosen to guarantee that the system has the great robustness to the outlier measurements.

9.5 Application to attitude control system with low precision sensors

According to transformed system discussed in Section 5.5, it is known that the PHVSF in Algorithm 9.1 and the HPHVSF in Algorithm 9.3 can be applied to estimate the states of microsatellite attitude control system with a low-precision

Algorithm 9.3: Sigma point HPHVSF.

Input Data: Initial estimation $\hat{\boldsymbol{x}}_0$, $\hat{\boldsymbol{y}}_0$, and measurement $\boldsymbol{y}_k$
Result: State estimation $\hat{\boldsymbol{x}}_k$
begin
  Initializing the state and the covariance matrix as
  $\hat{\boldsymbol{x}}_0 = \mathbb{E}(\boldsymbol{x}_0)$
  $\boldsymbol{P}_{\tilde{x},0} = \text{Var}(\boldsymbol{x}_0) = \mathbb{E}((\boldsymbol{x}_0 - \hat{\boldsymbol{x}}_0)(\boldsymbol{x}_0 - \hat{\boldsymbol{x}}_0)^{\text{T}})$
  for $k = 1, 2, \cdots$ **do**
    Applying the Cholesky decomposition to error covariance $\boldsymbol{P}_{\tilde{x},k}$
    $$\boldsymbol{P}_{\tilde{x},k} = \boldsymbol{S}_k \boldsymbol{S}_k^{\text{T}} = \sum_{i=1}^{n} \boldsymbol{\sigma}_{i,k} \boldsymbol{\sigma}_{i,k}^{\text{T}}$$
    Establishing sigma points
    $$\boldsymbol{\xi}_{i,k} = \hat{\boldsymbol{x}}_k + \sqrt{n+\kappa}\,\boldsymbol{\sigma}_{i,k}, i = 1, 2, \cdots, n$$
    $$\boldsymbol{\xi}_{i+n,k} = \hat{\boldsymbol{x}}_k - \sqrt{n+\kappa}\,\boldsymbol{\sigma}_{i,k}, i = 1, 2, \cdots, n$$
    Updating measurement
    $$\boldsymbol{e}_{i,y_{k+1/k}} = \boldsymbol{y}_{k+1} - \hat{\boldsymbol{y}}_{i,k} - \boldsymbol{Z}(\hat{\boldsymbol{x}}_{i,k})$$
    $$\hat{\boldsymbol{y}}_{i,k} = \boldsymbol{h}(\hat{\boldsymbol{x}}_{i,k})$$
    $$\boldsymbol{e}_{i,y_k} = \boldsymbol{y}_k - \hat{\boldsymbol{y}}_{i,k}$$
    $$\Delta \boldsymbol{e}_{i,y_k} = \boldsymbol{e}_{i,y_k} - \boldsymbol{e}_{y_{k-1}}$$
    $$\hat{\boldsymbol{d}}_{i,k} = (\boldsymbol{U}(\hat{\boldsymbol{x}}_{i,k}))^{-1}\left(\boldsymbol{e}_{i,y_{k+1/k}} - \frac{\boldsymbol{e}_{i,y_k}}{2} - \sqrt{\frac{\boldsymbol{e}_{i,y_k}^{\text{T}} \boldsymbol{e}_{i,y_k}}{4} + \frac{(\Delta \boldsymbol{e}_{i,y_k})^{\text{T}} \Delta \boldsymbol{e}_{i,y_k}}{2}\boldsymbol{\gamma}}\right)$$
    Updating the priori state covariance
    $$\hat{\boldsymbol{x}}_{i,k+1} = \boldsymbol{f}(\hat{\boldsymbol{x}}_{i,k}) + \hat{\boldsymbol{d}}_{i,k}$$
    $$\hat{\boldsymbol{x}}_{k+1} = \frac{1}{2n}\sum_{i=1}^{2n} \hat{\boldsymbol{x}}_{i,k+1}$$
    $$\hat{\boldsymbol{y}}_{i,k+1} = \boldsymbol{h}(\hat{\boldsymbol{x}}_{i,k+1})$$
    $$\hat{\boldsymbol{y}}_{k+1} = \frac{1}{2n}\sum_{i=1}^{2n} \hat{\boldsymbol{y}}_{i,k+1}$$
    $$\boldsymbol{P'}_{k+1} = \frac{1}{2n}\sum_{i=1}^{2n} (\hat{\boldsymbol{x}}_{i,k+1} - \hat{\boldsymbol{x}}_{k+1})(\hat{\boldsymbol{x}}_{i,k+1} - \hat{\boldsymbol{x}}_{k+1})^{\text{T}}$$
    $$\boldsymbol{P}_{k+1}^{\boldsymbol{xy}} = \frac{1}{2n}\sum_{i=1}^{2n} (\hat{\boldsymbol{y}}_{i,k+1} - \hat{\boldsymbol{y}}_{k+1})(\hat{\boldsymbol{y}}_{i,k+1} - \hat{\boldsymbol{y}}_{k+1})^{\text{T}}$$
    $$\boldsymbol{H}_{k+1} = (\boldsymbol{P'}^{-1}_{k+1} \boldsymbol{P}_{k+1}^{\boldsymbol{xy}})^{\text{T}}$$
    Calculating (9.30)
    Updating the state estimation and the state estimation error covariance
    $$\boldsymbol{x}_{k+1}^{(j+1)} = (\boldsymbol{\Gamma}_{k+1}^{\text{T}} \boldsymbol{\psi}^{(j)} \boldsymbol{\Gamma}_{k+1})^{-1} \boldsymbol{\Gamma}_{k+1}^{\text{T}} \boldsymbol{\psi}^{(j)} \boldsymbol{\Xi}_{k+1}$$
    $$\boldsymbol{P}_{\tilde{x},k+1} = (\boldsymbol{\Gamma}_{k+1}^{\text{T}} \boldsymbol{\psi}^{(final)} \boldsymbol{\Gamma}_{k+1})^{-1}$$
  end
end

three-axis magnetometer and a low-precision sun sensor as measurement unit. To validate this application, numerical simulation is done with all the simulation parameters and the modeling error chosen as Section 5.5.3. The estimation gains $\alpha_i = 0.1$ and $\lambda = 0.1$ are chosen for the PHVSF and HPHVSF, $i = 1, 2, \cdots, 6$. Once again, $\boldsymbol{\Theta}_e = [\phi_e \quad \theta_e \quad \psi_e]^{\mathrm{T}}$ and $\boldsymbol{\omega}_E = [\omega_{Ex} \quad \omega_{Ey} \quad \omega_{Ez}]^{\mathrm{T}} = \tilde{\boldsymbol{\omega}}_{bi}$ defined in Remark 4.5 are adopted in this section to represent the estimation error of the attitude and the angular velocity, respectively.

In simulation, the initial states of the microsatellite control system and the initial estimation of the attitude is chosen as the same as Section 4.6.2 with $\Theta_0 =$ 5 degrees. The initial estimation of angular velocity is set as $[0 \quad 0 \quad 0]^{\mathrm{T}}$ deg/sec. The microsatellite attitude control system is assumed to under the effect of the modeling error defined in (4.119) with $\Delta\boldsymbol{N}_c = [-0.003 \quad -0.004 \quad 0.003]^{\mathrm{T}}$ Nm, $\Delta\boldsymbol{N}_e = [0.003\cos(\omega_0 t) \quad 0.001\cos(\omega_0 t) \quad -0.002\cos(\omega_0 t)]^{\mathrm{T}}$ Nm, where $\omega_0 = 0.66$ deg/sec. Regarding measurement noise of sun sensor and the magnetometer, the following three cases are considered.

9.5.1 Simulation result in the presence of Gaussian white noise

In this case, the measurement noise $\Delta\Theta_{i\to j}$ in (7.62) is assumed to be Gaussian white noises. The state estimation results obtained from the PHVSF, the HPHVSF, and the standard CKF are shown in Figs. 9.1–9.2. It is seen in Fig. 9.1 that HPHVSF achieves better attitude estimation with its accuracy higher than the CKF and the PHVSF. Fig. 9.2 shows that the PHVSF results in overshoot in the angular velocity estimation, while the other two filters are very stable around zero. However, the CKF leads to periodic deviation of the angular velocity es-

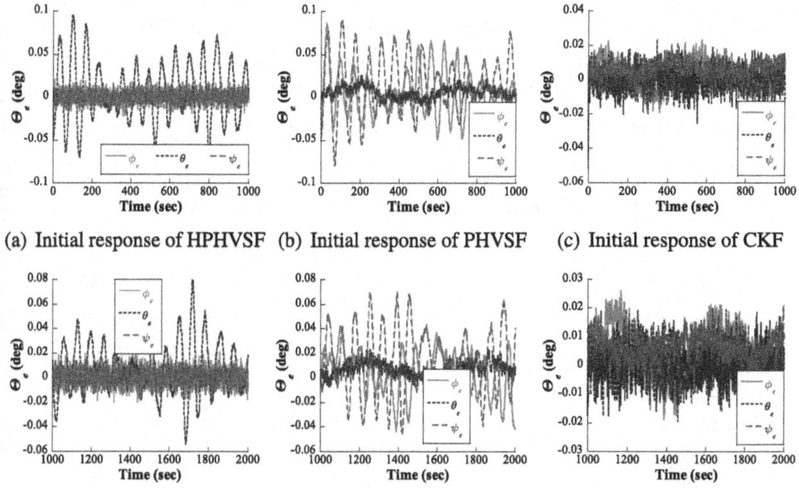

(a) Initial response of HPHVSF (b) Initial response of PHVSF (c) Initial response of CKF

(d) Stead error of HPHVSF (e) Stead error of PHVSF (f) Stead error of CKF

IGURE 9.1 Attitude estimation error using HPHVSF, PHVSF, and CKF in case of white noise.

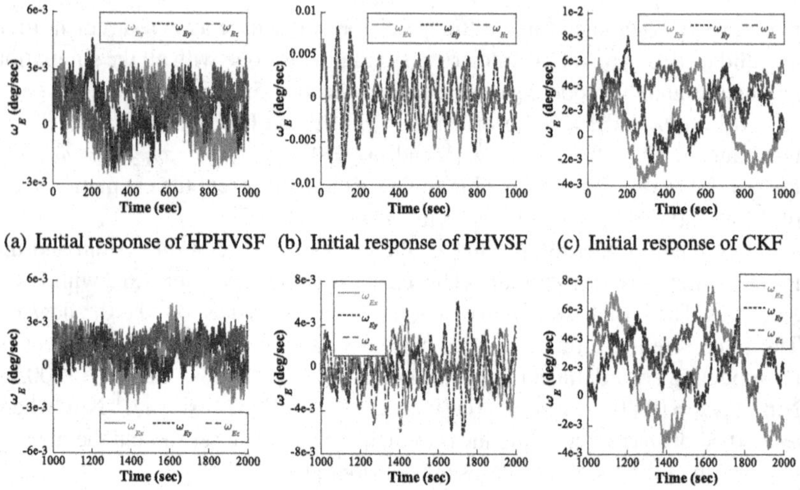

(a) Initial response of HPHVSF (b) Initial response of PHVSF (c) Initial response of CKF

(d) Stead error of HPHVSF (e) Stead error of PHVSF (f) Stead error of CKF

FIGURE 9.2 Angular velocity estimation error using HPHVSF, PHVSF, and CKF in case of white noise.

timation due to the modeling error. In contrast, the state estimation provided by the HPHVSF is more stable and can effectively attenuate thses fluctuations, because it has better capability of handling modeling error.

The estimated RSME comparison of the PHVSF, the HPHVSF, and the CKF is listed in Table 9.1. The PHVSF guarantees more accurate estimation than the CKF for the angular velocity state. On the contrary, it has worse performance in the attitude estimation compared with the CKF. However, the PHVSF demands less online calculation than the CKF, because it does not need to update the state estimate error covariance. In addition, the HPHVSF has better performance than the other two methods. The theoretical conclusions in Section 9.4 are validated

TABLE 9.1 RSME of state estimation error in case of white noise.

RSME	State estimation approaches		
	CKF	PSVSF	HPHVSF
ϕ_e (degrees)	0.0103	0.0181	0.0065
θ_e (degrees)	0.0090	0.0230	0.0065
ψ_e (degrees)	0.0077	0.0320	0.0053
ω_{Ex} (deg/sec)	0.0036	0.0014	0.0016
ω_{Ey} (deg/sec)	0.0031	0.0021	0.0014
ω_{Ez} (deg/sec)	0.0036	0.0023	0.0016

9.5.2 Simulation result in the presence of heavy-tailed noise

In this case, heavy-tailed measurement noise is assumed for $\Delta\Theta_{i\to j}$ in (7.62) with

$$\Delta\Theta_{i\to j} = \begin{cases} \mathcal{N}(\mathbf{0}, \boldsymbol{R}^0), \, w.p. \; 0.9 \\ \mathcal{N}(\mathbf{0}, 100\boldsymbol{R}^0), \, w.p. \; 0.1 \end{cases} \tag{9.36}$$

where $\boldsymbol{R}^0 = \frac{1}{8100}\boldsymbol{I}_3$. Moreover, the following heavy-tailed noise is considered into the attitude dynamics (3.1)–(3.2).

$$\boldsymbol{w}_f = \begin{cases} \mathcal{N}(\mathbf{0}, \boldsymbol{R}^1), \, w.p. \; 0.9 \\ \mathcal{N}(\mathbf{0}, 100\boldsymbol{R}^1), \, w.p. \; 0.1 \end{cases} \tag{9.37}$$

with $\boldsymbol{R}^1 = \mathbf{diag}([10^{-6} \quad 10^{-6} \quad 10^{-6} \quad 10^{-7} \quad 10^{-7} \quad 10^{-7}]^{\mathrm{T}})$.

The corresponding simulation results of this case are shown in Figs. 9.3–9.4. It is seen that the state estimation performance is inferior to the results in Subsection 9.5.1. This is induced by the heavy-tailed noise. However, the estimation accuracy of the HPHVSF is still better than that of the CKF and PHVSF, and its simulation trend is almost identical to that of the above white noise case. This is because the HPHVSF has the superior capability of updating the modeling error's estimation and applying the sigma-point extension strategy to improve the estimation accuracy. This endows the HPHVSF method with a significant advantage.

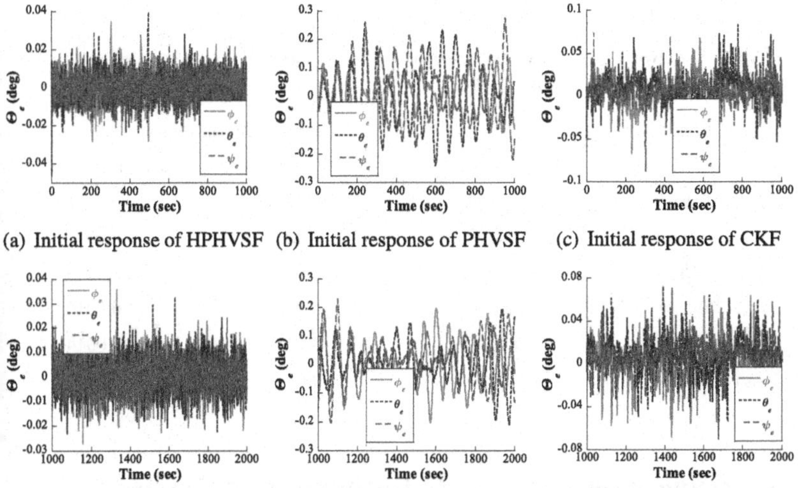

(a) Initial response of HPHVSF (b) Initial response of PHVSF (c) Initial response of CKF

(d) Stead error of HPHVSF (e) Stead error of PHVSF (f) Stead error of CKF

FIGURE 9.3 Attitude estimation error using HPHVSF, PHVSF, and CKF in the case of heavy-tailed noise case.

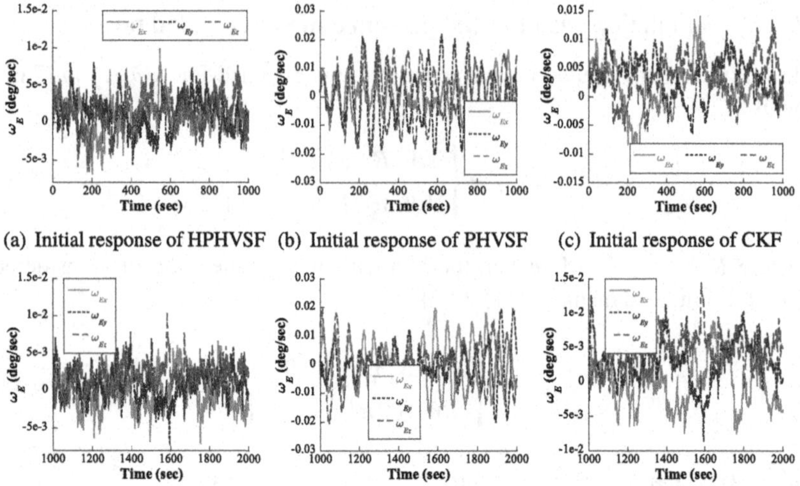

(a) Initial response of HPHVSF (b) Initial response of PHVSF (c) Initial response of CKF

(d) Stead error of HPHVSF (e) Stead error of PHVSF (f) Stead error of CKF

FIGURE 9.4 Angular velocity estimation error using HPHVSF, PHVSF, and CKF in the case of heavy-tailed noise case.

The estimation RSME comparison of this case is summarized in Table 9.2. It is seen that the accuracy of all the three filters is decreased due to the heavy-tailed noise, especially for the PHVSF and the CKF. The modeling error and the heavy-tailed noise are coupled to enlarge the system nonlinearity and uncertainty. Consequently, only the HPHVSF can still maintain the desirable convergence performance and guarantee high estimation accuracy.

TABLE 9.2 RSME of state estimation error in case of Gaussian white noise and the heavy-tailed noise case.

RSME	State estimation approaches		
	CKF	PSVSF	HPHVSF
ϕ_e (degrees)	0.0103	0.0181	0.0065
θ_e (degrees)	0.0090	0.0230	0.0065
ψ_e (degrees)	0.0077	0.0320	0.0053
ω_{Ex} (deg/sec)	0.0036	0.0014	0.0016
ω_{Ey} (deg/sec)	0.0031	0.0021	0.0014
ω_{Ez} (deg/sec)	0.0036	0.0023	0.0016

9.6 Application to distributed attitude control system

It is analyzed in Section 7.7 that the microsatellite distributed attitude system can be transformed into the nonlinear system (2.32)–(2.33). Hence, it is con cluded that the PHVSF in Algorithm 9.1 and the OPHVSF in Algorithm 9.2 are applicable to solve the state estimation problem of microsatellite distributed attitude system. Based on this conclusion, numerical simulations are conducted to demonstrate their effectiveness in comparison with the standard CKF. Except the measurement noise, other simulation parameters are chosen as same as Subsection 7.7.3. To validate the effectiveness and capability of the designed filters to handle different type of measurement noises, the following two noises cases are considered in the simulation.

1) Scenario #1: The measurement noise $\Delta\Theta_{i\to j}$ in (7.62) is assumed to be Gaussian white noises.

2) Scenario #2: Heavy-tailed measurement noise is considered with

$$\Delta\Theta_{i\to j} = \begin{cases} \mathcal{N}(\mathbf{0}, \boldsymbol{R}_{ij}^0), \, w.p. \ 0.9 \\ \mathcal{N}(\mathbf{0}, 100\boldsymbol{R}_{ij}^0), \, w.p. \ 0.1 \end{cases} \tag{9.38}$$

where $\boldsymbol{R}_{ij}^0 = \frac{\pi^2}{8100}\boldsymbol{I}_3$, $i, j = 0, 1, 2, \cdots, 3$. Moreover, the following heavy-tailed noise is considered into the distributed attitude dynamics (3.190)–(3.191):

$$\boldsymbol{w}_{i\to j} = \begin{cases} \mathcal{N}(\mathbf{0}, \boldsymbol{R}_k^1), \, w.p. \ 0.9 \\ \mathcal{N}(\mathbf{0}, 100\boldsymbol{R}_k^1), \, w.p. \ 0.1 \end{cases} \tag{9.39}$$

with $\boldsymbol{R}_k^1 = \mathbf{diag}([10^{-6} \quad 10^{-6} \quad 10^{-6} \quad 10^{-7} \quad 10^{-7} \quad 10^{-7}]^{\mathrm{T}})$ and $k = 1, 2, 3$.

Remark 9.5. As pointed out in Remark 7.6, $\boldsymbol{\Theta}_e = [\phi_{eij} \quad \theta_{eij} \quad \psi_{eij}]^{\mathrm{T}}$ and $\boldsymbol{\omega}_E = [\omega_{Exij} \quad \omega_{Eyij} \quad \omega_{Ezij}]^{\mathrm{T}} = \tilde{\boldsymbol{\omega}}_{i\to j}$ are used to denote the estimation error of the relative attitude state and the relative angular velocity state, respectively.

9.6.1 Simulation results

When implementing the PHVSF and the OPHVSF in simulations, the estimation gains of those two filters are chosen as $\alpha_i = 0.1$, $i = 1, 2, \cdots, 6$. For Scenario #1, the estimation error of the relative attitude is shown in Fig. 9.5. Some deviations and fluctuations are observed for all the filters due to modeling error. However, the fluctuation resulted from the CKF is more than the PHVSF and the OPHVSF. The OPHVSF ensures the smallest deviation. The relative angular velocity estimation error is illustrated in Fig. 9.6. Serious deviation and fluctuation in the relative angular velocity estimation are also witnessed for the CKF. In contrast, the estimates of the PHVSF and the OPHVSF converge to a small set of the true relative angular velocity. Regarding the relative angular velocity estimation, the proposed PHVSF and the OPHVSF can provider higher estimation accuracy than the CKF, since the PHVSF and the OPHVSF have better capability than the CKF to handle the modeling errors.

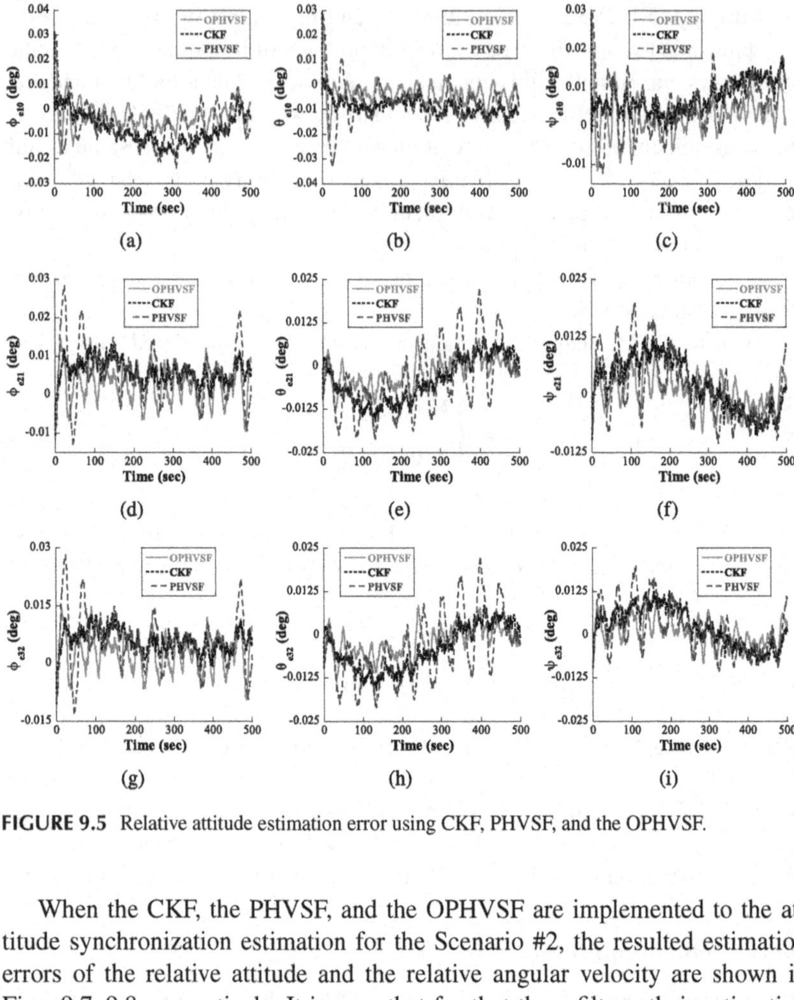

FIGURE 9.5 Relative attitude estimation error using CKF, PHVSF, and the OPHVSF.

When the CKF, the PHVSF, and the OPHVSF are implemented to the attitude synchronization estimation for the Scenario #2, the resulted estimation errors of the relative attitude and the relative angular velocity are shown in Figs. 9.7–9.8, respectively. It is seen that for that three filters, their estimation performance in Scenario #2 is inferior to the results in Scenario #1. This is due to the heavy-tailed noise considered in the Scenario #2. However, the estimation accuracy of the proposed PHVSF, and the OPHVSF is still better than the accuracy ensured by the CKF. That is because the PHVSF, and the OPHVSF has capability of handling the modeling error and heavy-tailed noise simultaneously while the CKF does not have this capability.

9.6.2 Quantitative analysis

To quantitatively evaluate the estimation accuracy obtained from the CKF, the PHVSF, and the OPHVSF, the root-square-mean-error (RSME) of both the relative attitude and the relative angular velocity estimation error. Moreover

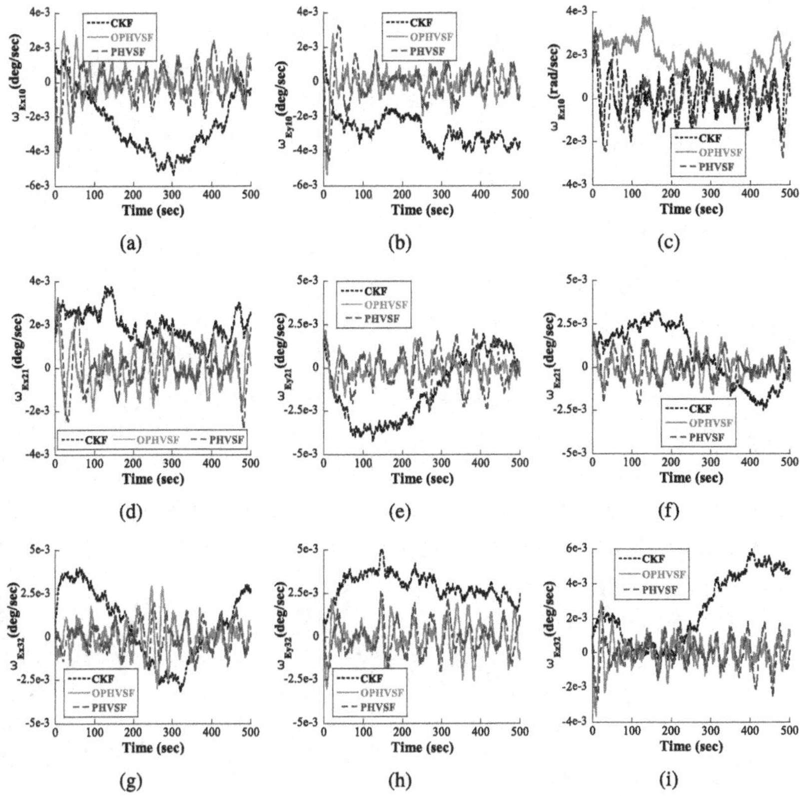

FIGURE 9.6 Relative angular velocity estimation error using CKF, PHVSF, and the OPHVSF.

comparison between the presented filters is also done by using the PSVSF in Chapter 8. The results of Scenarios #1–#2 are listed in Tables 9.3–9.4.

1) For Scenario #1, it is seen in Table 9.3 and Table 9.4 that the attitude es- timation accuracy ensured by the CKF, the PHVSF, and the OPHVSF is almost he same. The OPHVSF provides better attitude estimation accuracy than the PHVSF, the PSVSF, and the CKF. In Table 9.4, we see that the relative angular velocity estimation accuracy ensured by the PHVSF and the OPHVSF is almost he same, but higher than the CKF and the PSVSF; however, the angular velocity stimation accuracy ensured by PSVSF is better than the CKF. The conclusion hat the proposed two filters have better capability than the CKF and the PSVSF o handle modeling errors is hence verified.

2) For Scenario #2, the RSME of the relative attitude and the relative angu- ar velocity estimation error are listed in Table 9.5 and Table 9.6, respectively. The accuracy achieved by those four filters is reduced due to the heavy-tailed oise in Scenario #2. When the modeling error and the heavy-tailed noise are onsidered simultaneously, the system uncertainties will be increased. This will

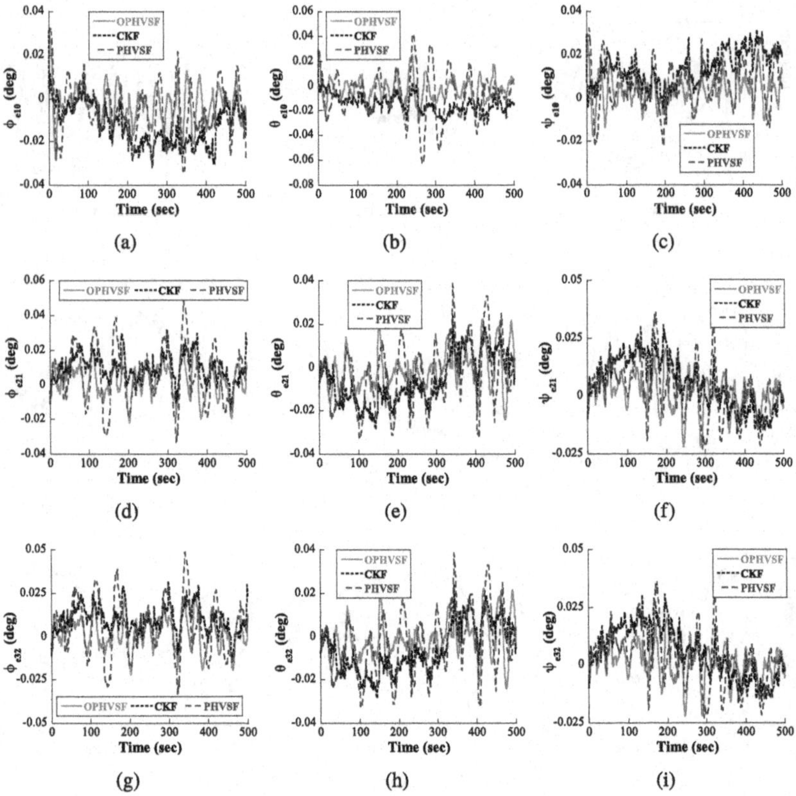

FIGURE 9.7 Relative attitude estimation error using CKF, PHVSF, and the OPHVSF.

deteriorate the filtering performance. Nevertheless, the OPHVSF and PHVSF maintain the desired estimation accuracy. Moreover, the OPHVSF still ensures higher estimation precision. The conclusion that the OPHVSF ensures good estimation result despite modeling error and heavy-tailed noise is validated.

Note that the Kalman filter requires the exact level of noise error including the modeling error and the measurement noise covariance to ensure high state estimation accuracy. This requirement may not be satisfied in practice. Moreover, when the conditions are abrupt or serious, the adaptive capacity and estimation performance of the KF will be seriously deteriorated. However, the proposed PHVSF and OPHVSF are independent of the exact information of modeling error or measurement noise. To this end, summarizing the above analysis, it can conclude that the developed filter has superior estimation performance than the existing approaches such as the KF and the CKF when the unknown modeling error and the heavy-tailed noise are considered. Moreover, the computation resource demanded by the PHVSF and OPHVSF is almost the same; however, they are less than the CKF.

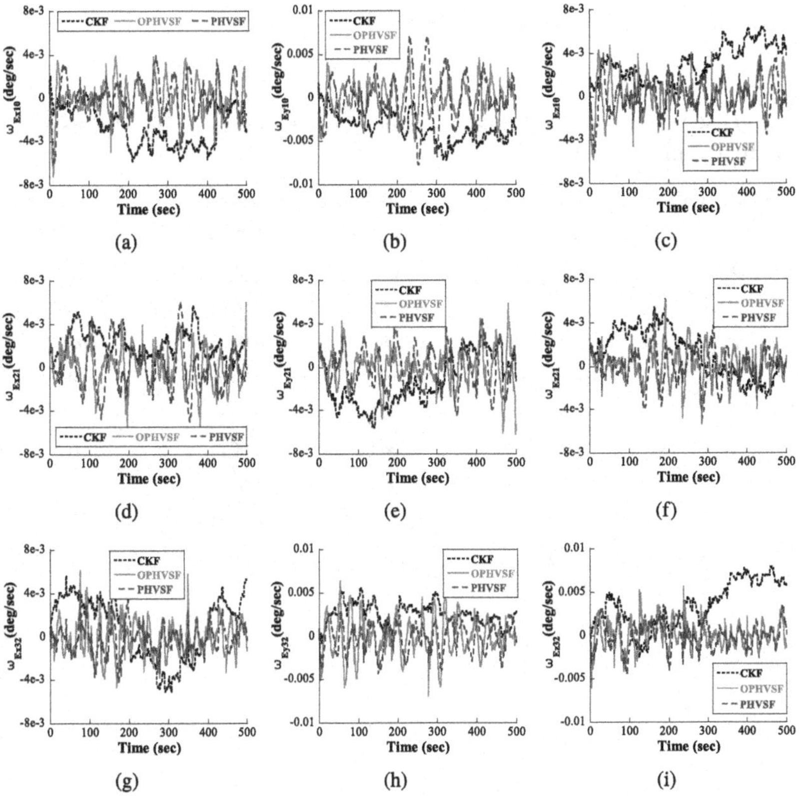

FIGURE 9.8 Relative angular velocity estimation error using CKF, PHVSF, and the OPHVSF.

9.7 Summary

This paper presented a novel approach to the state estimation problem of nonlinear systems subject to modeling error and heavy-tailed noise. A high-order variable structure predictive filtering scheme was developed. This method did not necessitate the assumption that the measurement noise should be the Gaussian white type. It was applicable to the state estimation problem of systems and the microsatellite control system with any measurement noise. Better estimation accuracy was ensured with the measurement error and its difference canceled simultaneously. In comparison with the variable structure-based predictive filters developed in Chapters 7–8, the prior estimation error was fully utilized to develop the high-order variable structure predictive filter and its robust version. Hence, they guaranteed higher estimation accuracy than the filters in Chapters 7–8. Moreover, the proposed filter and its robust version did not have chattering phenomenon. The shortcoming of the filters in Chapters 7–8 was avoided.

TABLE 9.3 RSME of the relative attitude estimation error from CKF, PSVSF, PHVSF, and OPHVSF in Scenario #1.

RSME (rad)	State estimation approaches			
	CKF	PSVSF	PHVSF	OPHVSF
ϕ_{e10}	0.0095	0.0098	0.0098	0.0057
θ_{e10}	0.0093	0.0093	0.0092	0.0053
ψ_{e10}	0.0095	0.0085	0.0083	0.0049
ϕ_{e21}	0.0070	0.0100	0.0100	0.0047
θ_{e21}	0.0073	0.0092	0.0089	0.0053
ψ_{e21}	0.0057	0.0064	0.0079	0.0047
ϕ_{e32}	0.0092	0.0090	0.0078	0.0045
θ_{e32}	0.0108	0.0107	0.0104	0.0059
ψ_{e32}	0.0102	0.0098	0.0094	0.0053

TABLE 9.4 RSME of relative angular velocity estimation error from CKF, PSVSF, PHVSF, and OPHVSF in Scenario #1.

RSME (rad/sec)	State estimation approaches			
	CKF	PSVSF	PHVSF	OPHVSF
ω_{Ex10}	0.002900	0.001005	0.000894	0.000787
ω_{Ey10}	0.002800	0.001200	0.000785	0.000679
ω_{Ez10}	0.002800	0.001400	0.000789	0.000653
ω_{Ex21}	0.002100	0.001800	0.001030	0.000730
ω_{Ey21}	0.002100	0.001600	0.000655	0.000632
ω_{Ez21}	0.001700	0.001000	0.000868	0.000747
ω_{Ex32}	0.002900	0.001500	0.000661	0.000638
ω_{Ey32}	0.003300	0.001400	0.000904	0.000812
ω_{Ez32}	0.003200	0.001200	0.001130	0.000760

TABLE 9.5 RSME of the relative attitude estimation error from CKF, PSVSF, PHVSF, and OPHVSF in Scenario #2.

RSME (rad)	State estimation approaches			
	CKF	PSVSF	PHVSF	OPHVSF
ϕ_{e10}	0.0175	0.0152	0.0139	0.0089
θ_{e10}	0.0166	0.0160	0.0157	0.0087
ψ_{e10}	0.0163	0.0152	0.0147	0.0094
ϕ_{e21}	0.0160	0.0150	0.0151	0.0077
θ_{e21}	0.0145	0.0140	0.0133	0.0077
ψ_{e21}	0.0213	0.0200	0.0182	0.0073
ϕ_{e32}	0.0196	0.0180	0.0171	0.0099
θ_{e32}	0.0184	0.0170	0.0157	0.0089
ψ_{e32}	0.0180	0.0160	0.0114	0.0076

TABLE 9.6 RSME of relative angular velocity estimation error from CKF, PSVSF, PHVSF, and OPHVSF in Scenario #2.

RSME (rad/sec)	State estimation approaches			
	CKF	PSVSF	PHVSF	OPHVSF
ω_{Ex10}	0.0040	0.0034	0.0017	0.0014
ω_{Ey10}	0.0038	0.0033	0.0019	0.0015
ω_{Ez10}	0.0030	0.0027	0.0021	0.0016
ω_{Ex21}	0.0032	0.0029	0.0021	0.0013
ω_{Ey21}	0.0031	0.0025	0.0014	0.0012
ω_{Ez21}	0.0019	0.0019	0.0020	0.0015
ω_{Ex32}	0.0035	0.0032	0.0023	0.0013
ω_{Ey32}	0.0041	0.0029	0.0018	0.0015
ω_{Ez32}	0.0042	0.0030	0.0019	0.0015

Chapter 10

Conclusion and future work

10.1 General conclusion

This book mainly investigated the state estimation problem of the microsatellite control system with severe modeling error and measurement noise. To achieve this objective, taking advantages of the predictive filtering approach, several advanced predictive filters were developed to estimate states of a general class of nonlinear discrete systems. Those filters were classified into two types. The first type was the sigma-point predictive filters in Chapters 4–6. The other type was the predictive variable structure filters in Chapters 7–9. In general, these two types of filters eliminated the drawbacks of the conventional Kalman filters and the classical predictive filters. Applying those new predictive filters to the microsatellite control system, they estimated its states with high-accuracy and great robustness to modeling error and sensor noises. The detailed features and advantages of those developed filters was highlighted as follows.

The sigma-point predictive filters in Chapters 4–6 were developed by using the unscented transformation, the Stirling polynomial interpolation technique, and the cubature rules, respectively. Those three filters guaranteed that for nonlinear Gaussian systems, the incorporated estimation of modeling error captured the posterior mean accurately up to the 3rd order; for nonlinear non-Gaussian systems, the estimation of modeling error captured the posterior mean accurately to the 2nd order. This estimation accuracy was higher than the classical predictive filter. It was proved that if initial estimation, measurement noise, and modeling error were bounded within a specific value, then the state estimation was governed to be stable and bounded. In comparison with the classical predictive filter and the Kalman filtering theory, the proposed sigma-point predictive filters achieved higher estimation accuracy for states and modeling error. They ensured more robustness to modeling error and measurement noise. Moreover, they were less dependent on the weighting matrix with more freedom to choose such a matrix.

The predictive variable structure filters in Chapters 7–9 were designed by using the variable structure control theory, the adaptive control theory, and the high-order variable structure control, respectively. They broke the limitation of the Kalman filtering theory in the assumption that the noise should be Gaussian white noise. Those predictive variable structure filtering approaches were applicable to achieve a high accuracy state estimation for nonlinear systems subject to any type of measurement noises. The estimation performance ensured was more

Predictive Filtering for Microsatellite Control System. https://doi.org/10.1016/B978-0-12-821865-5.00025-5

robust to modeling error, external disturbance, and unmodeled dynamics. Moreover, the drawback of the classical predictive filtering theory that its estimation performance partially depends on the weighting matrix was eliminated. The implementation of the presented predictive variable structure filtering method did not necessitate the computation of the state estimation error covariance. Hence, another main advantage of this method was that expensive computation was not involved.

Another main contribution of this book was that a high-precision mathematical model of the relative position control system of two formation flying microsatellites in an elliptical orbit was established. In comparison with the existing linear models, this model explicitly considered the effect of the J_2 and the atmospheric drag perturbations. Moreover, in comparison with the existing relative position control model, the relative position and velocity were the states of the presented model. This lets this model be more convenient and appropriate for state estimation and control system design.

10.2 Future work

All the filters in this book were developed based on the predictive filtering theory. It aimed to avoid the drawbacks of the classical predictive filters by using advanced mathematical tools. Although those filtering approaches have great robustness to modeling error and measurement noise, they also have some weaknesses. In the near future, new filters could be developed by integrating the designed predictive filters with the other filtering methods, which is capable of handling the weakness of the filters presented in this book.

The real-time property and the computation complexity of the predictive filter synthesized in this book were qualitatively analyzed only. In order to guarantee that they are applicable to solve the state estimation problem of the microsatellite control system in engineering, quantitative analysis should be carried out to evaluate the performance of those filters. Moreover, their practical implementation in embedded computer should be done and tested in real in-orbital microsatellite.

The sigma-point predictive filters in Part II and the predictive variable structure filters in Part III were developed based on system model. They were model-based and model-dependent. It is known that system model is currently and will never be precisely established. Hence, the limitation of the model-based filters will always exist. In the future, with the help of intelligent techniques, data-driven intelligent model-free state estimation approach could be developed to essentially solve this problem.

References

Abedor, J., Nagpal, K., Poolla, K., 1996. A linear matrix inequality approach to peak-to-peak gain minimization. International Journal of Robust and Nonlinear Control 6 (9–10), 899–927.

Adnane, A., Foitih, Z.A., Mohammed, M.A.S., Bellar, A., 2018. Real-time sensor fault detection and isolation for LEO satellite attitude estimation through magnetometer data. Advances in Space Research 61 (4), 1143–1157.

Agniel, R.G., Jury, E.I., 1971. Almost sure boundedness of randomly sampled systems. SIAM Journal on Control 9 (3), 372–384.

Ahmed-Ali, T., Lamnabhi-Lagarrigue, F., 1999. Sliding observer-controller design for uncertain triangular nonlinear systems. IEEE Transactions on Automatic Control 44 (6), 1244–1249.

Al-Shabi, M., Gadsden, S.A., Habibi, S.R., 2013. Kalman filtering strategies utilizing the chattering effects of the smooth variable structure filter. Signal Processing 93 (2), 420–431.

Ali, I., Radice, G., Kim, J., 2010. Backstepping control design with actuator torque bound for spacecraft attitude maneuver. Journal of Guidance, Control, and Dynamics 33 (1), 254–259.

Arasaratnam, I., Haykin, S., 2009. Cubature Kalman filters. IEEE Transactions on Automatic Control 54 (6), 1254–1269.

Banavar, R.N., Speyer, J.L., 1991. A linear-quadratic game approach to estimation and smoothing. In: American Control Conference, pp. 2818–2822.

Bar-Itzhack, I.Y., Deutschmann, J., Markley, F., 1991. Quaternion normalization in additive EKF for spacecraft attitude determination. In: Navigation and Control Conference, pp. 91–2706.

Bartolini, G., Ferrara, A., Usai, E., 1998. Chattering avoidance by second-order sliding mode control. IEEE Transactions on Automatic Control 43 (2), 241–246.

Battin, R.H., 1964. Astronautical Guidance. (Book on astronautical guidance covering celestial mechanics and navigation, two-body orbital transfer, perturbation methods and guidance theory) McGraw-Hill Book CO.

Bayes, T., 1763. An essay towards solving a problem in the doctrine of chances. Philosophical Transactions of the Royal Society of London 53, 370–418.

Baziw, E., 2005. Real-time seismic signal enhancement utilizing a hybrid Rao-Blackwellized particle filter and hidden Markov model filter. IEEE Geoscience and Remote Sensing Letters 2 (4), 418–422.

Berkane, S., Abdessameud, A., Tayebi, A., 2018. Hybrid output feedback for attitude tracking on SO(3). IEEE Transactions on Automatic Control 63 (11), 3956–3963.

Bhaumik, S., Swati, 2013. Cubature quadrature Kalman filter. IET Signal Processing 7 (7), 533–541.

Bierman, G.J., 1974. Sequential square root filtering and smoothing of discrete linear systems. Automatica 10 (2), 147–158.

Bucy, R.S., Senne, K.D., 1971. Digital synthesis of non-linear filters. Automatica 7 (3), 287–298.

Cao, L., Misra, A.K., 2015. Linearized J_2 and atmospheric drag model for satellite relative motion with small eccentricity. Proceedings of the Institution of Mechanical Engineers. Part G, Journal of Aerospace Engineering 229 (14), 2718–2736.

Carlson, N.A., 1973. Fast triangular formulation of the square root filter. AIAA Journal 11 (9), 1259–1265.

Carlson, N.A., 1990. Federated square root filter for decentralized parallel processors. IEEE Transactions on Aerospace and Electronic Systems 26 (3), 517–525.

Carmi, A., Oshman, Y., 2006. Vector observations-based gyroless spacecraft attitude/angular rate estimation using particle filtering. In: AIAA Guidance, Navigation, and Control Conference and Exhibit, pp. 21–24.

Carmi, A., Oshman, Y., 2007. Robust spacecraft angular rate estimation from vector observations using interlaced particle filtering. Journal of Guidance, Control, and Dynamics 30 (6), 1729–1741.

Carrington, C.K., Junkins, J.L., 1986. Optimal nonlinear feedback control for spacecraft attitude maneuvers. Journal of Guidance, Control, and Dynamics 9 (1), 99–107.

Carter, T., Humi, M., 2002a. Clohessy-Wiltshire equations modified to include quadratic drag. Journal of Guidance, Control, and Dynamics 25 (6), 1595–1607.

Carter, T., Humi, M., 2002b. Rendezvous equations in a central-force field with linear drag. Journal of Guidance, Control, and Dynamics 25 (1), 74–79.

Chalanga, A., Kamal, S., Fridman, L.M., Bandyopadhyay, B., Moreno, J.A., 2016. Implementation of super-twisting control: super-twisting and higher order sliding-mode observer-based approaches. IEEE Transactions on Industrial Electronics 63 (6), 3677–3685.

Chandra, K.R.B., Gu, D.W., 2019. Nonlinear Filtering Methods and Application. Springer.

Chang, L.B., Zha, F., Qin, F.J., 2017. Indirect Kalman filtering based attitude estimation for low-cost attitude and heading reference systems. IEEE/ASME Transactions on Mechatronics 22 (4), 1850–1858.

Chen, X., Liu, M., 2017. A two-stage extended Kalman filter method for fault estimation of satellite attitude control systems. Journal of the Franklin Institute 354 (2), 872–886.

Chen, Y.P., Lo, S.C., 1993. Sliding mode controller design for spacecraft attitude tracking maneuvers. IEEE Transactions on Aerospace and Electronic Systems 29 (4), 1328–1333.

Chen, B.S., Wu, C.S., Jan, Y.W., 2000. Adaptive fuzzy mixed $\mathcal{H}_2/\mathcal{H}_\infty$ attitude control of spacecraft. IEEE Transactions on Aerospace and Electronic Systems 36 (4), 1343–1359.

Cheng, Y., Crassidis, J.L., 2004. Particle filtering for sequential spacecraft attitude estimation. In: AIAA Guidance, Navigation, and Control Conference and Exhibit.

Cheng, Y., Crassidis, J.L., 2010. Particle filtering for attitude estimation using a minimal local-error representation. Journal of Guidance, Control, and Dynamics 33 (4), 1305–1310.

Cho, H.C., Park, S.Y., 2009. Analytic solution for fuel-optimal reconfiguration in relative motion Journal of Optimization Theory and Applications 141 (3), 495–512.

Cilden-Guler, D., Raitoharju, M., Piche, R., Hajiyev, C., 2019. Nanosatellite attitude estimation using Kalman-type filters with non-Gaussian noise. Aerospace Science and Technology 92 66–76.

Clohessy, W.H., Wiltshire, R.S., 1960. Terminal guidance system for satellite rendezvous. Journal of the Aerospace Sciences 27 (9), 653–658.

Costic, B.T., Dawson, D.M., Kapila, V., 2001. Quaternion-based adaptive attitude tracking controller without velocity measurements. Journal of Guidance, Control, and Dynamics 24 (6), 1214–1222.

Crassidis, J.L., 1999. Efficient and optimal attitude determination using model-error control synthesis. Journal of Guidance, Control, and Dynamics 22, 193–201.

Crassidis, J.L., Alonso, R., Junkins, J.L., 2000a. Optimal attitude and position determination from line-of-sight measurements. The Journal of the Astronautical Sciences 48 (2), 391–408.

Crassidis, J.L., Markley, F.J., 1996a. Sliding mode control using modified Rodrigues parameters Journal of Guidance, Control, and Dynamics 19 (6), 1381–1383.

Crassidis, J.L., Markley, F.L., 1996b. Attitude estimation using modified Rodrigues parameter Journal of Astronautics, 71–83.

Crassidis, J.L., Markley, F., 1997a. A minimum model error approach for attitude estimation. Journal of Guidance, Control, and Dynamics 20 (6), 1241–1247.

Crassidis, J.L., Markley, F.L., 1997b. Predictive filtering for attitude estimation without rate sensor Journal of Guidance, Control, and Dynamics 20 (3), 522–527.

Crassidis, J.L., Markley, F.L., 1997c. Predictive filtering for nonlinear systems. Journal of Guidance, Control, and Dynamics 20 (3), 566–572.

Crassidis, J.L., Markley, F.L., 2003. Unscented filtering for spacecraft attitude estimation. Journal of Guidance, Control, and Dynamics 26 (4), 536–542.

Crassidis, J.L., Mason, P.A.C., Mook, D.J., 1993. Riccati solution for the minimum model error algorithm. Journal of Guidance, Control, and Dynamics 16 (6), 1181–1183.

Crassidis, J.L., Vadali, S.R., Markley, F.L., 2000b. Optimal variable structure control tracking of spacecraft maneuvers. Journal of Guidance, Control, and Dynamics 23 (3), 564–566.

Crisan, D., Doucet, A., 2002. A survey of convergence results on particle filtering methods for practitioners. IEEE Transactions on Signal Processing 50 (3), 736–746.

Daly, J.M., Wang, D.W., 2009. Output feedback sliding mode control in the presence of unknown disturbances. Systems & Control Letters 58 (3), 188–193.

Dang, Z.H., Zhang, Y.L., 2012. Formation control using μ-synthesis for inner-formation gravity measurement satellite system. Advances in Space Research 49 (10), 1487–1505.

Davila, J., Fridman, L., Levant, A., 2005. Second-order sliding-mode observer for mechanical systems. IEEE Transactions on Automatic Control 50 (11), 1785–1789.

Doucet, A., Godsill, S., Andrieu, C., 2000. On sequential Monte Carlo sampling methods for Bayesian filtering. Statistics and Computing 10 (3), 197–208.

Du, H., Li, S., 2012. Finite-time attitude stabilization for a spacecraft using homogeneous method. Journal of Guidance, Control, and Dynamics 35 (3), 740–748.

Egeland, O., Godhavn, J.M., 1994. Passivity-based adaptive attitude control of a rigid spacecraft. IEEE Transactions on Automatic Control 39 (4), 842–846.

Elsayed, A., Grimble, M.J., 1989. A new approach to the H_∞ design of optimal digital linear filters. IMA Journal of Mathematical Control and Information 6 (2), 233–251.

Farza, M., M'Saad, M., Triki, T., Maatoug, T., 2011. High gain observer for a class of non-triangular systems. Systems & Control Letters 60 (1), 27–35.

Feng, Y., Yu, X., Han, F., 2013. High-order terminal sliding-mode observer for parameter estimation of a permanent-magnet synchronous motor. IEEE Transactions on Industrial Electronics 60 (10), 4272–4280.

Fialho, I.J., Georgiou, T.T., 1995. On the $\mathcal{L}_1$ norm of uncertain linear systems. In: Proceedings of the 1995 American Control Conference, vol. 1, pp. 939–943.

Fisher, R.A., 1912. On an absolute criterion for fitting frequency curves. Messenger of Mathematics 41, 44–155.

Floquet, T., Edwards, C., Spurgeon, S.K., 2007. On sliding mode observers for systems with unknown inputs. International Journal of Adaptive Control and Signal Processing 21 (8–9), 638–656.

Fox, D., 2002. KLD-sampling: adaptive particle filters. In: Advances in Neural Information Processing Systems, pp. 713–720.

Francis, B., 2012. Satellite Formation Maintenance Using Differential Atmospheric Drag. McGill University.

Fridman, E., Shaked, U., Xie, L., 2002. Robust $\mathcal{H}_2$ filtering of linear systems with time delays. In: Proceedings of the 41st IEEE Conference on Decision and Control, vol. 4, pp. 3877–3882.

Froberg, C.E., 1969. Introduction to Numerical Analysis. Addison-Wesley.

Fuh, C.C., 2008. Variable-thickness boundary layers for sliding mode control. Journal of Marine Science and Technology 16 (4), 288–294.

Gao, W., Liu, Y.L., Xu, B., 2014. Robust Huber-based iterated divided difference filtering with application to cooperative localization of autonomous underwater vehicles. Sensors 24 (12), 24523–24542.

Gao, H.J., Wang, C.H., 2003a. New approach to robust $\mathcal{L}_2$-$\mathcal{L}_\infty$ filter design for uncertain continuous-time systems. Acta Automatica Sinica 29, 809–814.

Gao, H.J., Wang, C.H., 2003b. New approaches to robust $\mathcal{L}_2$-$\mathcal{L}_\infty$ and $\mathcal{H}_\infty$ filtering for uncertain discrete-time systemsfiltering for uncertain discrete-time systems. Science in China. Series F. Information Sciences 46 (5), 355–358.

Gao, H.J., Wang, C.H., Li, Y.H., 2003. Robust $\mathcal{L}_2$-$\mathcal{L}_\infty$ filter design for uncertain discrete-time state-delayed systems. Acta Automatica Sinica 29 (5), 666–672.

Garcia, R.V., Pardal, P.C.P.M., Kuga, H.K., Zanardi, M.C., 2019. Nonlinear filtering for sequential spacecraft attitude estimation with real data: cubature Kalman filter, unscented Kalman filter and extended Kalman filter. Advances in Space Research 63 (2), 1038–1050.

Gauss, C.F., 1809. Theoria motus corporum coelestium in sectionibus conicis solem ambientium, vol. 7. Perthes et Besser.

Giap, V.N., Huang, S.C., 2020. Effectiveness of fuzzy sliding mode control boundary layer based on uncertainty and disturbance compensator on suspension active magnetic bearing system. Measurements & Control, 1–9.

Gim, D.W., Alfriend, K.T., 2003. State transition matrix of relative motion for the perturbed noncircular reference orbit. Journal of Guidance, Control, and Dynamics 26 (6), 956–971.

Giri, D.K., Sinha, M., 2016. Finite-time continuous sliding mode magneto-Coulombic satellite attitude control. IEEE Transactions on Aerospace and Electronic Systems 52 (5), 2397–2412.

Gonzalez, J.A., Barreiro, A., Dormido, S., Banos, A., 2017. Nonlinear adaptive sliding mode control with fast non-overshooting responses and chattering avoidance. Journal of the Franklin Institute 354 (7), 2788–2815.

Gordon, N.J., Salmond, D.J., Smith, A.F.M., 1993. Novel approach to nonlinear/non-Gaussian Bayesian state estimation. IEE Proceedings. Part F. Radar and Signal Processing 140, 107–113.

Grigoriadis, K.M., Watson, J.T., 1997. Reduced-order $\mathcal{H}_\infty$ and $\mathcal{L}_2$-$\mathcal{H}_\infty$ filtering via linear matrix inequalities. IEEE Transactions on Aerospace and Electronic Systems 33 (4), 1326–1338.

Grimble, M.J., 1988. $\mathcal{H}_\infty$ design of optimal linear filters. In: Linear Circuit Systems and Signal Processing: Theory and Application, pp. 533–540.

Gui, H., Vukovich, G., 2015. Adaptive integral sliding mode control for spacecraft attitude tracking with actuator uncertainty. Journal of the Franklin Institute 352 (12), 5832–5852.

Gui, H., Vukovich, G., 2017. Finite-time angular velocity observers for rigid-body attitude tracking with bounded inputs. International Journal of Robust and Nonlinear Control 27 (1), 15–38.

Gunnam, K.K., Hughes, D.C., Junkins, J.L., Kehtarnavaz, N., 2002. A vision-based DSP embedded navigation sensor. IEEE Sensors Journal 2 (5), 428–442.

Guo, Y., Huang, B., Song, S.M., Li, A.J., Wang, C.Q., 2019. Robust saturated finite-time attitude control for spacecraft using integral sliding mode. Journal of Guidance, Control, and Dynamics 42 (2), 440–446.

Habibi, S., 2006. The extended variable structure filter. Journal of Dynamic Systems, Measurement and Control 128 (2), 341–351.

Habibi, S., 2007. The smooth variable structure filter. Proceedings of the IEEE 95 (5), 1026–1059.

Habibi, S.R., Burton, R., 2003. The variable structure filter. Journal of Dynamic Systems, Measurement, and Control 125 (3), 157–165.

Habibi, S., Burton, R., Chinniah, Y., 2002. Estimation using a new variable structure filter. In: American Control Conference, pp. 2937–2942.

Haddad, W.M., Bernstein, D.S., Mustafa, D., 1991. Mixed-norm $\mathcal{H}_2$-H_∞ regulation and estimation the discrete-time case. Systems & Control Letters 16 (4), 235–247.

Hajiyev, C., Soken, H.E., 2014. Robust adaptive unscented Kalman filter for attitude estimation o pico satellites. International Journal of Adaptive Control and Signal Processing 28 (2), 107–12C

Hamel, J.F., Lafontaine, J., 2007. Linearized dynamics of formation flying spacecraft on J_2 perturbed elliptical orbit. Journal of Guidance, Control, and Dynamics 30 (6), 1649–1658.

Hammersley, J.M., Morton, K.W., 1954. Poor man's Monte Carlo. Journal of the Royal Statistica Society 16 (1), 23–38.

Hammouri, H., Bornard, G., Busawon, K., 2010. High gain observer for structured multi-outp nonlinear systems. IEEE Transactions on Automatic Control 55 (4), 987–992.

Handschin, J.E., Mayne, D.Q., 1969. Monte Carlo techniques to estimate the conditional expectatic in multi-stage non-linear filtering. International Journal of Control 9 (5), 547–559.

Haug, A.J., 2005. A tutorial on Bayesian estimation and tracking techniques applicable to nonline and non-Gaussian processes. MITRE Corporation.

Ho, Y.C., Lee, R.C.K.A., 1964. A Bayesian approach to problems in stochastic estimation and control. IEEE Transactions on Automatic Control 9 (4), 333–339.

Hu, Q.L., Jiang, B.Y., 2018. Continuous finite-time attitude control for rigid spacecraft based on angular velocity observer. IEEE Transactions on Aerospace and Electronic Systems 54 (3), 1082–1092.

Hu, J., Zhang, H., 2013. Bounded output feedback of rigid-body attitude via angular velocity observers. Journal of Guidance, Control, and Dynamics 36 (4), 1240–1248.

Hu, Q., Zhang, J., Zhang, Y.M., 2018. Velocity-free attitude coordinated tracking control for spacecraft formation flying. ISA Transactions 73, 54–65.

Huang, W., Xie, H., Shen, C., Li, J., 2016. A robust strong tracking cubature Kalman filter for spacecraft attitude estimation with quaternion constraint. Acta Astronautica 121, 153–163.

Huber, P.J., 1964. Robust estimation of a location parameter. The Annals of Mathematical Statistics 35, 73–101.

Huber, P.J., 2004. Robust Statistics, vol. 523. John Wiley.

Hussein, R., Shaban, K.B., El-Hag, A.H., 2007. Online wavelet denoising via a moving window. Acta Automatica Sinica 33 (9), 897–901.

Hussein, R., Shaban, K.B., El-Hag, A.H., 2015. Wavelet transform with histogram-based threshold estimation for online partial discharge signal denoising. IEEE Transactions on Instrumentation and Measurement 64 (12), 3601–3614.

Ibaraki, S., Suryanarayanan, S., Tomizuka, M., 2005. Design of Luenberger state observers using fixed-structure $\mathcal{H}_\infty$ optimization and its application to fault detection in lane-keeping control of automated vehicles. IEEE/ASME Transactions on Mechatronics 10 (1), 34–42.

Idan, M., 1996. Estimation of Rodrigues parameters from vector observations. IEEE Transactions on Aerospace and Electronic Systems 32 (2), 578–586.

Ito, K., Xiong, K., 2000. Gaussian filters for nonlinear filtering problems. IEEE Transactions on Automatic Control 45 (5), 910–927.

Jazwinski, A.H., 1970. Stochastic Processes and Filtering Theory. Academic Press.

Ji, H.X., Yang, J., 2010. Satellite attitude determination based on nonlinear predictive filter. Journal of System Simulation 22, 34–38.

Julier, S.J., 2002. The scaled unscented transformation. In: Proceedings of the 2002 American Control Conference, vol. 6, pp. 4555–4559.

Julier, S.J., 2003. The spherical simplex unscented transformation. In: Proceedings of the 2003 American Control Conference, vol. 3, pp. 2430–2434.

Julier, S.J., Uhlmann, J.K., 1997. New extension of the Kalman filter to nonlinear systems. In: The 11th International Symposium Aerospace/Defense Sensing, Simulation and Controls, pp. 54–65.

Julier, S.J., Uhlmann, J.K., 2002. Reduced sigma point filters for the propagation of means and covariances through nonlinear transformations. In: Proceedings of the 2002 American Control Conference, vol. 2, pp. 887–892.

Julier, S.J., Uhlmann, J.K., Durrant-Whyte, H.F., 2000. A new method for the nonlinear transformation of means and covariances in filters and estimators. IEEE Transactions on Automatic Control 45 (3), 477–482.

Junkins, J.L., Schaub, H., 2009. Analytical Mechanics of Space Systems. American Institute of Aeronautics and Astronautics.

Kalman, R.E., 1960. A new approach to linear filtering and prediction problems. Transactions of the ASME–Journal of Basic Engineering 82, 35–45.

Kalman, R.E., Bucy, R.S., 1961. New results in linear filtering and prediction theory. Transactions of the ASME–Journal of Basic Engineering 83, 95–107.

Karlgaard, C.D., Schaub, H., 2007. Huber-based divided difference filtering. Journal of Guidance, Control, and Dynamics 30 (3), 885–891.

Karlgaard, C.D., Schaub, H., 2011. Adaptive nonlinear Huber-based navigation for rendezvous in elliptical orbit. Journal of Guidance, Control, and Dynamics 24 (2), 388–402.

Kerr, T., 1987. Decentralized filtering and redundancy management for multisensor navigation. IEEE Transactions on Aerospace and Electronic Systems 23 (1), 83–119.

Kim, W., Won, S., 2000a. Robust induced $\mathcal{L}_\infty$-norm control for uncertain discrete-time systems: an LMI approach. IEICE Transactions on Fundamentals of Electronics, Communications and Computer Science 83 (3), 558–562.

Kim, W., Won, S., 2000b. Robust $\mathcal{L}_\infty$-gain filtering for structured uncertain systems. IEICE Transactions on Fundamentals of Electronics, Communications and Computer Science 83 (11), 2385–2389.

Kolmogoroff, A., 1941. Interpolation und Extrapolation von stationaren zufalligen Folgen. Izvestiya Rossiiskoi Akademii Nauk. Seriya Matematicheskaya 5 (1), 3–14.

Kotecha, J.H., Djuric, P.M., 2003a. Gaussian particle filtering. IEEE Transactions on Signal Processing 51 (10), 2592–2601.

Kotecha, J.H., Djuric, P.M., 2003b. Gaussian sum particle filtering. IEEE Transactions on Signal Processing 51 (10), 2602–2612.

Kristiansen, R., Loria, A., Chaillet, A., Nicklasson, P.J., 2009a. Spacecraft relative rotation tracking without angular velocity measurements. Automatica 45 (3), 750–756.

Kristiansen, R., Nicklasson, P.J., Gravdahl, J.T., 2009b. Satellite, attitude control by quaternion-based backstepping. IEEE Transactions on Control Systems Technology 17 (1), 227–232.

Krstic, M., Kanellakopoulos, I., Kokotovic, P.V., 1995. Nonlinear and Adaptive Control Design. John Wiley.

Kucera, V., 1980. Discrete Linear Control: The Polynomial Equation Approach. John Wiley.

Kurian, A.P., Puthusserypady, S., 2006. Performance analysis of nonlinear predictive filter based chaotic synchronization. IEEE Transactions on Circuits and Systems. II, Express Briefs 53 (9), 886–890.

Lefferts, E.J., Markley, F.L., Shuster, M.D., 1982. Kalman filtering for spacecraft attitude estimation. Journal of Guidance, Control, and Dynamics 5 (5), 417–429.

Leon-Garcia, A., 2008. Probability, Statistics, and Random Processes for Electrical Engineering. Pearson Prentice Hall.

Levant, A., 1993. Sliding order and sliding accuracy in sliding mode control. International Journal of Control 58 (6), 1247–1263.

Li, H.Z., Fu, M.Y., 1997. A linear matrix inequality approach to robust $\mathcal{H}_\infty$ filtering. IEEE Transactions on Signal Processing 45 (9), 2338–2350.

Li, W., Liu, M.H., Duan, D.P., 2014. Improved robust Huber-based divided difference filtering Proceedings of the Institution of Mechanical Engineers. Part G, Journal of Aerospace Engineering 228 (11), 2123–2129.

Li, J., Zhang, H.Y., 2006. Stochastic stability analysis of predictive filters. Journal of Guidance Control, and Dynamics 29 (3), 738–741.

Li, X., Zhu, F., Zhang, J., 2016. State estimation and simultaneous unknown input and measurement noise reconstruction based on adaptive $\mathcal{H}_\infty$ observer. International Journal of Control Automation, and Systems 14 (3), 647–654.

Lin, Y.W., Cheng, J.W.J., 2010. A high-gain observer for a class of cascade-feedback-connected nonlinear systems with application to injection molding. IEEE/ASME Transactions on Mechatronics 15 (5), 714–727.

Lin, Y.R., Deng, Z.L., 2002. Star-sensor-based predictive Kalman filter for satellite attitude estimation. Science in China. Series F. Information Sciences 45 (3), 189–195.

Liu, J., Laghrouche, S., Harmouche, M., Wack, M., 2014. Adaptive-gain second-order sliding mode observer design for switching power converters. Control Engineering Practice 30, 124–131.

Lu, P., 1994. Nonlinear predictive controllers for continuous systems. Journal of Guidance, Control and Dynamics 17 (3), 553–560.

Lu, P., 1995. Nonlinear predictive controllers for continuous nonlinear systems. Journal of Guidance, Control, and Dynamics 62 (3), 633–649.

Luenberger, D., 1966. Observers for multivariable systems. IEEE Transactions on Automatic Control 11 (2), 190–197.

Luo, W.C., Chu, Y.C., Ling, K.V., 2005. $\mathcal{H}_\infty$ inverse optimal attitude tracking control of rigid spacecraft. Journal of Guidance, Control, and Dynamics 28 (3), 481–494.

Macagnano, D., Abreu, G.T.F., 2012. Adaptive gating for multitarget tracking with Gaussian mixture filters. IEEE Transactions on Signal Processing 60 (3), 1533–1538.

Maganti, G.B., Singh, S.N., 2007. Simplified adaptive control of an orbiting flexible spacecraft. Acta Astronautica 61 (7–8), 575–589.

Mahmoud, M.S., Al-Muthairi, N.F., Bingulac, S., 1999. Robust Kalman filtering for continuous time-lag systems. Systems & Control Letters 38 (4–5), 309–319.

Markley, F.L., Berman, N., Shaked, U., 1993. $\mathcal{H}_\infty$-type filter for spacecraft attitude estimation. In: AAS/GSFC Proceedings of the American Control Conference.

Markley, F.L., Berman, N., Shaked, U., 1994. Deterministic EKF-like estimator for spacecraft attitude estimation. In: American Control Conference, pp. 247–251.

Maybeck, P.S., 1982. Stochastic Models, Estimation, and Control. Academic Press.

McGee, L., Schmidt, S., 1985. Discovery of the Kalman filter as a practical tool for aerospace and industry. In: NASA. Technical Memo, vol. 86847.

Mercorelli, P., 2015. A two-stage sliding-mode high-gain observer to reduce uncertainties and disturbances effects for sensorless control in automotive applications. IEEE Transactions on Industrial Electronics 62 (9), 5929–5940.

Merwe, R.V., Wan, E.A., 2004. Sigma-point Kalman filters for integrated navigation. In: Proceedings of the 60th Annual Meeting of the Institute of Navigation, pp. 641–654.

Mishne, D., 2004. Formation control of satellite subject to drag variations and J_2 perturbations. Journal of Guidance, Control, and Dynamics 27 (4), 685–692.

Mook, D.J., Junkins, J.L., 1988. Minimum model error estimation for poorly modeled dynamic systems. Journal of Guidance, Control, and Dynamics 11 (3), 256–261.

Morgan, D., Chung, S., 2012. Swarm-keeping strategies for spacecraft under J_2 and atmospheric drag perturbations. Journal of Guidance, Control, and Dynamics 35 (5), 1492–1505.

Musso, C., Oudjane, N., Gland, F.L., 2001. Improving regularised particle filters. In: Sequential Monte Carlo Methods in Practice, pp. 247–271.

Nagashio, T., Kida, T., Ohtani, T., Hamada, Y., 2011. Design and implementation of robust symmetric attitude controller for ETS-VIII spacecraft. Control Engineering Practice 18 (12), 1440–1451.

Nagpal, K.M., Khargonekar, P.P., 1991. Filtering and smoothing in an H_∞ setting. IEEE Transactions on Automatic Control 36 (2), 152–166.

Nørgaard, M., Poulsen, N.K., Ravn, O., 2000. New developments in state estimation for nonlinear systems. Automatica 36 (11), 1627–1638.

Oppenheim, A.V., Schafer, R.W., Buck, J.R., 1999. Discrete-Time Signal Processing. Prentice-Hall.

Orlov, Y., Aoustin, Y., Chevallereau, C., 2011. Finite time stabilization of a perturbed double integrator—part I: continuous sliding mode-based output feedback synthesis. IEEE Transactions on Automatic Control 56 (3), 614–618.

Oshman, Y., Bar-Itzhack, I.Y., 1986. Square root filtering via covariance and information eigenfactors. Automatica 22 (5), 599–604.

Oshman, Y., Carmi, A., 2004. Estimating attitude from vector observations using a genetic algorithm-embedded quaternion particle filter. In: AIAA Guidance, Navigation, and Control Conference and Exhibit.

Oshman, Y., Carmi, A., 2006. Attitude estimation from vector observations using a genetic-algorithm-embedded quaternion particle filter. Journal of Guidance, Control, and Dynamics 29 (4), 879–891.

Palhares, R.M., Peres, P.L.D., 2000. Robust filtering with guaranteed energy-to-peak performance—an LMI approach. Automatica 36 (6), 851–858.

Papoulis, A., Pillai, S.U., 2002. Probability, Random Variables, and Stochastic Processes. Tata McGraw-Hill Education.

Park, Y., 2013. Inverse optimal and robust nonlinear attitude control of rigid spacecraft. Aerospace Science and Technology 28 (1), 257–265.

Parlos, A.G., Sunkel, J.W., 1992. Adaptive attitude control and momentum management for large-angle spacecraft maneuvers. Journal of Guidance, Control, and Dynamics 15 (4), 1018–1028.

Pertew, A., Marquez, H., Zhao, Q., 2006. $\mathcal{H}_\infty$ observer design for Lipschitz nonlinear systems. IEEE Transactions on Automatic Control 51 (7), 1211–1216.

Pisu, P., Serrani, A., 2007. Attitude tracking with adaptive rejection of rate gyro disturbances. IEEE Transactions on Automatic Control 52 (12), 2374–2379.

Pittelkau, M.E., 2003. Rotation vector in attitude estimation. Journal of Guidance, Control, and Dynamics 26 (6), 855–860.

Plestan, F., Shtessel, Y., Bregeault, V., Poznyak, A., 2010. New methodologies for adaptive sliding mode control. International Journal of Control 83 (9), 1907–1919.

Porat, B., 1997. A Course in Digital Signal Processing. John Wiley.

Psiaki, M.L., 2004. Global magnetometer-based spacecraft attitude and rate estimation. Journal of Guidance, Control, and Dynamics 27 (2), 240–250.

Psiaki, M.L., 2005. Backward-smoothing extended Kalman filter. Journal of Guidance, Control, and Dynamics 28 (5), 885–894.

Pukdeboon, C., 2013. Optimal output feedback controllers for spacecraft attitude tracking. Asian Journal of Control 15 (5), 1284–1294.

Qiu, Z., Qian, H., Wang, G., 2018. Adaptive robust cubature Kalman filtering for satellite attitude estimation. Chinese Journal of Aeronautics 31 (4), 806–819.

Ran, D., Tao, S., Cao, L., Chen, X., Zhao, Y., 2014. Attitude control system design and on-orbit performance analysis of nano-satellite—"Tian Tuo 1". Chinese Journal of Aeronautics 27 (3), 593–601.

Reid, T., Misra, A.K., 2011. Formation flight of satellites in the presence of atmospheric drag. Journal of Aerospace Engineering 30 (1), 64–91.

Reif, K., Gunther, S., Yaz, E., Unbehauen, R., 1999. Stochastic stability of the discrete-time extended Kalman filter. IEEE Transactions on Automatic Control 44 (4), 714–728.

Roberts, J.A., Roberts, P.C.E., 2004. The development of high fidelity linearized J_2 models for satellite formation flying control. In: Advances in the Astronautical Sciences, vol. 119. American Astronautical Society.

Rotea, M.A., 1993. The generalized $\mathcal{H}_2$ control problem. Automatica 29 (2), 373–385.

Sabatini, M., Palmerini, G.B., 2008. Linearized formation flying dynamics in a perturbed orbital environment. In: IEEE Aerospace Conference, pp. 1–13.

Sage, A.P., Andrew, G.W., 1969. Adaptive filtering with unknown prior statistics. In: Joint Automatic Control Conference, number 7, pp. 760–769.

Schaub, H., Alfriend, K.T., 2001. J_2 invariant relative orbits for spacecraft formations. Celestia Mechanics & Dynamical Astronomy 79 (2), 77–95.

Schlanbusch, R., Loria, A., Kristiansen, R., Nicklasson, P.J., 2012. Pd+ based output feedback atti tude control of rigid bodies. IEEE Transactions on Automatic Control 57 (8), 2146–2152.

Schmidt, S.F., 1970. Computational techniques in Kalman filtering theory. NATO Advisory Grou for Aerospace Research and Development 139, 219–292.

Schmidt, S.F., 2012. The Kalman filter: its recognition and development for aerospace applications Journal of Guidance, Control, and Dynamics 4 (1), 4–7.

Schon, T., Gustafsson, F., Nordlund, P.J., 2005. Marginalized particle filters for mixed linear/non linear state-space models. IEEE Transactions on Signal Processing 53 (7), 2279–2289.

Schweigart, S.A., Sedwick, R.J., 2002. High-fidelity linearized J_2 model for satellite formatio flight. Journal of Guidance, Control, and Dynamics 25 (6), 1073–1080.

Sengupta, P., Vadali, S.R., 2007. Relative motion and the geometry of formations in Keplerian el liptic orbits with arbitrary eccentricity. Journal of Guidance, Control, and Dynamics 30 (4 953–964.

Shaked, U., 1990. $\mathcal{H}_\infty$-minimum error state estimation of linear stationary processes. IEEE Trans actions on Automatic Control 35 (5), 554–558.

Shaked, U., Souza, C.E., 1994. Robust $\mathcal{H}_2$ filtering. Proceedings of the IFAC Symposium On Robu Control Design 22 (2), 335–340.

Shaked, U., Theodor, Y., 1992. $\mathcal{H}_\infty$-optimal estimation: a tutorial. In: Proceedings of the 31st IEEE Conference on Decision and Control, pp. 2276–2286.

Sharma, R., Tewari, A., 2004. Optimal nonlinear tracking of spacecraft attitude maneuvers. IEEE Transactions on Control Systems Technology 12 (5), 677–682.

Shi, X., Zhou, Z., Zhou, D., 2017. Finite-time attitude trajectory tracking control of rigid spacecraft. IEEE Transactions on Aerospace and Electronic Systems 53 (6), 2913–2923.

Shtessel, Y., Edwards, C., Fridman, L., Levant, A., 2014. Sliding Mode Control and Observation. Springer.

Shtessel, Y., Taleb, M., Plestan, F., 2012. A novel adaptive-gain super-twisting sliding mode controller: methodology and application. Automatica 48 (5), 759–769.

Shuster, M.D., 1990. Kalman filtering of spacecraft attitude and the quest model. The Journal of the Astronautical Sciences 38 (3), 377–393.

Sidi, M.J., 1997. Spacecraft Dynamics and Control: A Practical Engineering Approach. Cambridge university press.

Simandl, M., Dunik, J., 2009. Derivative-free estimation methods: new results and performance analysis. Automatica 45 (7), 1749–1757.

Singh, S.N., Yim, W., 2005. Nonlinear adaptive spacecraft attitude control using solar radiation pressure. IEEE Transactions on Aerospace and Electronic Systems 41 (3), 770–779.

Singh, S.N., Zhang, R., 2004. Adaptive output feedback control of spacecraft with flexible appendages by modeling error compensation. Acta Astronautica 54 (4), 229–243.

Smith, R.H., 1995. An $\mathcal{H}_\infty$-type filter for GPS-based attitude estimation. In: Spaceflight Mechanics, pp. 559–567.

Soken, H.E., Hajiyev, C., 2010. Pico satellite attitude estimation via robust unscented Kalman filter in the presence of measurement faults. ISA Transactions 49 (3), 249–256.

Soken, H.E., Hajiyev, C., 2013. Adaptive fading UKF with Q-adaptation: application to picosatellite attitude estimation. Journal of Aerospace Engineering 26 (3), 628–636.

Soken, H.E., Hajiyev, C., Sakai, S.I., 2014. Robust Kalman filtering for small satellite attitude estimation in the presence of measurement faults. European Journal of Control 20 (2), 64–72.

Souza, C.E., Shaked, U., 1999. Robust $\mathcal{H}_2$ filtering for uncertain systems with measurable inputs. IEEE Transactions on Signal Processing 47 (8), 2286–2292.

Souza, C.E., Shaked, U., Fu, M., 1995. Robust $\mathcal{H}_\infty$ filtering for continuous time varying uncertain systems with deterministic input signals. IEEE Transactions on Signal Processing 43 (3), 709–719.

Speyer, J., 1979. Computation and transmission requirements for a decentralized linear-quadratic-Gaussian control problem. IEEE Transactions on Automatic Control 24 (2), 266–269.

Subrahmanya, N., Shin, Y.C., 2009. Adaptive divided difference filtering for simultaneous state and parameter estimation. Automatica 45 (7), 1686–1693.

Sun, D., Crassidis, J.L., 2002. Observability analysis of six-degree-of-freedom configuration determination using vector observations. Journal of Guidance, Control, and Dynamics 25, 1149–1157.

Sun, L., Zheng, Z.W., 2017. Disturbance-observer-based robust backstepping attitude stabilization of spacecraft under input saturation and measurement uncertainty. IEEE Transactions on Industrial Electronics 64 (10), 7994–8002.

Sunahara, Y., 1969. An approximate method of state estimation for nonlinear dynamical systems. In: Joint Automatic Control Conference, number 7, pp. 161–172.

Tarn, T.J., Rasis, Y., 1976. Observers for nonlinear stochastic systems. IEEE Transactions on Automatic Control 21 (4), 441–448.

Tayebi, A., 2008. Unit quaternion-based output feedback for the attitude tracking problem. IEEE Transactions on Automatic Control 53 (6), 1516–1520.

Theodor, U., Shaked, U., Souza, C.E., 1994. A game theory approach to robust discrete-time $\mathcal{H}_\infty$-estimation. IEEE Transactions on Signal Processing 42 (6), 1486–1495.

Tsao, T., Liu, D., 2009. Gyroless stellar acquisition algorithm for spacecraft closed-loop control and its accuracy. In: AIAA Guidance, Navigation, and Control Conference and Exhibit, pp. 10–13.

Tschauner, J., Hempel, P., 1965. Rendezvous with a target in an elliptical orbit. Astronautica Acta 11 (2), 104–109.

Vallado, D.A., 2005. Fundamentals of Astrodynamics and Applications, vol. 12. Springer.

Varma, S., Kumar, K.D., 2009. Fault tolerant satellite attitude control using solar radiation pressure based on nonlinear adaptive sliding mode. Acta Astronautica 66 (3–4), 486–500.

Veluvolu, K.C., Soh, Y.C., 2009. High-gain observers with sliding mode for state and unknown input estimations. IEEE Transactions on Industrial Electronics 56 (9), 3386–3393.

Vincent, T., Abedor, J., Nagpal, K., Khargonekar, P.P., 1996. Discrete-time estimators with guaranteed peak-to-peak performance. IFAC Proceedings Volumes 29 (1), 4470–4475.

Voulgaris, P.G., 1996. Optimal $\mathcal{L}_\infty$ to $\mathcal{L}_\infty$ estimation for periodic systems. IEEE Transactions on Automatic Control 41 (9), 1392–1396.

Wang, Z.D., Guo, Z., Unbehauen, H., 1997. Robust $\mathcal{H}_2/\mathcal{H}_\infty$-state estimation for discrete-time systems with error variance constraints. IEEE Transactions on Automatic Control 42 (10), 1431–1435.

Wang, Z.D., Huang, B., 2000. Robust $\mathcal{H}_2/\mathcal{H}_\infty$ filtering for linear systems with error variance constraints. IEEE Transactions on Signal Processing 48 (8), 2463–2467.

Wang, Z.D., Huang, B., Unbehauen, H., 1999. Robust $\mathcal{H}_\infty$ observer design of linear state delayed systems with parametric uncertainty: the discrete-time case. Automatica 35 (6), 1161–1167.

Wang, Z.D., Unbehauen, H., 1999. Robust $\mathcal{H}_2/\mathcal{H}_\infty$-state estimation for systems with error variance constraints: the continuous-time case. IEEE Transactions on Automatic Control 44 (5), 1061–1065.

Watson, J.T., Grigoriadis, K.M., 1998. Optimal unbiased filtering via linear matrix inequalities. Systems & Control Letters 35 (2), 111–118.

Wei, C., Park, S.Y., Park, C., 2013. Linearized dynamics model for relative motion under a J_2 perturbed elliptical reference orbit. International Journal of Non-Linear Mechanics 55, 55–69.

Werve, R.V., 2004. Sigma-point Kalman filters for probabilistic inference in dynamic state-space models. PhD Thesis. OGI School of Science & Engineering at OHSU.

Werve, R.V., Wan, E.A., 2001. The square-root unscented Kalman filter for state and parameter-estimation. In: 2001 IEEE International Conference on Acoustics, Speech, and Signal Processing, vol. 6, pp. 3461–3464.

White, R.L., Adams, M.B., Geisler, E.G., Grant, F.D., 1975. Attitude and orbit estimation using stars and landmarks. IEEE Transactions on Aerospace and Electronic Systems 11 (2), 195–203

Wiener, N., 1942. The Extrapolation, Interpolation and Smoothing of Stationary Time Series. Massachusetts Institute of Technology.

Wilson, D.A., 1989. Convolution and Hankel operator norms for linear systems. IEEE Transaction on Automatic Control 34 (1), 94–97.

Wu, Y.X., Hu, D.W., Hu, X.P., 2007. Comments on "Performance evaluation of UKF-based nonlin ear filtering". Automatica 43 (3), 567–568.

Wu, Z.W., Sun, Z.G., Zhang, W.Z., Chen, Q., 2016. A novel approach for attitude estimation base on MEMS inertial sensors using nonlinear complementary filters. IEEE Sensors Journal 16 (10) 3856–3864.

Xiao, B., Hu, Q.L., Ma, G.F., 2010. Adaptive sliding mode backstepping control for attitude trackin of flexible spacecraft under input saturation and singularity. Proceedings of the Institution c Mechanical Engineers. Part G, Journal of Aerospace Engineering 224 (G2), 199–214.

Xiao, B., Hu, Q.L., William, S., Huo, X., 2013. Reaction wheel fault compensation and disturbanc rejection for spacecraft attitude tracking with finite-time convergence. Journal of Guidanc Control, and Dynamics 36 (6), 1565–1575.

Xie, L.H., Soh, Y.C., 1994. Robust Kalman filtering for uncertain systems. Systems & Contr Letters 22 (2), 123–129.

Xie, L.H., Soh, Y.C., Souza, C.E., 1994. Robust Kalman filtering for uncertain discrete-time sy tems. IEEE Transactions on Automatic Control 39 (6), 1310–1314.

Xing, Y.J., Cao, X.B., Zhang, S.J., Guo, H.B., Wang, F., 2010. Relative position and attitud estimation for satellite formation with coupled translational and rotational dynamics. Acta A tronautica 67 (3–4), 455–467.

Xing, Y., Zhang, S., Zhang, J., Cao, X., 2012. Robust-extended Kalman filter for small satellite attitude estimation in the presence of measurement uncertainties and faults. Proceedings of the Institution of Mechanical Engineers. Part G, Journal of Aerospace Engineering 226 (G1), 30–41.

Xiong, K., Zhang, H.Y., Chan, C.W., 2006. Performance evaluation of UKF-based nonlinear filtering. Automatica 42 (2), 261–270.

Xiong, K., Zhang, H.Y., Chan, C.W., 2007. Author's reply to "Comments on 'Performance evaluation of UKF-based nonlinear filtering'". Automatica 43 (3), 569–570.

Xiong, K., Zhang, H., Liu, L., 2008. Adaptive robust extended Kalman filter for nonlinear stochastic systems. IET Control Theory & Applications 2 (3), 239–250.

Yaesh, I., Shaked, U., 1989. Game theory approach to optimal linear estimation in the minimum $\mathcal{H}_\infty$-norm sense. In: Proceedings of the 28th IEEE Conference on Decision and Control, pp. 421–425.

Yang, C.D., Sun, Y.P., 2002. Mixed $\mathcal{H}_2/\mathcal{H}_\infty$ state-feedback design for microsatellite attitude control. Control Engineering Practice 10 (9), 951–970.

Yang, C.C., Wu, C.J., 2007. Optimal large-angle attitude control of rigid spacecraft by momentum transfer. IET Control Theory & Applications 1 (3), 657–664.

Yokoyama, N., 2011. Parameter estimation of aircraft dynamics via unscented smoother with expectation-maximization algorithm. Journal of Guidance, Control, and Dynamics 34 (2), 426–436.

Yoshimura, T., Soeda, T., 1978. A technique for compensating the filter performance by a fictitious noise. Journal of Dynamic Systems, Measurement, and Control 100 (2), 154–156.

Zhang, B., Liu, K., Xiang, J., 2013. A stabilized optimal nonlinear feedback control for satellite attitude tracking. Aerospace Science and Technology 27 (1), 17–24.

Zhang, M., Xie, H.W., Shi, D.H., 1999. Robust $\mathcal{H}_\infty$ filtering for uncertain parameter system. IFAC Proceedings Volumes 32 (2), 484–494.

Zhao, L., Yu, J., Yu, H., 2018. Adaptive finite-time attitude tracking control for spacecraft with disturbances. IEEE Transactions on Aerospace and Electronic Systems 54 (3), 1297–1305.

Zheng, Q., Wu, F., 2009. Nonlinear $\mathcal{H}_\infty$ control designs with axisymmetric spacecraft control. Journal of Guidance, Control, and Dynamics 32 (3), 850–859.

Zhou, D.H., Park, P.M., 1996. Strong tracking Kalman filter for nonlinear time-varying stochastic systems with colored noise: application to parameter estimation and empirical robustness analysis. International Journal of Control 65 (2), 295–307.

Zou, A., de Ruiter, A.H., Kumar, K.D., 2017. Finite-time attitude tracking control for rigid spacecraft with control input constraints. IET Control Theory & Applications 11 (7), 931–940.

Index

R

S

Printed in the United States
By Bookmasters